Graph Spectral Image Processing

SCIENCES

Image, Field Director – Laure Blanc-Feraud

Compression, Coding and Protection of Images and Videos,
Subject Head – Christine Guillemot

Graph Spectral Image Processing

Coordinated by

Gene Cheung
Enrico Magli

iSTE

WILEY

First published 2021 in Great Britain and the United States by ISTE Ltd and John Wiley & Sons, Inc.

ISTE Ltd
27-37 St George's Road
London SW19 4EU
UK

www.iste.co.uk

John Wiley & Sons, Inc.
111 River Street
Hoboken, NJ 07030
USA

www.wiley.com

Library of Congress Control Number: 2021932054

British Library Cataloguing-in-Publication Data
A CIP record for this book is available from the British Library
ISBN 978-1-78945-028-6

ERC code:
PE7 Systems and Communication Engineering
PE7_7 Signal processing

MIX
Paper from responsible sources
FSC® C013604

Contents

Introduction to Graph Spectral Image Processing

Gene CHEUNG[1] and Enrico MAGLI[2]
[1] *York University, Toronto, Canada*
[2] *Politecnico di Torino, Turin, Italy*

I.1. Introduction

Image processing is a mature research topic. The first specification of Joint Photographic Experts Group (JPEG), now the predominant image coding standard on the Internet, was published in 1992. MPEG1, the first digital video compression standard by ISO, was standardized in 1993. The IEEE International Conference on Image Processing (ICIP), the flagship image processing conference held annually for the IEEE Signal Processing Society (SPS), was also started in 1993 and has been in existence for 27 years, making it older than many image processing researchers now studying in graduate schools! Given the topic's maturity, it is a legitimate question to ask if yet another book on image processing is warranted. As co-editors of this book, we emphatically answer this question with a resounding "Yes". We will first discuss the following recent technological trends, which also serve as motivations for the creation of this publication.

1) *Sensing and Display Technologies*: The advent of image sensing technologies, such as active depth sensors and display technologies like head-mounted displays (HMD), in the last decade alone, means that the nature of a digital image has drastically changed. Beyond higher spatial resolution and bit-depth per pixel, a modern imaging sensor can also acquire scene depth, hyper-spectral properties, etc. Further, often acquired image data is not represented as a traditional 2D array of pixel information, but in an alternative form, such as light fields and 3D point clouds. This means that the processing tools must flexibly adapt to richer and evolving imaging contents and formats.

2) *Graph Signal Processing*: In the last eight years, we have also witnessed the birth of a new signal processing topic – called *graph signal processing* (GSP) – that generalizes traditional mathematical tools like transforms and wavelets, to process signals residing on irregular data kernels described by graphs (Shuman *et al.* 2013). Central to GSP is the notion of *graph frequencies*: orthogonal components, computed from a graph variation operator like the graph Laplacian matrix, that generalize the notion of Fourier modes to the graph domain, spanning a graph signal space. Because of its inherent powerful generality, one can easily adopt or design GSP tools for different imaging applications, where a node in a graph represents a pixel, and the graph connectivity is chosen to reflect inter-pixel similarities or correlations. For an example of the GSP tool being used for image restoration, see Figure I.1 for an illustration of a graph spectral method called left eigenvectors of the random walk graph Laplacian (LeRAG) for JPEG image dequantization (Liu *et al.* 2017). GSP tools can also easily adapt to the aforementioned modern imaging modalities, such as light field images and 3D point clouds, that do not reside on regular 2D grids.

(a) BM3D (23.91) (b) KSVD (24.55) (c) LERaG (25.82)

Figure I.1. *Visual comparison of JPEG dequantization methods for a butterfly at QF = 5. The corresponding PSNR values are also given. For a color version of this figure, see www.iste.co.uk/cheung/graph.zip*

3) *Deep Neural Networks*: Without a doubt, the singular seismic paradigm shift in data science in the last decade is *deep learning*. Using layers of convolutional filters, pointwise nonlinearities and pooling functions, deep neural network (DNN) architectures like *convolutional neural networks* (CNN) have demonstrated superior performance in a wide range of imaging tasks from denoising to classification, when a large volume of labeled data is available for training (Vemulapalli *et al.* 2016; Zhang *et al.* 2017). When labeled training data is scarce, or when the underlying data kernel is irregular (thus complicating the training of convolutional filters and the selection of pooling operators), how to best design and construct DNN for a targeted image application is a challenging problem. Moreover, a CNN purely trained from labeled data often remains a "black box", i.e. the learned operators like filtering remain unexplainable.

Motivated by these technological trends, we have focused this book on the theory and applications of GSP tools for image processing, covering conventional images and videos, new modalities like light fields and 3D point clouds, and hybrid GSP/deep learning approaches. Different from other graph-based image processing books (Lezoray and Grady 2012), we concentrate on *spectral* processing techniques with frequency interpretations such as graph Fourier transforms (GFT) and graph wavelets, drawing inspiration from the long history of frequency analysis tools in traditional signal processing. *Graph frequency analysis* enables the definition of familiar signal processing notions, such as graph Fourier modes, bandlimitedness, and signal smoothness, using graph spectral methods that can be designed.

Specifically, the content of this book is structured into two parts:

1) The first part of the book discusses the fundamental GSP theories. Chapter 1, titled "Graph Spectral Filtering" by Y. Tanaka, reviews the basics of graph filtering such as graph transforms and wavelets. Chapter 2, titled "Graph Learning" by X. Dong, D. Thanou, M. Rabbat and P. Frossard, reviews recent techniques to learn an underlying graph structure given a set of observable data. Chapter 3, titled "Graph Neural Networks" by G. Fracastoro and D. Valsesia, overviews recent works generalizing DNN architectures to the graph data domain, where input signals reside on irregular graph structures.

2) The second part of the book reviews different imaging applications of GSP. Chapters 4 and 5, titled "Graph Specral Image and Video Compression" by H.E. Egilmez, Y.-H. Chao and A. Ortega and "Graph Spectral 3D Image Compression" by T. Maugey, M. Rizkallah, N. M. Bidgoli, A. Roumy and C. Guillemot, focus on the design and applications of GSP tools for the compression of traditional images/videos and 3D images, respectively. Chapter 6, titled "Graph Spectral Image Restoration" by J. Pang and J. Zeng, focuses on the general recovery of corrupted images, e.g. image denoising and deblurring. As a new imaging modality, Chapter 7, titled "Graph Spectral Point Cloud Processing" by W. Hu, S. Chen and D. Tian, focuses on the processing of 3D point clouds for applications, such as low-level restoration and high-level unsupervised feature learning. Chapters 8 and 9, titled "Graph Spectral Image Segmentation" by M. Ng and "Graph Spectral Image Classification" by M. Ye, V. Stankovic, L. Stankovic and G. Cheung, narrow the discussion specifically to segmentation and classification, respectively, two popular research topics in the computer vision community. Finally, Chapter 10, titled "Graph Neural Networks for Image Processing" by G. Fracastoro and D. Valsesia, reviews the growing efforts to employ recent GNN architectures for conventional imaging tasks such as denoising.

Before we jump into the various chapters, we begin with the basic definitions in GSP that will be used throughout the book. Specifically, we formally define a graph, graph spectrum, variation operators and graph signal smoothness priors in the following sections.

I.2. Graph definition

A graph $\mathcal{G}(\mathcal{V}, \mathcal{E}, \mathbf{W})$ contains a set $\mathcal{V}$ of N nodes and a set $\mathcal{E}$ of M edges. While directed graphs are also possible, in this book we more commonly assume an undirected graph, where each existing edge $(i, j) \in \mathcal{E}$ is undirected and contains an edge weight $w_{i,j} \in \mathbb{R}$, which is typically positive. A large positive edge weight $w_{i,j}$ would mean that samples at nodes i and j are expected to be similar/correlated.

There are many ways to compute appropriate edge weights. Especially common for images, edge weight $w_{i,j}$ can be computed using a Gaussian kernel, as done in the *bilateral filter* (Tomasi and Manduchi 1998):

$$w_{i,j} = \exp\left(-\frac{\|\mathbf{l}_i - \mathbf{l}_j\|_2^2}{\sigma_l^2}\right) \exp\left(-\frac{\|x_i - x_j\|_2^2}{\sigma_x^2}\right) \qquad \text{[I.1]}$$

where $\mathbf{l}_i \in \mathbb{R}^2$ is the location of pixel i on the 2D image grid, $x_i \in \mathbb{R}$ is the intensity of pixel i, and σ_l^2 and σ_x^2 are two parameters. Hence, $0 \leq w_{i,j} \leq 1$. Larger geometric and/or photometric distances between pixels i and j would mean a smaller weight $w_{i,j}$. Edge weights can alternatively be defined based on local pixel patches, features, etc. (Milanfar 2013b). To a large extent, the appropriate definition of edge weight is application dependent, as will be discussed in various forthcoming chapters.

A *graph signal* $\mathbf{x}$ on $\mathcal{G}$ is a discrete signal of dimension N – one sample $x_i \in \mathbb{R}$ for each node[1] i in $\mathcal{V}$. Assuming that nodes are appropriately labeled from 1 to N, we can simply treat a graph signal as a vector $\mathbf{x} \in \mathbb{R}^N$.

I.3. Graph spectrum

Denote by $\mathbf{W} \in \mathbb{R}^{N \times N}$ an *adjacency matrix*, where the (i, j)th entry is $W_{i,j} = w_{i,j}$. Next, denote by $\mathbf{D} \in \mathbb{R}^{N \times N}$ a diagonal *degree matrix*, where the (i, i)th entry is $D_{i,i} = \sum_j W_{i,j}$. A *combinatorial graph Laplacian matrix* $\mathbf{L}$ is $\mathbf{L} = \mathbf{D} - \mathbf{W}$ (Shuman *et al.* 2013). Because $\mathbf{L}$ is real and symmetric, one can show, via the spectral theorem, that it can be eigen-decomposed into:

$$\mathbf{L} = \mathbf{U}\mathbf{\Lambda}\mathbf{U}^\top \qquad \text{[I.2]}$$

where $\mathbf{\Lambda}$ is a diagonal matrix containing real eigenvalues λ_k along the diagonal, and $\mathbf{U}$ is an eigen-matrix composed of orthogonal eigenvectors $\mathbf{u}_i$ as columns. If all edge

1 If a graph node represents a pixel in an image, each pixel would typically have three color components: red, green and blue. For simplicity, one can treat each color component separately as a different graph signal.

weights $w_{i,j}$ are restricted to be positive, then graph Laplacian $\mathbf{L}$ can be proven to be *positive semi-definite* (PSD) (Chung 1997)[2], meaning that $\lambda_k \geq 0, \forall k$ and $\mathbf{x}^\top \mathbf{L}\mathbf{x} \geq 0, \forall \mathbf{x}$. Non-negative eigenvalues λ_k can be interpreted as *graph frequencies*, and eigenvectors $\mathbf{U}$ can be interpreted as corresponding graph Fourier modes. Together they define the *graph spectrum* for graph $\mathcal{G}$.

The set of eigenvectors $\mathbf{U}$ for $\mathbf{L}$ collectively form the GFT (Shuman *et al.* 2013), which can be used to decompose a graph signal $\mathbf{x}$ into its frequency components via $\alpha = \mathbf{U}^\top \mathbf{x}$. In fact, one can interpret GFT as a generalization of known discrete transforms like the *Discrete Cosine Transform* (DCT) (see Shuman *et al.* 2013 for details).

Note that if the multiplicity m_k of an eigenvalue λ_k is larger than 1, then the set of eigenvectors that span the corresponding eigen-subspace of dimension m_k is non-unique. In this case, it is necessary to specify the graph spectrum as the collection of eigenvectors $\mathbf{U}$ themselves.

If we also consider negative edge weights $w_{i,j}$ that reflect inter-pixel dissimilarity/anti-correlation, then graph Laplacian $\mathbf{L}$ can be indefinite. We will discuss a few recent works (Su *et al.* 2017; Cheung *et al.* 2018) that employ negative edges in later chapters.

I.4. Graph variation operators

Closely related to the combinatorial graph, Laplacian $\mathbf{L}$, are other variants of Laplacian operators, each with their own unique spectral properties. A *normalized graph Laplacian* $\mathbf{L}_n = \mathbf{D}^{-1/2}\mathbf{L}\mathbf{D}^{-1/2}$ is a symmetric normalized variant of $\mathbf{L}$. In contrast, a *random walk graph Laplacian* $\mathbf{L}_r = \mathbf{D}^{-1}\mathbf{L}$ is an asymmetric normalized variant of $\mathbf{L}$. A *generalized graph Laplacian* $\mathbf{L}_g = \mathbf{L} + \text{diag}(\mathbf{D})$ is a graph Laplacian with self-loops $d_{i,i}$ at nodes i – called the *loopy graph Laplacian* in Dörfler and Bullo (2013) – resulting in a general symmetric matrix with non-positive off-diagonal entries for a positive graph (Biyikoglu *et al.* 2005). Eigen-decomposition can also be performed on these operators to acquire a set of graph frequencies and graph Fourier modes. For example, normalized variants $\mathbf{L}_n$ and $\mathbf{L}_r$ (which are similarity transforms of each other) share the same eigenvalues between 0 and 2. While $\mathbf{L}$ and $\mathbf{L}_n$ are both symmetric, $\mathbf{L}_n$ does not have the constant vector as an eigenvector. Asymmetric $\mathbf{L}_r$ can be symmetrized via left and right diagonal matrix multiplications (Milanfar 2013a). Different variation operators will be used throughout the book for different applications.

---

2 One can prove that a graph $\mathcal{G}$ with positive edge weights has PSD graph Laplacian $\mathbf{L}$ via the Gershgorin circle theorem: each Gershgorin disc corresponding to a row in $\mathbf{L}$ is located in the non-negative half-space, and since all eigenvalues reside inside the union of all discs, they are non-negative.

I.5. Graph signal smoothness priors

Traditionally, for graph $\mathcal{G}$ with positive edge weights, signal $\mathbf{x}$ is considered *smooth* if each sample x_i on node i is similar to samples x_j on neighboring nodes j with large $w_{i,j}$. In the graph frequency domain, it means that $\mathbf{x}$ mostly contains low graph frequency components, i.e. coefficients $\boldsymbol{\alpha} = \mathbf{U}^\top \mathbf{x}$ are zeros (or mostly zeros) for high frequencies. The smoothest signal is the constant vector – the first eigenvector $\mathbf{u}_1$ for $\mathbf{L}$, corresponding to the smallest eigenvalue $\lambda_1 = 0$.

Mathematically, we can declare that a signal $\mathbf{x}$ is smooth if its *graph Laplacian regularizer* (GLR) $\mathbf{x}^\top \mathbf{L} \mathbf{x}$ is small (Pang and Cheung 2017). GLR can be expressed as:

$$\mathbf{x}^\top \mathbf{L} \mathbf{x} = \sum_{(i,j)\in\mathcal{E}} w_{i,j} \left(x_i - x_j\right)^2 = \sum_k \lambda_k \, \alpha_k^2 \qquad \text{[I.3]}$$

Because $\mathbf{L}$ is PSD, $\mathbf{x}^\top \mathbf{L} \mathbf{x}$ is lower bounded by 0 and achieved when $\mathbf{x} = c\mathbf{u}_1$ for some scalar constant c. One can also define GLR using the normalized graph Laplacian $\mathbf{L}_n$ instead of $\mathbf{L}$, resulting in $\mathbf{x}^\top \mathbf{L}_n \mathbf{x}$. The caveats is that the constant vector $\mathbf{u}_1$ – typically the most common signal in imaging – is no longer the first eigenvector, and thus $\mathbf{u}_1^\top \mathbf{L}_n \mathbf{u}_1 \neq 0$.

In Chen *et al.* (2015), the adjacency matrix $\mathbf{W}$ is interpreted as a shift operator, and thus, graph signal smoothness is instead defined as the difference between a signal $\mathbf{x}$ and its shifted version $\mathbf{W}\mathbf{x}$. Specifically, *graph total variation* (GTV) based on l_p-norm is:

$$\mathrm{TV}_{\mathbf{W}}(\mathbf{x}) = \left\| \mathbf{x} - \frac{1}{|\lambda_{\max}|} \mathbf{W} \mathbf{x} \right\|_p^p \qquad \text{[I.4]}$$

where $\lambda_{\max}$ is the eigenvalue of $\mathbf{W}$ with the largest magnitude (also called the *spectral radius*), and p is a chosen integer. As a variant to equation [I.4], a quadratic smoothness prior is defined in Romano *et al.* (2017), using a row-stochastic version $\mathbf{W}_n = \mathbf{D}^{-1}\mathbf{W}$ of the adjacency matrix $\mathbf{W}$:

$$S_2(\mathbf{x}) = \frac{1}{2} \|\mathbf{x} - \mathbf{W}_n \mathbf{x}\|_2^2 \qquad \text{[I.5]}$$

To avoid confusion, we will call equation [I.5] the *graph shift variation* (GSV) prior. GSV is easier to use in practice than GTV, since the computation of $\lambda_{\max}$ is required for GTV. Note that GSV, as defined in equation [I.5], can also be used for signals on directed graphs.

I.6. References

Biyikoglu, T., Leydold, J., Stadler, P.F. (2005). Nodal domain theorems and bipartite subgraphs. *Electronic Journal of Linear Algebra*, 13, 344–351.

Chen, S., Sandryhaila, A., Moura, J., Kovacevic, J. (2015). Signal recovery on graphs: Variation minimization. *IEEE Transactions on Signal Processing*, 63(17), 4609–4624.

Cheung, G., Su, W.-T., Mao, Y., Lin, C.-W. (2018). Robust semisupervised graph classifier learning with negative edge weights. *IEEE Transactions on Signal and Information Processing over Networks*, 4(4), 712–726.

Chung, F. (1997). Spectral graph theory. *CBMS Regional Conference Series in Mathematics*, 92.

Dörfler, F. and Bullo, F. (2013). Kron reduction of graphs with applications to electrical networks. *IEEE Transactions on Circuits and Systems I: Regular Papers*, 60(1), 150–163.

Lezoray, O. and Grady, L. (2012). *Image Processing and Analysis with Graphs: Theory and Practice*, CRC Press, Boca Raton, Florida.

Liu, X., Cheung, G., Wu, X., Zhao, D. (2017). Random walk graph Laplacian based smoothness prior for soft decoding of JPEG images. *IEEE Transactions on Image Processing*, 26(2), 509–524.

Milanfar, P. (2013a). Symmetrizing smoothing filters. *SIAM Journal on Imaging Sciences*, 6(1), 263–284.

Milanfar, P. (2013b). A tour of modern image filtering. *IEEE Signal Processing Magazine*, 30(1), 106–128.

Pang, J. and Cheung, G. (2017). Graph Laplacian regularization for image denoising: Analysis in the continuous domain. *IEEE Transactions on Image Processing*, 26(4), 1770–1785.

Romano, Y., Elad, M., Milanfar, P. (2017). The little engine that could: Regularization by denoising (RED). *SIAM Journal on Imaging Sciences*, 10(4), 1804–1844.

Shuman, D.I., Narang, S.K., Frossard, P., Ortega, A., Vandergheynst, P. (2013), The emerging field of signal processing on graphs: Extending high-dimensional data analysis to networks and other irregular domains. *IEEE Signal Processing Magazine*, 30(3), 83–98.

Su, W.-T., Cheung, G., Lin, C.-W. (2017). Graph Fourier transform with negative edges for depth image coding. *IEEE International Conference on Image Processing*, Beijing.

Tomasi, C. and Manduchi, R. (1998), Bilateral filtering for gray and color images. *IEEE International Conference on Computer Vision*, 839–846.

Vemulapalli, R., Tuzel, O., Liu, M.-Y. (2016). Deep Gaussian conditional random field network: A model-based deep network for discriminative denoising. *Proceedings of the IEEE Conference on Computer Vision and Pattern Recognition*, 4801–4809.

Zhang, K., Zuo, W., Chen, Y., Meng, D., Zhang, L. (2017). Beyond a Gaussian denoiser: Residual learning of deep CNN for image denoising. *IEEE Transactions on Image Processing*, 26(7), 3142–3155.

PART 1

Fundamentals of Graph Signal Processing

1

Graph Spectral Filtering

Yuichi TANAKA
Tokyo University of Agriculture and Technology, Japan

1.1. Introduction

The filtering of time- and spatial-domain signals is one of the fundamental techniques for image processing and has been studied extensively to date. GSP can treat signals with irregular structures that are mathematically represented as graphs. Theories and methodologies for the filtering of graph signals are studied using spectral graph theory. In image processing, graphs are strong tools for representing structures formed by pixels, like edges and textures.

The filtering of graph signals is not only an extension of that for standard time- and spatial-domain signals, but it also has its own interesting properties. For example, GSP can represent traditional pixel-dependent image filtering methods as graph spectral domain filters. Furthermore, theory and design methods for wavelets and filter banks, which are studied extensively in signal and image processing, are also updated to treat graph signals.

In this chapter, the spectral-domain filtering of graph signals is introduced. In section 1.2, the filtering of time-domain signals is briefly described as a starting point. The filtering of graph signals, both in the vertex and spectral domains, is detailed in section 1.3, in addition to its relationship with classical filtering. Edge-preserving image smoothing is represented as a graph filter in section 1.4. Furthermore, a framework of filtering by multiple graph filters, i.e. graph wavelets and filter banks, is presented in section 1.5. Eventually, section 1.6 introduces several fast computation methods of graph filtering. Finally, the concluding remarks of this chapter are discussed in section 1.7.

Graph Spectral Image Processing,
coordinated by Gene CHEUNG and Enrico MAGLI.

1.2. Review: filtering of time-domain signals

We start by reviewing the filtering in discrete-time linear time-invariant (LTI) systems, which has been extensively studied in literature. Suppose that a one-dimensional discrete-time signal x_n is obtained by sampling its continuous-time counterpart $x(t)$, with a fixed sampling period T, i.e. $x_n = x(nT)$. A two-dimensional image signal can be similarly obtained by performing sampling in both the horizontal and vertical directions. In this case, the spatial sampling period usually corresponds to the spacing between an array of photosensors.

Suppose that an impulse response of a filter h_n is given *a priori*. The discrete-time filtered signal y_n in the LTI system is calculated from x_n and h_n by convolution as follows:

$$y_n = h_n * x_n := \sum_{k=-\infty}^{\infty} h_{n-k} x_k = \sum_{k=-\infty}^{\infty} x_{n-k} h_k \qquad [1.1]$$

This equation is based on the *shift* of the signal or impulse response. In LTI systems, we (implicitly) assume that the shift of a discrete-time signal is well defined, i.e. x_{n-k} is unique and time invariant. Therefore, equation [1.1] is equivalently represented as

$$\mathbf{y} = \begin{bmatrix} \ddots & & & & & \\ \dots & h_{-1} & h_0 & h_1 & \dots & \\ & \dots & h_{-1} & h_0 & h_1 & \dots \\ & & & & & \ddots \end{bmatrix} \mathbf{x}, \qquad [1.2]$$

where $\mathbf{x} := [\dots, x_{-1}, x_0, x_1, \dots]^\top$ and $\mathbf{y} := [\dots, y_{-1}, y_0, y_1, \dots]^\top$.[1]

In equation [1.2], the impulse response h_k is invariant for n, i.e. the same filter is used for different values of n. Instead, we can use different filters for different values of n to yield y_n, whose impulse response $h_k[n]$ is often defined in a signal-dependent manner, i.e. $h_k[n] \neq h_k[m]$ for $m \neq n$. It is formulated as

$$y_n := \sum_{k=-\infty}^{\infty} h_{n-k}[n] x_k \qquad [1.3]$$

1 Here, we assume both $\mathbf{x}$ and $\mathbf{y}$ are finite length signals and their boundaries are extended or filtered by a boundary filter to ensure that the equation is valid.

and its matrix form representation is

$$\mathbf{y} = \begin{bmatrix} \ddots & & & & & \\ \dots & h_{-1}[n-1] & h_0[n-1] & h_1[n-1] & \dots & \\ & \dots & h_{-1}[n] & h_0[n] & h_1[n] & \dots \\ & & & & & \ddots \end{bmatrix} \mathbf{x}. \qquad [1.4]$$

Famous image processing filters in this category include the bilateral filter (Tomasi and Manduchi 1998; Barash 2002; Durand and Dorsey 2002; Fleishman *et al.* 2003), anisotropic diffusion (Weickert 1998; Desbrun *et al.* 1999), adaptive directional wavelets (Chang and Girod 2007; Ding *et al.* 2007; Tanaka *et al.* 2010) and their variants.

It is well known that convolution in the time domain equation [1.1] has an equivalent expression in the frequency (i.e. Fourier) domain as follows:

$$\hat{y}(\omega) = \hat{h}(\omega)\hat{x}(\omega), \qquad [1.5]$$

where

$$\hat{x}(\omega) := \sum_{n=-\infty}^{\infty} x_n e^{-j\omega n}. \qquad [1.6]$$

Here, $\hat{x}(\omega)$ is the discrete-time Fourier transform (DTFT) of x_n. We utilize the fact that convolution in the time domain is identical to multiplication in the frequency domain. Note that the fixed filter has a corresponding fixed frequency response, and thus, we can intuitively understand the filter characteristics from the frequency response. In contrast, the frequency response of a signal-dependent filter is not always clear in general. Fortunately, this drawback can be partially solved with a graph spectral domain perspective, which is described further.

1.3. Filtering of graph signals

In this chapter, we consider *linear* graph filters. Readers can find nonlinear graph filters, like one used in deep learning, in the following chapters, specifically Chapter 10.

Let us denote a graph filter as $\mathbf{H} \in \mathbb{R}^{N \times N}$, where its elements are typically derived from $\mathcal{G}$ and $\mathbf{x}$. As in the LTI system, the filtered signal is represented as

$$\mathbf{y} = \mathbf{H}\mathbf{x}. \qquad [1.7]$$

The representation of its element y_n is similar to that observed in equation [1.3], i.e.

$$y_n := \sum_{k=0}^{N-1} [\mathbf{H}]_{n,k} x_k, \qquad [1.8]$$

where $[\cdot]_{n,k}$ is the n, k-element in the matrix. Similar to discrete-time signals, graph signal filtering may be defined in the vertex and graph frequency domains. These are described in the following.

1.3.1. *Vertex domain filtering*

Vertex domain filtering is an analog of filtering in the time domain. However, GSP systems are not shift-invariant: This means node indices do not have any physical meaning, in general. Therefore, the shift of graph signals based on the indices of nodes, similar to that used for discrete-time signals, would be inappropriate. Moreover, the underlying graph will exhibit a highly irregular connectivity, i.e. the degree in each node will vary significantly. For example, the star graph shown in Figure 1.1 has one center node and $N-1$ surrounding nodes. It is clear that $N-1$ edges are connected to the center node, i.e. the center node has degree $N-1$, whereas all of the surrounding nodes have degree 1. In an image processing perspective, such an irregularity comes from the edge and texture regions. An example is provided on the right side of Figure 1.1. Suppose that we construct a graph based on pixel intensity, i.e. pixels are nodes, and they are connected by edges with higher weights when their pixel values are closer. In this situation, pixels *along* edge/texture directions will be connected to each other strongly with a large degree, whereas those *across* edge/texture directions may have weaker edges, or may even be disconnected. Filtering based on such a graph will therefore reflect structures in the vertex (i.e. pixel) domain.

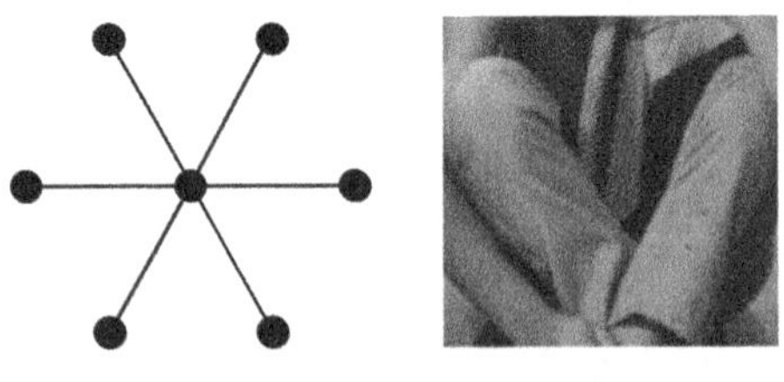

Figure 1.1. *Left: Star graph with $N = 7$. Right: Textured region of Barbara*

Vertex domain filtering can be defined more formally as follows. Let $\mathcal{N}_{n,p}$ be a set of p-hop neighborhood nodes of the nth node. Clearly $|\mathcal{N}_{n,p}|$ varies according to n.

Vertex domain filtering may be typically defined as a local linear combination of the neighborhood samples

$$y_n := \sum_{k \in \mathcal{N}_{n,p}} [\mathbf{H}]_{n,k} x_k. \qquad [1.9]$$

Since $\mathcal{N}_{n,p}$ varies according to n, $[\mathbf{H}]_{n,k}$ should be appropriately determined for all n. The matrix form of equation [1.9] may be represented as

$$\mathbf{y} = (\mathrm{diag}([\mathbf{H}]_{0,0}, \ldots, [\mathbf{H}]_{N-1,N-1}) + h(\mathbf{W}))\mathbf{x}, \qquad [1.10]$$

where $h(\mathbf{W})$ is a matrix containing filter coefficients $h[n,k]$ $(n \neq k)$ as a function of the adjacency matrix $\mathbf{W}$, in which $[h(\mathbf{W})]_{n,k} = 0$ if $k \notin \mathcal{N}_{n,p}$.

The vertex domain filtering in equations [1.9] and [1.10] requires the determination of $\sum_n |\mathcal{N}_{n,p}|$ filter coefficients, in general; moreover, it sometimes needs increased computational complexity. Typically, $[\mathbf{H}]_{n,k}$ may be parameterized in the following form:

$$\mathbf{y} = \left(\sum_{p=0}^{P-1} h_p \mathbf{W}_p \right) \mathbf{x}, \qquad [1.11]$$

where h_p is a real value and $\mathbf{W}_p \in \mathbb{R}^{N \times N}$ is a masked adjacency matrix that only contains p-hop neighborhood elements of $\mathbf{W}$. It is formulated as

$$[\mathbf{W}_p]_{n,k} = \begin{cases} [\mathbf{W}]_{n,k} & \text{if } k \in \mathcal{N}_{n,p}, \\ 0 & \text{otherwise.} \end{cases} \qquad [1.12]$$

The number of parameters required in equation [1.12] is P, which is significantly smaller than that required in equation [1.10].

One may find a similarity between the time domain filtering in equation [1.2] and the parameterized vertex domain filtering in equation [1.11]. In fact, if the underlying graph is a cycle graph, equation [1.11] coincides with equation [1.2] with a proper definition of $\mathbf{W}_p$. However, they do not coincide in general cases: It is easily confirmed that the sum of each row of the filter coefficient matrix in equation [1.11] is not constant due to the irregular nature of the graph, whereas $\sum_k h_k$ is a constant in time-domain filtering. Therefore, the parameters of equation [1.11] should be determined carefully.

1.3.2. *Spectral domain filtering*

The vertex domain filtering introduced above intuitively parallels time-domain filtering. However, it has a major drawback in a frequency perspective. As mentioned in section 1.2, time-domain filtering and frequency domain filtering are identical up to the DTFT. Unfortunately, in general, such a simple relationship does not hold in GSP. As a result, the naïve implementation of the vertex domain filtering equation [1.10] does not always have a diagonal response in the graph frequency domain. In other words, the filter coefficient matrix $\mathbf{H}$ is not always diagonalizable by the GFT matrix $\mathbf{U}$, i.e. $\mathbf{U}^\top\mathbf{HU}$ is not diagonal in general. Therefore, the graph frequency response of $\mathbf{H}$ is not always clear when filtering is performed in the vertex domain. This is a clear difference between the filtering of discrete-time signals and that of the graph signals.

From the above description, we can come up with another possibility for the filtering of graph signals: graph signal filtering defined in the graph frequency domain. This is an analog of filtering in the Fourier domain in equation [1.5]. This spectral domain definition of graph signal filtering has many desirable properties listed as follows:

– diagonal graph frequency response;

– fast computation;

– interpretability of pixel-dependent image filtering as graph spectral filtering.

These properties are described further.

As shown in equation [1.5], the convolution of h_n and x_n in the time domain is a multiplication of $\hat{h}(\omega)$ and $\hat{x}(\omega)$ in the Fourier domain. Filtering in the graph frequency domain utilizes such an analog to define generalized convolution (Shuman *et al.* 2016b):

$$\hat{y}_i := \hat{h}(\lambda_i)\hat{x}_i \qquad [1.13]$$

where $\hat{x}_i = \langle \mathbf{u}_i, \mathbf{x} \rangle$ is the ith GFT coefficient of $\mathbf{x}$ and the GFT basis $\mathbf{u}_i$ is given by the eigendecomposition of the chosen graph operator equation [I.2]. Furthermore, $\hat{h}(\lambda_i)$ is the graph frequency response of the graph filter. The filtered signal in the vertex domain, $y[n]$, can be easily obtained by transforming $\hat{\mathbf{y}}$ back to $y_n = \sum_{i=0}^{N-1} \hat{y}_i [\mathbf{u}_i]_n$, where $[\mathbf{u}_i]_n$ is the nth element of $\mathbf{u}_i$. This is equivalently written in the matrix form as

$$\begin{aligned}\mathbf{y} &= \mathbf{U}\begin{bmatrix}\hat{h}(\lambda_0) & & \\ & \ddots & \\ & & \hat{h}(\lambda_{N-1})\end{bmatrix}\mathbf{U}^\top\mathbf{x} \qquad [1.14]\\ &= \left(\sum_{i=0}^{N-1}\hat{h}(\lambda_i)\mathbf{P}_{\lambda_i}\right)\mathbf{x},\end{aligned}$$

where

$$\mathbf{P}_\lambda := \sum_{k \in \sigma(\lambda)} \mathbf{u}_k \mathbf{u}_k^\top \quad [1.15]$$

is a projection matrix in which $\sigma(\lambda)$ is a set of indices for repeated eigenvalues, i.e. a set of indices such that $\mathbf{L}\mathbf{u}_k = \lambda \mathbf{u}_k$.

For simplicity, let us assume that all eigenvalues are distinct. Under a given GFT basis $\mathbf{U}$, graph frequency domain filtering in equation [1.13] is realized by specifying N graph frequency responses in $\hat{h}(\lambda_i)$. Since this is a diagonal matrix, as shown in equation [1.14], its frequency characteristic becomes considerably clear in contrast to that observed in vertex domain filtering. Note that the naïve realization of equation [1.13] requires specific values of λ_i, i.e. graph frequency values. Therefore, the eigenvalues of the graph operator must be given prior to the filtering. Instead, in this case, we can parameterize a continuous spectral response $\hat{h}(\lambda)$ for the range $\lambda \in [\lambda_{\min}, \lambda_{\max}]$. This graph-independent design procedure has been widely implemented in many spectral graph filters, since the eigenvalues often vary significantly in different graphs.

For the classical Fourier domain filtering, it is enough to consider the frequency range $\omega \in [-\pi, \pi]$ (or an arbitrary 2π interval). However, graph frequency varies according to an underlying graph and/or the chosen graph operator. For example, symmetric normalized graph Laplacians have eigenvalues within $[0, 2]$, whereas combinatorial graph Laplacians do not have such a graph-independent maximum bound. The simple maximum bound of combinatorial graph Laplacian is, for example, given as (Anderson Jr and Morley 1985)

$$\lambda_{N-1} \leq \max\{d_u + d_v | (u, v) \in \mathcal{E}\}, \quad [1.16]$$

where d_u is the degree of the vertex u. Several other improvements on the bound are also found in literature. Although the graph Laplacians mentioned above have a bound of the largest eigenvalue, such bounds are not applicable to the adjacency matrix. Considering this, appropriate care of the graph frequency range must be taken while designing graph filters.

As mentioned, graph frequency domain filtering is an analog of Fourier domain filtering. However, this does not mean we always obtain a vertex domain expression of this similar to equation [1.9]. Hence, we need to compute the GFT of the input signal, which raises a computational issue described as follows. For the GFT, the eigenvector matrix $\mathbf{U}$ has to be calculated from the graph operator. The eigendecomposition requires $\mathcal{O}(N^3)$ complexity for a dense matrix[2]. This

2 While the computation cost for eigendecomposition of a sparse matrix is generally lower than $\mathcal{O}(N^3)$, it still requires a high computational complexity, especially for large graphs.

calculation often becomes increasingly complex, especially for big data applications, including image processing.

Typically, graph spectral image processing vectorizes image pixels. Let us assume that we have a grayscale image of size $W \times H$ pixels. Its vectorized version is $\mathbf{x} \in \mathbb{R}^{WH}$ and its corresponding graph operator would be $\mathbb{R}^{WH \times WH}$. For example, 4K ultra-high-definition resolution corresponds to $W = 3,840$ and $H = 2,160$, which leads to $WH > 8 \times 10^6$: this is too large to perform eigendecomposition, even for a recent high-spec computer. In section 1.6, several fast computation methods of graph spectral filtering will be discussed to alleviate this problem.

1.3.3. *Relationship between graph spectral filtering and classical filtering*

Filtering in the graph frequency domain seems to be an intuitive extension of Fourier domain filtering into the graph setting. In fact, it coincides with time-domain filtering in a special case, beyond the intuition.

Suppose that the underlying graph is a cycle graph with length N, and its graph Laplacian $\mathbf{L}_{\text{cycle}}$ is assumed as follows:

$$\mathbf{L}_{\text{cycle}} = \begin{bmatrix} 2 & -1 & & & -1 \\ -1 & 2 & -1 & & \\ & \ddots & \ddots & \ddots & \\ & & -1 & 2 & -1 \\ -1 & & & -1 & 2 \end{bmatrix}, \qquad [1.17]$$

where its blank elements are zero. It is well known that the eigenvector matrix of $\mathbf{L}_{\text{cycle}}$ is the DFT (Strang 1999), i.e.

$$\mathbf{L}_{\text{cycle}} = \mathbf{U}\boldsymbol{\Lambda}_{\text{cycle}}\mathbf{U}^* \qquad [1.18]$$

in which

$$\mathbf{u}_k = [1, w^k, w^{2k}, \dots, w^{(N-1)k}]^\top, \quad w = \exp(2\pi/N). \qquad [1.19]$$

In other words, when we consider a cycle graph and assume its associated graph Laplacian is $\mathbf{L}_{\text{cycle}}$, its GFT is the DFT. Therefore, graph spectral filtering in equation [1.13] is identical to the time-domain filtering. Note that, while $\mathbf{U}$ is the DFT, the interval of its eigenvalues is not equal to $2\pi k/N$. Specificallly, the kth eigenvalue of $\mathbf{L}_{\text{cycle}}$ is $\lambda_k = 2 - 2\cos(2\pi k/N)$.

1.4. Edge-preserving smoothing of images as graph spectral filters

This book (especially this chapter) focuses on graph spectral domain operations for image processing. Here, we describe interconnections between well-studied edge preserving filters and their GSP-based representations. As previously mentioned in this section, pixel-dependent filters do not have frequency domain expressions in a classical sense. This is because the impulse responses vary for different pixel index values n. In the following, we show that such a pixel-dependent filter can be viewed as a graph spectral filter, i.e. it presents a diagonal graph frequency response. Roughly speaking, GSP-based image processing considers the *pixel structure* and the *filter kernel* independently. Therefore, the pixel-dependent processing can be performed with a fixed filter kernel, owing to the underlying graph.

1.4.1. *Early works*

Let us begin with the history before the GSP era. In the mid-1990s, Taubin proposed seminal works on smoothing using graph spectral analysis for 3D mesh processing (Taubin 1995; Taubin *et al.* 1996)[3]. He determined the edge weights of polygon meshes using the Euclidean (geometric) distance between nodes. Assuming $\mathbf{p}_i \in \mathbb{R}^3$ as a 3-D coordinate of the ith node, the edge weight is then defined as

$$W_{i,j} = \eta \cdot \phi(\mathbf{p}_i, \mathbf{p}_j), \qquad [1.20]$$

where η is the normalizing factor and $\phi(\mathbf{p}_i, \mathbf{p}_j)$ is a non-negative function, which assigns a large weight if $\mathbf{p}_i$ and $\mathbf{p}_j$ are close. The typical choice of $\phi(\mathbf{p}_i, \mathbf{p}_j)$ will be $\|\mathbf{p}_i - \mathbf{p}_j\|^{-1}$.

The matrix $\mathbf{W}$ is symmetric. If we choose $\phi(\mathbf{p}_i, \mathbf{p}_i) = 0$, its diagonal elements would become zero, and as a result, $\mathbf{W}$ could be viewed as a normalized adjacency matrix. The coordinates are then smoothed by a graph low-pass filter, after computing the GFT basis $\mathbf{U}$. Similar approaches to this method have been used in several computer graphics/vision tasks (Zhang *et al.* 2010; Vallet and Lévy 2008; Desbrun *et al.* 1999; Fleishman *et al.* 2003; Kim and Rossignac 2005).

For image smoothing, filtering with a heat kernel represented in the graph frequency domain has also been proposed by Zhang and Hancock (2008). In this work, the edge weights of the pixel graph are computed according to photometric distance, i.e. large weights are assigned to the edges whose ends have similar pixel

3 The term "graph signal" was first introduced in Taubin *et al.* (1996), to the best of our knowledge.

values and vice versa. Additionally, the graph spectral filter is defined as a solution for the heat equation on the graph, and is expressed as follows:

$$\hat{h}(\lambda) = e^{-t\lambda}, \qquad [1.21]$$

where $t > 0$ is an arbitrary parameter that helps control the spreading speed caused by diffusion. Note that this method still needs eigendecomposition of the graph Laplacian if we decide to implement equation [1.21] naïvely. Instead, (Zhang and Hancock 2008) represent equation [1.21] using the Taylor series around the origin as follows:

$$e^{-t\lambda} = \sum_{k=0}^{\infty} \frac{t^k}{k!}(-\lambda)^k. \qquad [1.22]$$

By truncating the above equation with an arbitrary order K, we can approximate the heat kernel as a finite-order polynomial (Hammond *et al.* 2011; Shuman *et al.* 2013). In Zhang and Hancock (2008), the Krylov subspace method is used, along with equation [1.22] to approximate the graph filter. The polynomial method for graph spectral smoothing is detailed in section 1.6.5.

Figure 1.2 depicts the approximation error of the heat kernel using the Taylor series. Clearly, its approximation accuracy gets significantly worse when λ is away from 0. Since the maximum eigenvalue $\lambda_{\max}$ highly depends on the graph used, it is better to use different approximation methods like the Chebyshev approximation, which is introduced in section 1.6.

1.4.2. *Edge-preserving smoothing*

Edge-preserving image smoothing is widely used for various tasks, as well as for image restoration (Nagao and Matsuyama 1979; Pomalaza-Raez and McGillem 1984; Weickert 1998; Tomasi and Manduchi 1998; Barash 2002; Durand and Dorsey 2002; Farbman *et al.* 2008; Xu *et al.* 2011; He *et al.* 2013). Image restoration aims to approximate an unknown ground-truth image from its degraded version(s). In contrast, edge-preserving smoothing is typically used to yield a user-desired image from the original one. The resulting image is not necessarily close to the original one.

In the graph setting, we need to define pixel-wise or patch-wise relationships as a distance between pixels or patches, and it is used to construct a graph. The following distances are often considered (Milanfar 2013b), where i and j are pixel or patch indices and $\phi(\cdot)$ is some nonnegative function:

1) Geometric distance: $d_g(i,j) = \phi(\|\mathbf{p}_i - \mathbf{p}_j\|)$, where $\mathbf{p}_i$ is the ith pixel coordinate.

2) Photometric distance: $d_p(i,j) = \phi(\|[\mathbf{x}]_i - [\mathbf{x}]_j\|)$, where $[\mathbf{x}]_i$ is the pixel value (often three dimensional) of the ith pixel/patch.

3) Saliency distance: $d_s(i,j) = \phi(\|s_i - s_j\|)$, where s_i is the ith saliency value.

4) Combinations of the above.

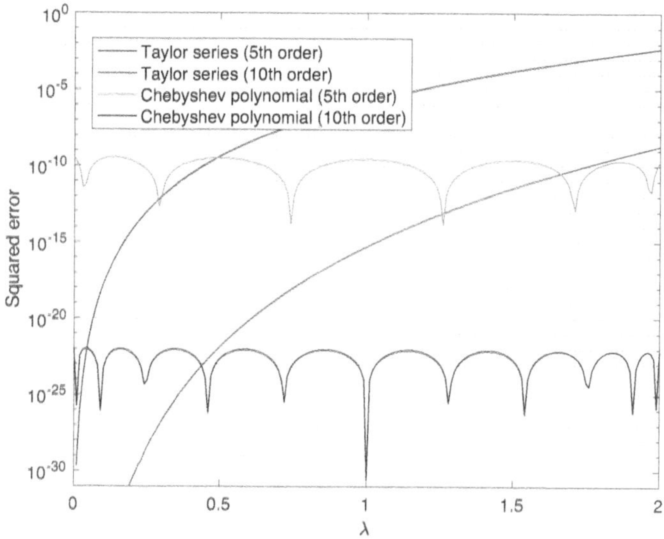

Figure 1.2. *Comparison of approximation errors in $\hat{h}(\lambda) = e^{-\lambda}$. For a color version of this figure, see www.iste.co.uk/cheung/graph.zip*

Saliency of the image/region/pixel is designed to simulate perceptual behavior (Itti *et al.* 1998; Harel *et al.* 2006). A popular choice of $\phi(\cdot)$ is the Gaussian weight

$$\phi(x) = \frac{1}{\sqrt{2\pi}\sigma} \exp\left(-\frac{x^2}{2\sigma^2}\right), \qquad [1.23]$$

where σ controls the spread of the filter kernel.

Suppose that the filter coefficients are determined based on the above features, and that they are symmetric, i.e. the output pixel value y_i is represented as

$$y_i := \frac{1}{D_i} \sum_j W_{i,j} x_j, \qquad [1.24]$$

where

$$W_{i,j} := \prod_{k=1}^{K} d_k(i,j). \qquad [1.25]$$

Here, $d_k(\cdot,\cdot)$ is one of the distance metrics mentioned earlier and K is the number of features we considered. The scaling factor D_i normalizes the filter weights as $D_i = \sum_j W_{i,j}$. For example, the bilateral filter is $K = 2$ for $d_g(\cdot,\cdot)$ and $d_p(\cdot,\cdot)$.

The Fourier domain representation of such pixel-dependent filters cannot be calculated in a classical sense because it is no longer shift-invariant: the filter matrix $\mathbf{W}$ cannot be diagonalized by the DFT matrix. In contrast, GSP provides a frequency-like notion in the graph frequency domain. In general, the weight matrix $\mathbf{W}$ in equation [1.24] can be regarded as an adjacency matrix because all $d_k(\cdot,\cdot)$ are assumed to be distances between pixels. Suppose that there is no self-loop in $\mathbf{W}$, for simplicity. In general, the smoothed image in equation [1.24] is represented in the following matrix form:

$$\mathbf{y} = \mathbf{D}^{-1}\mathbf{W}\mathbf{x} \qquad [1.26]$$

where $\mathbf{D} = \mathrm{diag}(D_0, \ldots, D_{N-1})$. This can be rewritten by using the relationship $\mathbf{L} = \mathbf{D} - \mathbf{W}$ as (Gadde *et al.* 2013):

$$\begin{aligned} \mathbf{y} &= \mathbf{D}^{-1}(\mathbf{D} - \mathbf{L})\mathbf{x} && [1.27] \\ &= \mathbf{D}^{-1/2}(\mathbf{I} - \mathbf{D}^{-1/2}\mathbf{L}\mathbf{D}^{-1/2})\mathbf{D}^{1/2}\mathbf{x} && [1.28] \\ &= \mathbf{D}^{-1/2}(\mathbf{I} - \mathbf{L}_n)\underline{\mathbf{x}} && [1.29] \end{aligned}$$

where $\underline{\mathbf{x}} := \mathbf{D}^{1/2}\mathbf{x}$ is a degree-normalized signal. Let us denote the eigendecomposition of $\mathbf{L}_n$ as $\mathbf{L}_n := \underline{\mathbf{U}\boldsymbol{\Lambda}\mathbf{U}}^\top$. The above filtering in equation [1.29] is further rewritten as:

$$\begin{aligned} \underline{\mathbf{y}} &= \underline{\mathbf{U}}(\mathbf{I} - \underline{\boldsymbol{\Lambda}})\underline{\mathbf{U}}^\top\underline{\mathbf{x}} && [1.30] \\ &= \underline{\mathbf{U}}\hat{h}(\underline{\boldsymbol{\Lambda}})\underline{\mathbf{U}}^\top\underline{\mathbf{x}} && [1.31] \\ &= h(\mathbf{L}_n)\underline{\mathbf{x}}, && [1.32] \end{aligned}$$

where $\underline{\mathbf{y}} := \mathbf{D}^{1/2}\mathbf{y}$ and the graph spectral filter is defined as $\hat{h}(\underline{\lambda}) := 1 - \underline{\lambda}$. Moreover, $\underline{\lambda} \in [0, 2]$ for the symmetric normalized graph Laplacian; therefore, it acts as a linear decay low-pass filter in the graph frequency domain.

This graph spectral representation of a pixel-dependent filter suggests that *the pixel-dependent filter* $\mathbf{W}$ *implicitly and simultaneously designs the underlying graph (and therefore, the GFT basis) and the spectral response of the graph filter*. In other words, the GSP expression of the pixel-dependent filter is free to design the spectral response $\hat{h}(\underline{\lambda})$, apart from the linear decay one, once we determine $\mathbf{W}$. For example, let us consider the following spectral response:

$$\hat{h}(\underline{\lambda}) = \frac{1}{1 + \eta\hat{h}_{\mathrm{HPF}}(\underline{\lambda})}, \qquad [1.33]$$

where $\hat{h}_{\text{HPF}}(\underline{\lambda})$ is an arbitrary graph high-pass filter and $\eta > 0$ is a parameter. In this case, $\hat{h}(\underline{\lambda})$ works as a graph low-pass filter and its spectral shape is controlled by $\hat{h}_{\text{HPF}}(\underline{\lambda})$. In fact, Gadde *et al.* (2013) show that equation [1.33] is the optimal solution for the following signal restoration problem:

$$\arg\min_{\mathbf{x}} \|\mathbf{z} - \mathbf{x}\|_2^2 + \eta\|\mathbf{H}_{\text{HPF}}\mathbf{x}\|_2^2, \tag{1.34}$$

where $\mathbf{z} = \mathbf{x} + \mathbf{n}$ with additive noise $\mathbf{n}$ and $\mathbf{H}_{\text{HPF}} = \mathbf{U}\hat{h}_{\text{HPF}}(\underline{\mathbf{\Lambda}})\mathbf{U}^\top$.

Image filtering sometimes needs numerous iterations to smooth out the details, in case of textured and/or noisy images. Therefore, to boost up the smoothing effect, the trilateral filter method (Choudhury and Tumblin 2003) first smooths the gradients of the image, and subsequently, the smoothed gradient is utilized to smooth the intensities. Its counterpart in the graph spectral domain is also proposed in Onuki *et al.* (2016) with the parameter optimization method for ρ in equation [1.33], which minimizes MSE after denoising it.

Figure 1.3. *Image denoising example using bilateral filters. From left to right: Original, noisy (PSNR: 20.02 dB), bilateral filter in the pixel domain (PSNR: 26.23 dB), and bilateral filter in the graph frequency domain (PSNR: 27.14 dB). Both bilateral filters use the same parameters*

Figure 1.3 depicts an example of image denoising by the bilateral filter in the graph frequency domain (Gadde *et al.* 2013). The image is degraded by additive white Gaussian noise. The bilateral filter in the graph frequency domain uses the spectral filter parameterized in equation [1.33], with $\hat{h}_{\text{HPF}} = \lambda$ and $\eta = 5$. It is clear that the graph spectral version efficiently removes noise while preserving image edges.

1.5. Multiple graph filters: graph filter banks

In the previous sections, we only considered the case where a single graph spectral filter was applied. Several image processing applications, such as compression and restoration, often require multiple filters that have different passbands (typically low-pass and high-pass). This signal processing system – so-called filter banks – is also important for GSP. In this section, the spectral domain design of graph filter banks is briefly introduced.

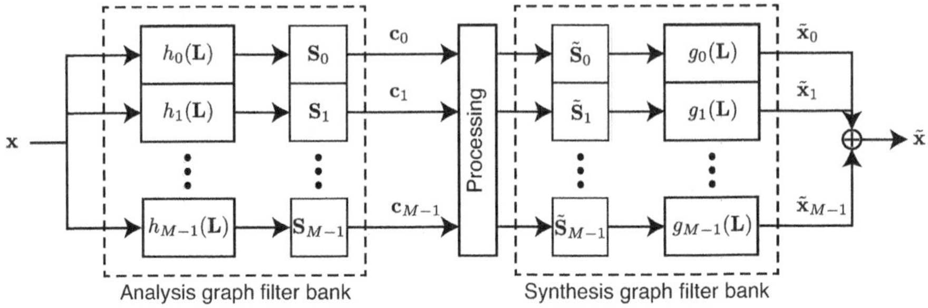

Figure 1.4. *Framework of graph filter bank*

1.5.1. *Framework*

A typical framework of a graph filter bank is illustrated in Figure 1.4. The *analysis transform* decomposes the input signal into some graph frequency components using a set of graph filters $\{h_k(\mathbf{L})\}$ $(k = 0, \dots, M-1)$. We assume that the graph operator is a graph Laplacian $\mathbf{L}$; however, in general, any graph operator can be applied. The decomposed coefficients (called *transformed coefficients*) are often downsampled by the sampling matrix $\mathbf{S}_k \in \mathbb{R}^{M_k \times N}$, where M_k is the sampling ratio, to reduce the number of coefficients. As a result, the transformed coefficients in each subband are represented as

$$\mathbf{c}_k = \mathbf{S}_k h_k(\mathbf{L})\mathbf{x}. \qquad [1.35]$$

The entire analysis transform is given as follows:

$$\begin{aligned}[\mathbf{c}_0^\top, \dots, \mathbf{c}_{M-1}^\top]^\top &:= \mathbf{E}\mathbf{x} \qquad [1.36]\\ &= \text{diag}(\mathbf{S}_0, \dots, \mathbf{S}_{M-1})[h_0^\top(\mathbf{L}), \dots, h_{M-1}^\top(\mathbf{L})]^\top \mathbf{x}.\end{aligned}$$

The size of $\mathbf{E}$ is $(\sum_k M_k) \times N$ and $\rho := (\sum_k M_k)/N$ is often called the *redundancy* of the transform. The redundancies of transforms are classified as follows:

– $\rho = 1$: *critically sampled* transform. The number of transformed coefficients is the same as N, i.e. the number of elements in $\mathbf{x}$.

– $\rho > 1$: *oversampled* transform. The number of transformed coefficients is larger than N.

– $\rho < 1$: *undersampled* transform. The number of transformed coefficients is smaller than N.

If $\mathbf{S}_k = \mathbf{I}_N$, i.e. no sampling is performed, $\rho = M$, and the transform is called an *undecimated* transform. In general, undersampled transforms will lose the information of the original signal. They cannot recover the original signal $\mathbf{x}$ from the transformed coefficients.

After the analysis transformation, an arbitrary linear and nonlinear operation is performed to $\mathbf{c}_k$ for a target application. For example, small magnitude elements in $\mathbf{c}_k$ are thresholded to denoise or compress the signal. Let us denote $\tilde{\mathbf{c}}_k$ as processed coefficients.

The *synthesis transform* combines $\tilde{\mathbf{c}}_k$ to reconstruct the signal. This is represented as

$$\tilde{\mathbf{x}} := \mathbf{R}[\tilde{\mathbf{c}}_0^\top, \ldots, \tilde{\mathbf{c}}_{M-1}^\top]^\top, \tag{1.37}$$

where $\mathbf{R} \in \mathbb{R}^{N\times(\sum_k M_k)}$ is the synthesis transform matrix. The *perfect reconstruction* transform is defined as the transform that recovers the original signal perfectly, when no processing is performed between the analysis and synthesis transforms. Formally, it satisfies the following condition:

$$\mathbf{R}\mathbf{E}\mathbf{x} = \mathbf{x}. \tag{1.38}$$

The details of perfect reconstruction graph filter banks are provided in the next section.

While $\mathbf{R}$ can be arbitrary, one may need a *symmetric* structure: the synthesis transform represented by multiple filters and upsampling as a counterpart of the analysis transform. In classical signal processing, most filter banks are designed to be symmetric, which, in contrast, is difficult for the graph versions, mainly due to the sampling operations. Several design methods make it possible to design perfect reconstruction graph transforms with a symmetric structure (Narang and Ortega 2012; Narang and Ortega 2013; Shuman *et al.* 2015; Leonardi and Van De Ville 2013; Tanaka and Sakiyama 2014; Sakiyama and Tanaka 2014; Sakiyama *et al.* 2016; Sakiyama *et al.* 2019a; Teke and Vaidyanathan 2016; Sakiyama *et al.* 2019b).

1.5.2. *Perfect reconstruction condition*

Suppose that the redundancy is $\rho \geq 1$ and the columns of $\mathbf{E}$ are linearly independent. The perfect reconstruction condition equation [1.38] is clearly rewritten as

$$\mathbf{R}\mathbf{E} = \mathbf{I}_N. \tag{1.39}$$

The critically sampled system constrains that $\mathbf{E}$ is a square matrix; therefore, $\mathbf{R}$ must be $\mathbf{E}^{-1}$ for perfect reconstruction. For the oversampled system, we generally have an infinite number of $\mathbf{R}$ satisfying the condition in equation [1.39]. The most simple and well-known solution is the least squares solution, which is expressed as

$$\mathbf{R} = (\mathbf{E}^\top \mathbf{E})^{-1}\mathbf{E}^\top \qquad [1.40]$$

This is nothing but the Moore–Penrose pseudo inverse of $\mathbf{E}$[4]. This GSP system is generally asymmetric: while the analysis transform has graph filters and possible sampling, the synthesis transform does not have such a clear notion of filtering and upsampling. In general, the asymmetric structure requires a matrix inversion. Additionally, the $N \times N$ matrix $\mathbf{E}^\top\mathbf{E}$ is usually dense, which leads to $\mathcal{O}(N^3)$ complexity.

Therefore, symmetric structures are often desired instead, and they are similar to those that are widely used in classical signal processing. The synthesis transform with a symmetric structure has the following form:

$$\mathbf{R} = [\mathbf{I}_N, \cdots, \mathbf{I}_N]\text{diag}(g_0(\mathbf{L}), \ldots, g_{M-1}(\mathbf{L}))\text{diag}(\tilde{\mathbf{S}}_0, \ldots, \tilde{\mathbf{S}}_{M-1}), \qquad [1.41]$$

where $g_k(\mathbf{L})$ is the kth synthesis filter and $\tilde{\mathbf{S}}_k$ is an upsampling matrix. As a result, each subband has the following input–output relationship:

$$\tilde{\mathbf{x}}_k = g_k(\mathbf{L})\tilde{\mathbf{S}}_k\mathbf{S}_k h_k(\mathbf{L})\mathbf{x}. \qquad [1.42]$$

The resulting output is therefore represented as $\sum_k \tilde{\mathbf{x}}_k$ and for perfect reconstruction, it must be $\mathbf{x}$.

1.5.2.1. *Design of perfect reconstruction transforms: undecimated case*

There are various methods available for designing perfect reconstruction graph transforms. First, let us consider undecimated transforms that exhibit symmetrical structure.

An undecimated transform has no sampling, i.e. $\mathbf{S}_k = \mathbf{I}_N$ for all k. Therefore, the analysis and synthesis transforms, respectively, are represented in the following simple forms:

$$\mathbf{E}_{\text{UD}} = [h_0^\top(\mathbf{L}), \ldots, h_{M-1}^\top(\mathbf{L})]^\top \qquad [1.43]$$

$$\mathbf{R}_{\text{UD}} = [g_0(\mathbf{L}), \ldots, g_{M-1}(\mathbf{L}))]. \qquad [1.44]$$

4 In fact, this $\mathbf{R}$ can also be used for the reconstruction of the undersampled systems.

Accordingly, the perfect reconstruction condition can also be simple. The input–output relationship in equation [1.42] is reduced to

$$\tilde{\mathbf{x}}_k = g_k(\mathbf{L})h_k(\mathbf{L})\mathbf{x}. \tag{1.45}$$

Assuming $p_k(\mathbf{L}) := g_k(\mathbf{L})h_k(\mathbf{L})$ as the kth product filter, the output signal is thus given by

$$\tilde{\mathbf{x}} = \sum_{k=0}^{M-1} p_k(\mathbf{L})\mathbf{x}. \tag{1.46}$$

Therefore, the product filters must satisfy the following condition for perfect reconstruction:

$$\sum_{k=0}^{M-1} p_k(\mathbf{L}) = c\mathbf{I}, \tag{1.47}$$

where c is some constant.

Suppose that $h_k(\mathbf{L})$ and $g_k(\mathbf{L})$ are parameterized as $h_k(\mathbf{L}) = \mathbf{U}\hat{h}_k(\mathbf{\Lambda})\mathbf{U}^\top$ and $g_k(\mathbf{L}) = \mathbf{U}\hat{g}_k(\mathbf{\Lambda})\mathbf{U}^\top$, respectively. In this case, equation [1.47] can be further reduced to

$$\sum_{k=0}^{M-1} \hat{p}_k(\lambda) = c \text{ for all } \lambda \in [\lambda_{\text{min}}, \lambda_{\text{max}}], \tag{1.48}$$

where $\hat{p}_k(\lambda) := \hat{g}_k(\lambda)\hat{h}_k(\lambda)$. This condition is similar to that considered in biorthogonal FIR filter banks in classical signal processing (Vaidyanathan 1993; Vetterli and Kovacevic 1995; Strang and Nguyen 1996). When $\hat{h}_k(\lambda) = \hat{g}_k(\lambda)$ and the filter set satisfies equation [1.48], the filter bank is called a *tight frame* because the perfect reconstruction condition can be rewritten as

$$\sum_{k=0}^{M-1} |\hat{h}_k(\lambda)|^2 = c. \tag{1.49}$$

If $c = 1$, the frame is called a *Parseval frame*. In this case, it conserves the energy of the original signal in the transformed domain. Tight spectral graph filter banks can be constructed by employing the design methods of tight frames in classical signal processing. Examples can be found in Leonardi and Van De Ville (2013); Shuman *et al.* (2015); Sakiyama *et al.* (2016).

1.5.2.2. *Design of perfect reconstruction transforms: decimated case*

Constructing perfect reconstruction graph transforms with sampling is much more difficult than the undecimated version. However, it is required as the storage cost can be increased tremendously for the undecimated versions, especially for signals on a very large graph. Though the general condition is given by equation [1.42], the challenges are designing and choosing the appropriate sampling operator $\mathbf{S}_k$ and the appropriate filters $h_k(\mathbf{L})$ and $g_k(\mathbf{L})$. The perfect reconstruction condition can be satisfied with proper sets of these.

Various methodologies have been proposed in literature with different strategies. The recent methods of such transforms are summarized in Sakiyama *et al.* (2019c). We have omitted these details because they are beyond the scope of this chapter. Instead, some design guidelines are listed as follows:

– Sampling operator: In GSP, two different definitions of sampling operators exist (Tanaka *et al.* 2020; Tanaka 2018; Tanaka and Eldar 2020). One is vertex domain sampling, which is an analog of time-domain sampling. The other is graph frequency domain sampling, which is a counterpart of the Fourier domain expression of sampling. Both are not interchangeable in general, and have their own advantages and disadvantages.

– Localized transform: As mentioned later in section 1.6.5, polynomial filters are localized in the vertex domain. Furthermore, if all filters are polynomials, the entire transform can be eigendecomposition free, and thus, its computation decreases significantly.

– Flexible design adapting various spectra: Different graphs have unique eigenvalue distributions, and different graph operators have unique characteristics. Additionally, we sometimes encounter repeated eigenvalues. A unified design strategy is required for representing various graph signals sparsely.

1.6. Fast computation

As we mentioned in the previous sections, a naïve implementation of spectral graph filters often requires a high computational complexity. Since we often need to process high-resolution images, reducing such a complexity would be a high priority. Moreover, spectral filtering is often iteratively employed for image restoration, like an internal algorithm for convex optimization problems. In this section, we describe a workaround to alleviate computational burden for graph spectral filtering.

1.6.1. *Subdivision*

Digital image processing has a long history, and the subdivision of images has been widely used for various image processing tasks. For example, JPEG and MPEG image/video compression standards still use block-based predictions and transforms, even in their most recent standards. Moreover, graph-based image processing also uses such a subdivision as preprocessing. It can also be combined with the following fast computation approaches.

A simple solution for image subdivision is the block-based approach. It divides the input image into an equal-sized subblocks (these blocks can be overlapped with an appropriate window function), after which the favorite image processing tasks can be performed. The advantage it provides is simplicity: we only have to consider the size of subblocks to make a trade-off between performance and complexity. Sizes of the consistent image regions vary significantly; as a result, a recursive subdivision, called quadtree decomposition, provides a good trade-off.

More complex image subdivisions are also possible by utilizing an image segmentation. Although these segmentated sub-images are not rectangular in general, we can directly perform graph-based image processing in such non-rectangular regions by using appropriate graphs.

1.6.2. *Downsampling*

Downsampling is another typical approach for reducing the computation cost, and this has been used everywhere in image processing applications. A challenge for graph-based image processing is to find a good low-resolution graph, which properly reflects the original pixel structure.

Reducing the size of a graph is called *graph reduction* or *graph coarsening*. It is divided into two phases:

1) Phase 1: reducing the number of nodes;

2) Phase 2: reconnecting edges for the downsampled pixels.

In image processing, Phase 1 is relatively straightforward. We can assume the original image pixels are nodes on a uniform grid. As in usual image processing, picking up every other node (when the image signal is downsampled by two) will be reasonable.

For more general graphs like those used in point cloud processing, we need to select the "best" set of nodes. This problem is called *sampling set selection*. Although this is beyond our scope in this chapter, please refer to (Tanaka *et al.* 2020; Sakiyama *et al.* 2019c) and references therein.

In contrast to Phase 1, Phase 2 is not very straightforward. If pixels in the original image/block are associated with a graph, the downsampled one should be reconnected by edges, such that the reduced-size graph reflects the original structure. In the general

GSP study, desiderata for the reduced-size graphs have been suggested in Shuman *et al.* (2016a) as follows:

1) the reduced-size graph has non-negative edge weights;

2) the connectivity of the original graph is preserved in the reduced-size graphs;

3) the spectrum of the reduced-size graph is representative of the original graph;

4) the reduced-size graph preserves the original structural properties;

5) if two nodes are connected in the original graph, they should have a similar edge weights in the reduced-size graph;

6) it is tractable in terms of implementation and computational complexity;

7) the reduced-size graph preserves the sparsity (i.e. the ratio between the number of nonzero edges and that of pixels) of the original one.

Existing reconnection methods do not always satisfy all of these simultaneously; however, they do exhibit some of these properties. The order of the desired properties depends on applications considered. Major approaches have been summarized in Shuman *et al.* (2016a).

1.6.3. *Precomputing GFT*

If an image or subblock has several typical patterns, i.e. graphs, precomputing GFT bases for these graphs may be a reasonable choice to decrease the computational burden. This is because when we use them off the shelf, the computation cost reduces significantly as a result of decreased $\mathcal{O}(N^3)$ complexity, with a sacrifice of the storage cost for the GFT matrices and the cost of searching the optimal precomputed GFT from a given image. This precomputing strategy is popular in standard image processing. For example, in modern image/video coding standards, some precomputed transforms, such as DCT and discrete sine transform (DST) with various sizes, are utilized to represent image blocks as sparsely as possible.

Some precomputing methods have been proposed by Hu *et al.* (2015) and Zhang and Liang (2017), and they are mainly used for image compression. As expected, the GFT yields sparse transformed coefficients for piecewise smooth images/blocks. For those without such piecewise regions, conventional transforms like the DCT and DST are basically included as a set of precomputed bases.

1.6.4. *Partial eigendecomposition*

To emphasize, the eigendecomposition of the graph operator will need $\mathcal{O}(N^3)$ complexity in general. In other words, we can reduce the complexity if we can assume graph signals on the underlying graph are bandlimited. Suppose that the signal is K-bandlimited, which is typically defined as

$$||\hat{\mathbf{x}}||_0 \leq K, \qquad [1.50]$$

where $||\cdot||_0$ represents the number of non-zero elements, i.e. ℓ_0 pseudo-norm. Here, without loss of generality, we can assume the first K GFT coefficients are non-zero:

$$\hat{x}_i = 0 \text{ for } i > K. \qquad [1.51]$$

With the GFT basis $\mathbf{U}$, it is equivalently represented as

$$\hat{\mathbf{x}} = \mathbf{U}^\top \mathbf{x} = [\underbrace{\times \quad \dots \quad \times}_{K} \underbrace{0 \quad \dots \quad 0}_{N-K}]^\top = [(\mathbf{U}_K^\top \mathbf{x})^\top, \mathbf{0}^\top]^\top, \qquad [1.52]$$

where $\times$ represents some possible non-zero elements and

$$\mathbf{U}_K := \begin{bmatrix} \mathbf{u}_0 & \dots & \mathbf{u}_{K-1} \end{bmatrix}. \qquad [1.53]$$

A partial eigendecomposition proposed in literature gives the following approximation of $\mathbf{L}$:

$$\tilde{\mathbf{L}} := \mathbf{U}_K \text{diag}(\lambda_0, \dots, \lambda_{K-1}) \mathbf{U}_K^\top. \qquad [1.54]$$

Evaluating $h(\tilde{\mathbf{L}})\mathbf{x}$ only requires K ($< N$) eigenvectors and eigenvalues, which is significantly less than that obtained using the full eigendecomposition. In general, its computational complexity will be $\mathcal{O}(KN^2)$.

1.6.5. *Polynomial approximation*

The previous subsection proposes that we can alleviate the heavy computational burden by assuming the bandlimitedness of the graph signal. However, this requires the assumption on the signal model prior to filtering, but the signal is *not* bandlimited in general.

In many application scenarios, we often only need the *evaluation* of $\mathbf{x}$ with a given (linear) matrix function $h(\mathbf{L})$. That is, the *eigenvalues and eigenvectors themselves are often unnecessary*. The polynomial approximation methods introduced here enable us to calculate an approximation of $\mathbf{y} = h(\mathbf{L})\mathbf{x}$ without the (partial) decomposition of the variation operator.

Another advantage of filtering using a polynomial filter function is the vertex localization. The local filtering could capture local variations of pixel values, which are generally preferable. In contrast, filtering in the graph frequency domain (equation [1.13]) is usually not localized in the vertex domain, because eigenvectors often have global support on the graph. Therefore, localizing graph filter response, both in the vertex and graph frequency domains, has been studied extensively (Shuman *et al.* 2013; Shuman *et al.* 2016b; Sakiyama *et al.* 2016). In fact, the localization of graph spectral filters can be controlled using polynomial filtering.

Polynomial graph filters are defined as follows:

$$h(\mathbf{L}) := \sum_{k=0}^{K} c_k \mathbf{L}^k, \qquad [1.55]$$

where c_k is the kth order coefficient of the polynomial. It is known that each row of $\mathbf{L}^k$ collects its k-hop neighborhood; therefore, equation [1.55] is exactly the K-hop localized in the vertex domain. Note that $\mathbf{L}^k$ can be represented as

$$\mathbf{L}^k = (\mathbf{U}\boldsymbol{\Lambda}\mathbf{U}^\top)^k = \mathbf{U}\boldsymbol{\Lambda}^k\mathbf{U}^\top = \mathbf{U}\begin{bmatrix} \lambda_0^k & & \\ & \ddots & \\ & & \lambda_{N-1}^k \end{bmatrix}\mathbf{U}^\top. \qquad [1.56]$$

Here, we utilized the orthogonality of $\mathbf{U}$. We can rewrite equation [1.55] by using equation [1.56] as:

$$h(\mathbf{L}) = \sum_{k=0}^{K} c_k \mathbf{L}^k = \mathbf{U}\begin{bmatrix} \sum_k c_k \lambda_0^k & & \\ & \ddots & \\ & & \sum_k c_k \lambda_{N-1}^k \end{bmatrix}\mathbf{U}^\top. \qquad [1.57]$$

Consequently, the polynomial graph filter has the following graph frequency response:

$$\hat{h}(\lambda) = c_0 + c_1\lambda + c_2\lambda^2 + \cdots + c_K\lambda^K = \sum_{k=0}^{K} c_k \lambda^k. \qquad [1.58]$$

Especially, the output signal in the vertex domain is given by

$$\mathbf{y} = h(\mathbf{L})\mathbf{x} = \left(\sum_{k=0}^{K} c_k \mathbf{L}^k\right)\mathbf{x}. \qquad [1.59]$$

This indicates that we do not need to compute specific eigenvalues and eigenvectors for just calculating $\mathbf{y}$. Specifically, we need to evaluate $\mathbf{L}\mathbf{x}, \mathbf{L}^2\mathbf{x}, \ldots, \mathbf{L}^K\mathbf{x}$. Calculating $\mathbf{L}\mathbf{z}$, where $\mathbf{z}$ is an arbitrary vector, requires $\mathcal{O}(|\mathcal{E}|)$

complexity. Additionally, $\mathcal{O}(N)$ is required for computing $c_k \mathbf{L}^k \mathbf{x}$ (and it is repeated K times). As a result, the entire complexity will be $\mathcal{O}(K(|\mathcal{E}| + N))$. It is usually much lower than the partial eigendecomposition. In general, $K \ll |\mathcal{E}|$; therefore, the number of edges is a dominant factor affecting the complexity.

Suppose that a fast computation is required for the spectral response of a graph filter $\hat{h}(\lambda)$, which is not a polynomial. Based on equation [1.59], we can approximate the output $\mathbf{y}$ if $\hat{h}(\lambda)$ is satisfactorily approximated by a polynomial.

Any polynomial approximation methods, e.g. Taylor expansion, are possible for the above-mentioned polynomial filtering. In GSP, Chebyshev polynomial approximation is implemented frequently. The Chebyshev expansion gives an approximate minimax polynomial, i.e. the maximum approximation error can be reduced.

The approximated version of $h(\mathbf{L})\mathbf{x}$ by the Kth order shifted Chebyshev polynomial, $h_{\text{Cheb}}(\mathbf{L})\mathbf{x}$, is given by Shuman *et al.* (2013); Hammond *et al.* (2011)

$$h_{\text{Cheb}}(\mathbf{L})\mathbf{x} = \left\{ \frac{1}{2} c_0 + \sum_{k=1}^{K} c_i \bar{T}_k(\mathbf{L}) \right\} \mathbf{x} \qquad [1.60]$$

and it has the recurrence property:

$$\bar{T}_k(\mathbf{L}) = \frac{4}{\lambda_{\max}} (\mathbf{L} - \mathbf{I}) \bar{T}_{k-1}(\mathbf{L}) - \bar{T}_{k-2}(\mathbf{L}) \qquad [1.61]$$

with $\bar{T}_0(\mathbf{L}) = \mathbf{I}$ and $\bar{T}_1(\mathbf{L}) = 2(\mathbf{L} - \mathbf{I})/\lambda_{\max}$. The kth Chebyshev coefficient c_k is defined as

$$c_k = \frac{2}{P} \sum_{p=1}^{P} \cos\left(\frac{k\left(p - \frac{1}{2}\right)\pi}{P} \right) \hat{h}\left(\frac{\lambda_{\max}}{2} \left(\cos\left(\frac{\left(p - \frac{1}{2}\right)\pi}{P} \right) + 1 \right) \right) \qquad [1.62]$$

for $k = 0, \ldots, K$, where P is the number of sampling points used to compute the Chebyshev coefficients and is usually set to $P = K + 1$. The approximated filter in equation [1.60] is clearly a Kth order polynomial of λ. As a result, it is K-hop localized in the vertex domain, as previously mentioned (Shuman *et al.* 2011; Hammond *et al.* 2011).

The approximation error for the Chebyshev polynomial has been well studied in the context of numerical computation (Vetterli *et al.* 2014; Phillips 2003):

THEOREM 1.1.– Let K be the polynomial degree of the Chebyshev polynomial and assume that $\hat{h}(\xi)$ has $(K+1)$ continuous derivatives on $[-1, 1]$. In this case, the upper bound of the error is given as follows:

$$|E_{\max,K}| \leq \frac{1}{2^K (K+1)!} \max \left| \frac{d^{(K+1)}}{d\xi^{(K+1)}} \hat{h}(\xi) \right|. \qquad [1.63]$$

1.6.6. *Krylov subspace method*

The Krylov subspace $\mathcal{K}_K$ of an arbitrary square matrix $\mathbf{W} \in \mathbb{R}^{N \times N}$ and a vector $\mathbf{x}$ is defined as follows:

$$\mathcal{K}_K := \text{span}\{\mathbf{x}, \mathbf{W}\mathbf{x}, \mathbf{W}^2\mathbf{x}, \ldots, \mathbf{W}^{K-1}\mathbf{x}\} \qquad [1.64]$$

The Krylov subspace method, in terms of GSP, refers to filtering, i.e. evaluating an arbitrary filtered response $h(\mathbf{W})\mathbf{x}$, realized in a Krylov subspace $K \ll N$. Many methods to evaluate $h(\mathbf{W})\mathbf{x}$ in a Krylov subspace have been proposed, mainly in computational linear algebra and numerical computation (Golub and Van Loan 1996). A famous approximation method is the Arnoldi approximation, which is given by

$$\mathbf{y} \sim \beta \mathbf{V}_K h(\mathbf{H}_K)[1, 0, \ldots, 0]^\top, \qquad [1.65]$$

where $h(\mathbf{H}_K)$ is evaluating $h(\cdot)$ for the upper Heisenberg matrix $\mathbf{H}_K$, which is obtained by using the Arnoldi process. Furthermore, $\mathbf{H}_K$ is expected to be much smaller than the original matrix; therefore, evaluating $h(\mathbf{H}_K)$ using full eigendecomposition will be feasible and light-weighted.

1.7. Conclusion

This chapter introduces the filtering of graph signals performed in the graph frequency domain. This is a key ingredient of graph spectral image processing presented in the following chapters. The design methods of efficient and fast graph filters and filter banks, along with *fast GFT* (such attempts can be found in Girault *et al.* (2018); Lu and Ortega (2019)), are still a vibrant area of GSP: the chosen graph filters directly affect the quality of processed images. This chapter only provided a brief overview of graph spectral filtering. Please refer to the references for more details.

1.8. References

Anderson Jr. W.N. and Morley, T.D. (1985). Eigenvalues of the Laplacian of a graph. *Linear and Multilinear Algebra*, 18(2), 141–145.

Barash, D. (2002). Fundamental relationship between bilateral filtering, adaptive smoothing, and the nonlinear diffusion equation. *IEEE Trans. Pattern Anal. Mach. Intell.*, 24(6), 844–847.

Chang, C.-L. and Girod, B. (2007). Direction-adaptive discrete wavelet transform for image compression. *IEEE Trans. Image Process.*, 16(5), 1289–1302.

Choudhury, P. and Tumblin, J. (2003). The trilateral filter for high contrast images and meshes. *Eurographics Rendering Symposium*, 186–196.

Desbrun, M., Meyer, M., Schröder, P., Barr, A.H. (1999). Implicit fairing of irregular meshes using diffusion and curvature flow. *Proceedings of the 26th Annual Conference on Computer Graphics and Interactive Techniques*, 317–324.

Ding, W., Wu, F., Wu, X., Li, S., Li, H. (2007). Adaptive directional lifting-based wavelet transform for image coding. *IEEE Trans. Image Process*, 16(2), 416–427.

Durand, F. and Dorsey, J. (2002). Fast bilateral filtering for the display of high-dynamic-range images. *ACM Transactions on Graphics (TOG)*, 21, 257–266.

Farbman, Z., Fattal, R., Lischinski, D., Szeliski, R. (2008). Edge-preserving decompositions for multi-scale tone and detail manipulation. *ACM Transactions on Graphics (TOG)*, 27, 67.

Fleishman, S., Drori, I., Cohen-Or, D. (2003). Bilateral mesh denoising. *ACM Transactions on Graphics (TOG)*, 22, 950–953.

Gadde, A., Narang, S.K., Ortega, A. (2013). Bilateral filter: Graph spectral interpretation and extensions. *IEEE International Conference on Image Processing*, 1222–1226.

Girault, B., Ortega, A., Narayanan, S. (2018). Irregularity-aware graph Fourier transforms. *IEEE Transactions on Signal Processing*, 66(21), 5746–5761.

Golub, G.H. and Van Loan, C.F. (1996). *Matrix Computations*, Johns Hopkins University Press, Maryland.

Hammond, D.K., Vandergheynst, P., Gribonval, R. (2011). Wavelets on graphs via spectral graph theory. *Applied and Computational Harmonic Analysis*, 30(2), 129–150.

Harel, J., Koch, C., Perona, P. (2006). Graph-based visual saliency. *Proceedings of the 19th International Conference on Neural Information Processing Systems*, 545–552.

He, K., Sun, J., Tang, X. (2013). Guided image filtering. *IEEE Transactions on Pattern Analysis and Machine Intelligence*, 35(6), 1397–1409.

Hu, W., Cheung, G., Ortega, A., Au, O.C. (2015). Multiresolution graph Fourier transform for compression of piecewise smooth images. *IEEE Transactions on Image Processing*, 24(1), 419–433.

Itti, L., Koch, C., Niebur, E. (1998). A model of saliency-based visual attention for rapid scene analysis. *IEEE Trans. Pattern Anal. Mach. Intell.*, 20(11), 1254–1259.

Kim, B.-M. and Rossignac, J. (2005). Geofilter: Geometric selection of mesh filter parameters. *Computer Graphics Forum*, 24, 295–302.

Leonardi, N. and Van De Ville, D. (2013). Tight wavelet frames on multislice graphs. *IEEE Trans. Signal Process.*, 16(13), 3357–3367.

Lu, K.-S. and Ortega, A. (2019). Fast graph Fourier transforms based on graph symmetry and bipartition. *IEEE Trans. Signal Process.*, 67(18), 4855–4869.

Milanfar, P. (2013b). A tour of modern image filtering. *IEEE Signal Processing Magazine*, 30(1), 106–128.

Nagao, M. and Matsuyama, T. (1979). Edge preserving smoothing. *Computer Graphics and Image Processing*, 9(4), 394–407.

Narang, S.K. and Ortega, A. (2012). Perfect reconstruction two-channel wavelet filter banks for graph structured data. *IEEE Trans. Signal Process*, 60(6), 2786–2799.

Narang, S.K. and Ortega, A. (2013). Compact support biorthogonal wavelet filterbanks for arbitrary undirected graphs. *IEEE Transactions on Signal Processing*, 61(19), 4673–4685.

Onuki, M., Ono, S., Yamagishi, M., Tanaka, Y. (2016). Graph signal denoising via trilateral filter on graph spectral domain. *IEEE Trans. Signal Inf. Process. Netw.*, 2(2), 137–148.

Phillips, G.M. (2003). *Interpolation and Approximation by Polynomials*. Springer, New York.

Pomalaza-Raez, C. and McGillem, C. (1984). An adaptative, nonlinear edge-preserving filter. *IEEE Trans. Acoust., Speech, Signal Process*, 32(3), 571–576.

Sakiyama, A. and Tanaka, Y. (2014). Oversampled graph Laplacian matrix for graph filter banks. *IEEE Trans. Signal Process*, 62(24), 6425–6437.

Sakiyama, A., Watanabe, K., Tanaka, Y. (2016). Spectral graph wavelets and filter banks with low approximation error. *IEEE Trans. Signal Inf. Process. Netw.*, 2(3), 230–245.

Sakiyama, A., Tanaka, Y., Tanaka, T., Ortega, A. (2019a). Eigendecomposition-free sampling set selection for graph signals. *IEEE Trans. Signal Process*, 67(10), 2679–2692.

Sakiyama, A., Watanabe, K., Tanaka, Y. (2019b). m-channel critically sampled spectral graph filter banks with symmetric structure. *IEEE Signal Processing Letters*, 26(5), 665–669.

Sakiyama, A., Watanabe, K., Tanaka, Y., Ortega, A. (2019c). Two-channel critically-sampled graph filter banks with spectral domain sampling. *IEEE Trans. Signal Process*, 67(6), 1447–1460.

Shuman, D.I., Vandergheynst, P., Frossard, P. (2011). Chebyshev polynomial approximation for distributed signal processing. *Proc. DCOSS'11*, 1–8.

Shuman, D.I., Narang, S.K., Frossard, P., Ortega, A., Vandergheynst, P. (2013). The emerging field of signal processing on graphs: Extending high-dimensional data analysis to networks and other irregular domains. *IEEE Signal Processing Magazine*, 30(3), 83–98.

Shuman, D.I., Wiesmeyr, C., Holighaus, N., Vandergheynst, P. (2015). Spectrum-adapted tight graph wavelet and vertex-frequency frames. *IEEE Trans. Signal Process*, 63(16), 4223–4235.

Shuman, D.I., Faraji, M.J., Vandergheynst, P. (2016a). A multiscale pyramid transform for graph signals. *IEEE Trans. Signal Process*, 64(8), 2119–2134.

Shuman, D.I., Ricaud, B., Vandergheynst, P. (2016b). Vertex-frequency analysis on graphs. *Applied and Computational Harmonic Analysis*, 40(2), 260–291.

Strang, G. (1999). The discrete cosine transform. *SIAM Rev.*, 41(1), 135–147.

Strang, G. and Nguyen, T.Q. (1996). *Wavelets and Filter Banks*. Wellesley-Cambridge, Massachusetts.

Tanaka, Y. (2018). Spectral domain sampling of graph signals. *IEEE Trans. Signal Process*, 66(14), 3752–3767.

Tanaka, Y. and Eldar, Y.C. (2020). Generalized sampling on graphs with subspace and smoothness priors. *IEEE Transactions on Signal Processing*, 68, 2272–2286.

Tanaka, Y. and Sakiyama, A. (2014). M-channel oversampled graph filter banks. *IEEE Trans. Image Process*, 62(14), 3578–3590.

Tanaka, Y., Hasegawa, M., Kato, S., Ikehara, M., Nguyen, T.Q. (2010). Adaptive directional wavelet transform based on directional prefiltering. *IEEE Trans. Image Process.*, 19(4), 934–945.

Tanaka, Y., Eldar, Y.C., Ortega, A., Cheung, G. (2020). Sampling signals on graphs: From theory to applications. *IEEE Signal Processing Magazine*, 37(6), 14–30.

Taubin, G. (1995). A signal processing approach to fair surface design. *Proc. SIGGRAPH'95*, 351–358.

Taubin, G., Zhang, T., Golub, G.H. (1996). Optimal surface smoothing as filter design. *Proc. ECCV'96*, 283–292.

Teke, O. and Vaidyanathan, P.P. (2016). Extending classical multirate signal processing theory to graphs – Part II: M-channel filter banks. *IEEE Trans. Image Process*, 65(2), 423–437.

Tomasi, C. and Manduchi, R. (1998), Bilateral filtering for gray and color images. *IEEE International Conference on Computer Vision*, 839–846.

Vaidyanathan, P.P. (1993). *Multirate Systems and Filter Banks*. Prentice Hall, New Jersey.

Vallet, B. and Lévy, B. (2008). Spectral geometry processing with manifold harmonics. *Computer Graphics Forum*, 27, 251–260.

Vetterli, M. and Kovacevic, J. (1995). *Wavelets and Subband Coding*. Prentice Hall, New Jersey.

Vetterli, M., Kovacevi, J., Goyal, V.K. (2014). *Foundations of Signal Processing*. Cambridge University Press, Cambridge.

Weickert, J. (1998). *Anisotropic Diffusion in Image Processing 1*, Teubner, Stuttgart.

Xu, L., Lu, C., Xu, Y., Jia, J. (2011). Image smoothing via $l0$ gradient minimization. *ACM Transactions on Graphics (TOG)*, 30, 174.

Zhang, F. and Hancock, E.R. (2008). Graph spectral image smoothing using the heat kernel. *Pattern Recognition*, 41(11), 3328–3342.

Zhang, D. and Liang, J. (2017). Graph-based transform for 2D piecewise smooth signals with random discontinuity locations. *IEEE Transactions on Image Processing*, 26(4), 1679–1693.

Zhang, H., Van Kaick, O., Dyer, R. (2010). Spectral mesh processing. *Computer Graphics Forum*, 29, 1865–1894.

2

Graph Learning

Xiaowen DONG[1], Dorina THANOU[2], Michael RABBAT[3], Pascal FROSSARD[2]

[1] *University of Oxford, UK*

[2] *École polytechnique fédérale de Lausanne, Switzerland*

[3] *Facebook, Montreal, Canada*

2.1. Introduction

Modern data analysis and processing tasks typically involve large sets of structured data, where the structure carries critical information about the nature of the data. Numerous examples of such data sets can be found in a wide diversity of application domains, including transportation networks, social networks, computer networks and brain networks. An image, which consists of a regular array of pixels, is also a special form of structured data. Typically, graphs are used as mathematical tools to describe the underlying data structure, as they provide a flexible way of representing relationships between data entities. Numerous signal processing and machine learning algorithms have been introduced in the past decade for analyzing structured data on *a priori* known graphs (Zhu 2005; Fortunato 2010). However, there are often settings where the graph is not readily available, and the structure of the data has to be estimated in order to permit effective representation, processing, analysis or visualization of graph data. Furthermore, the pairwise relationships between data entities encoded in the format of graphs are often the goal of analysis. In both cases, a crucial task is to infer a graph topology that describes the characteristics of the data observations, hence capturing the underlying relationships between these entities. For example, an area of significant interest in neuroscience is to infer, from the blood-oxygen-level-dependent (BOLD) signals measured in

Graph Spectral Image Processing,
coordinated by Gene CHEUNG and Enrico MAGLI.

different regions of the brain, the functional connectivity between these regions, which may help to reveal the underpinnings of some neurodegenerative diseases.

Formally, the problem of graph learning is as follows: given M observations on N variables or data entities, represented in a data matrix $\mathbf{X} \in \mathbb{R}^{N\times M}$, and given some prior knowledge (e.g. distribution, data model, etc.) about the data, we would like to build or infer the relationship or a distance metric between these variables that take the form of a graph $\mathcal{G}$. As a result, each column of the data matrix $\mathbf{X}$ becomes a graph signal defined on the node or vertex set of the estimated graph, and the observations can be represented as $\mathbf{X} = \mathcal{F}(\mathcal{G})$, where $\mathcal{F}$ represents a certain generative process or function on the graph. The graph learning problem is an important one because (1) a graph may capture the actual geometry of structured data, which is essential to efficient processing, analysis and visualization; (2) learning relationships between data entities benefits numerous application domains, such as understanding functional connectivity between brain regions or behavioral influence between a group of people; (3) the inferred graph can help in predicting data evolution in the future.

Generally speaking, learning graph topologies from observations is an ill-posed problem, and there are many ways of associating a topology with the observed data samples. A main challenge in this problem is to define a model for the generative process or function $\mathcal{F}$, so that it can capture the relationship between the observed data $\mathbf{X}$ and the learned graph topology $\mathcal{G}$. Historically, there have been two general approaches for learning graphs from data, one based on statistical models and one based on physically motivated models. From the statistical perspective, $\mathcal{F}(\mathcal{G})$ is modeled as a function that draws a realization from a probability distribution over the variables, that is determined by the structure of $\mathcal{G}$. One prominent example is found in probabilistic graphical models (Koller and Friedman 2009), where the graph structure encodes conditional independence relationships among random variables that are represented by the vertices. Therefore, learning the graph structure is equivalent to learning a factorization of a *joint probability distribution* of these random variables. For physically motivated models, $\mathcal{F}(\mathcal{G})$ is defined based on the assumption of an underlying physical phenomenon or process on the graph. One popular process is *network diffusion or cascades* (Gomez-Rodriguez *et al.* 2010), where $\mathcal{F}(\mathcal{G})$ dictates the diffusion behavior on $\mathcal{G}$, which leads to the observation of $\mathbf{X}$, possibly at different time steps. In this case, the problem is equivalent to learning a graph structure on which the generative process of the observed signals may be explained.

The problem of graph learning presents both challenges and opportunities for the fast-growing field of graph signal processing (GSP) (Shuman *et al.* 2013; Sandryhaila and Moura 2013; Ortega *et al.* 2018). On one hand, the construction of a meaningful graph topology plays a crucial role in the effectiveness of GSP analysis and algorithms. On the other hand, GSP offers a new perspective on the problem of graph learning by utilizing signal processing tools. In the GSP setting, the columns of the observation matrix $\mathbf{X}$ are explicitly considered as signals that are defined on the vertex set of a weighted graph $\mathcal{G}$. The learning problem can then be cast to learn a graph $\mathcal{G}$ such that $\mathcal{F}(\mathcal{G})$ is permitted to make certain properties or characteristics of the observations $\mathbf{X}$ explicit, e.g. smoothness with respect to $\mathcal{G}$ or sparsity related to $\mathcal{G}$. This *signal representation* perspective is particularly interesting as it puts a strong and explicit emphasis on the relationship between the signal representation and the graph topology, where $\mathcal{F}(\mathcal{G})$ often comes with an interpretation of frequency-domain analysis or filtering operation of signals on the graph.

In addition to its general applications, graph learning has also received attention from the image processing community. As mentioned at the beginning of the chapter, an image can naturally be thought of as a signal defined on a regular grid graph, with, for example, four neighbor connections (edges) between the image pixels (nodes). The graph learning problem, in this case, may consist of finding the optimal weights of the edges such that the resulting graph-based transforms, having been adapted to the actual image structure, may lead to efficient transform coding of the image.

In this chapter, we first review well-established solutions to the problem of graph learning that adopt a statistical or physical perspective. Next, we survey a series of recent GSP-based approaches and show how signal processing tools and concepts can be utilized to provide novel solutions to the graph learning problem. Finally, we showcase the application of GSP-based graph learning methods in imaging problems and conclude with open questions and challenges in the design of future signal processing and machine learning algorithms, for learning graphs from data.

2.2. Literature review

Two general approaches to the problem of graph learning have been proposed in literature, one based on statistical models and the other based on physically motivated models. We provide a brief review of the two approaches in this section.

2.2.1. *Statistical models*

The general philosophy behind the statistical view is that a graph $\mathcal{G}$ exists, whose structure determines the joint probability distribution of the observations on the data entities, i.e. columns of the data matrix $\mathbf{X}$. In this case, the function $\mathcal{F}(\mathcal{G})$ in the problem formulation is one that draws a collection of realizations, i.e. the columns of $\mathbf{X}$, from the distribution governed by $\mathcal{G}$. Such models are known as probabilistic graphical models (Koller and Friedman 2009; Meinshausen and Bühlmann 2006; Banerjee *et al.* 2008; Friedman *et al.* 2008; Hsieh *et al.* 2011), where the edges (or lack thereof) in the graph encode conditional independence relationships among the random variables represented by the vertices.

In this section, a type of undirected graphical model is of particular interest, i.e. the so-called Gaussian graphical models or Gaussian Markov Random Fields (GMRFs). Suppose we have N random variables, i.e. $\{x_i\}_{i=1}^N$. In the case of a (zero-mean) GMRF, their joint probability can be written as follows:

$$p(\mathbf{x}|\boldsymbol{\Theta}) = \frac{|\boldsymbol{\Theta}|^{1/2}}{(2\pi)^{N/2}}\exp\big(-\frac{1}{2}\mathbf{x}^\top\boldsymbol{\Theta}\mathbf{x}\big), \qquad [2.1]$$

where $\mathbf{x} = [x_1, x_2, ..., x_N]^\top$, $\boldsymbol{\Theta} \in \mathbb{R}^{N\times N}$ is the inverse covariance or *precision* matrix and $|\cdot|$ represents the determinant operator. The matrix $\boldsymbol{\Theta}$ can be interpreted as corresponding to a graph structure, where the random variables $\{x_i\}_{i=1}^N$ are represented by the vertices, and x_i and x_j are conditionally independent, given the remaining variables if there is no edge between the corresponding vertices in the graph, i.e. $\theta_{ij} = 0$ where θ_{ij} is the ijth entry of $\boldsymbol{\Theta}$. Given their generality, GMRFs have been used to model data in a wide range of applications; in particular, they have been used as image models for transform coding (Zhang and Florencio 2013). In the context of graph learning, learning the structure boils down to learning the matrix $\boldsymbol{\Theta}$ that encodes pairwise conditional independence between the variables. Typical applications include inferring interactions between genes using gene expression profiles, or relationships between politicians given their voting behavior (Banerjee *et al.* 2008).

Early approaches for learning GMRFs include the work of Dempster (1972), where the author has proposed to learn $\boldsymbol{\Theta}$ by sequentially pruning the smallest elements in the inverse of the sample covariance matrix. More recently, *neighborhood selection* (Meinshausen and Bühlmann 2006) and *graphical Lasso* (Yuan and Lin 2007; Banerjee *et al.* 2008; Friedman *et al.* 2008) have been proposed to solve the problem. In the former, the problem of graph learning boils down to learning local connections (i.e. a neighborhood) for each vertex, by assuming that the observation at a particular vertex may be approximated by a sparse linear combination of the observations at other vertices. The overall graph structure (the locations of non-zero entries of $\boldsymbol{\Theta}$ in case of (Meinshausen and Bühlmann 2006)) is then determined by combining the individual neighborhoods. In the latter, graph learning is formulated as an optimization problem, whose objective is an explicit function of $\boldsymbol{\Theta}$:

$$\max_{\boldsymbol{\Theta}} \log|\boldsymbol{\Theta}| - \mathrm{tr}(\widehat{\boldsymbol{\Sigma}}\boldsymbol{\Theta}) - \rho||\boldsymbol{\Theta}||_1, \qquad [2.2]$$

where $\widehat{\boldsymbol{\Sigma}}$ is the sample covariance matrix, and $|\cdot|$ and $\mathrm{tr}(\cdot)$ represent the determinant and trace operators, respectively. The first two terms can together be interpreted as the log-likelihood under a GMRF, and the entry-wise ℓ_1-norm of $\boldsymbol{\Theta}$ is added to enforce sparsity of the connections with a regularization parameter ρ. The main difference between this approach and the neighborhood selection method (Meinshausen and Bühlmann 2006) is that the optimization in the latter is decoupled for each vertex, while the one in graphical Lasso is coupled, which can be essential for stability under noise. Other notable approaches inspired by or related to the graphical Lasso include the works in Hsieh *et al.* (2011); Cai *et al.* (2011).

Most previous approaches for learning GMRFs recover a precision matrix $\mathbf{\Theta}$ with both positive and negative entries. A positive off-diagonal entry in the precision matrix implies a negative partial correlation between the two random variables, which is difficult to interpret in some contexts, such as a social friendship or road traffic network. For such application scenarios, it is therefore desirable to learn a graph with non-negative weights. To this end, a few studies have considered learning a precision matrix from the family of so-called M-matrices (Poole and Boullion 1974), which are symmetric and positive definite matrices with non-positive off-diagonal entries, leading to the *attractive* GMRFs. One typical choice is the graph Laplacian matrix, which is a (singular) M-matrix that uniquely determines the adjacency matrix of the graph. As we shall see in the following section, this is similar to the setting that has been adopted in some GSP-based models for graph learning.

2.2.2. *Physically motivated models*

A second general approach tackles the graph learning problem by taking on a physically motivated perspective. In this case, the observations $\mathbf{X}$ are considered as outcomes of some physical phenomena on the graph, specified by the function $\mathcal{F}(\mathcal{G})$ in the problem formulation, and the learning problem consists of capturing the structure inherent to the physical generative process of the observed data. For example, the field of *network tomography* concerns methods for inferring properties of networks from indirect observations (Castro *et al.* 2004). These methods can be interpreted by choosing the function $\mathcal{F}(\mathcal{G})$ to measure network responses by exhaustively sending probes between all possible pairs of end-hosts. More recently, an increasing number of studies have focused on epidemic or *information propagation* models, where the physical process represents a disease spreading over a contact network, or a meme spreading over social media. These models have been applied to infer latent biological, social and financial networks based on observations of epidemics, memes or other signals diffusing over them (e.g. (Gomez-Rodriguez *et al.* 2010; Myers and Leskovec 2010; Gomez-Rodriguez *et al.* 2014; Du *et al.* 2012)).

Adopting the terminology of epidemiology, information propagation models are characterized by three main components: (a) the *nodes*, (b) an *infection process* (i.e. the change in the state of the node that is transferred by neighboring nodes in the network) and (c) the *causality* (i.e. the underlying graph structure, based on which the infection is propagated). Given a known graph structure, epidemic processes over graphs have been well-studied through popular models in which nodes may be susceptible, infected and possibly recovered (Pastor-Satorras *et al.* 2015). On the other hand, when the structure is not known beforehand, it may be inferred by considering the propagation of contagions over the edges of an unknown network. Usually, only the time steps when the nodes became infected are given.

A (fully observed) cascade may be represented by the sequence of triples $\{(v'_p, v_p, t_p)\}_{p=0}^{P}$, where $P \leq N$, representing that node v'_p infected its neighbor v_p at time t_p. In many applications, one may observe when a node becomes infected, but not which neighbor infected it. Then, the task is to recover a graph $\mathcal{G}$, given the (partial) observations $\{(v_p, t_p)\}_{p=0}^{P}$, usually for a number of such cascades. In this case, the set of nodes is given and the goal is to recover the edge structure. The common convention is to shift the infection times so that the initial infection in each cascade always occurs at time $t_0 = 0$. Equivalently, let $\mathbf{x}$ denote a length-N vector, where x_i is the time when v_i is infected, using the convention that $x_i = \infty$ if v_i is not infected in this cascade. The observations from M cascades can then be represented in a N-by-M matrix $\mathbf{X} = \mathcal{F}(\mathcal{G})$.

The methods for inferring networks from information cascades can generally be divided into two main categories, depending on whether they are based on homogeneous or heterogeneous models. Methods based on *homogeneous* models assume that cascades propagate in a statistically identical manner across all edges, such as in the work of (Myers and Leskovec 2010; Gomez-Rodriguez *et al.* 2010). In comparison, methods based on *heterogeneous* models relax these requirements and allow cascades to propagate at different rates across different edges, such as the NETRATE algorithm (Gomez-Rodriguez *et al.* 2014). For a more detailed description of these methods, please refer to Dong *et al.* (2019).

Many physically motivated approaches generally fall under the bigger umbrella of probabilistic inference of the network of diffusion or epidemic data. Notice, however, that despite its probabilistic nature, such inference is carried out with a specific model of the physical phenomenon in mind, instead of using a general probability distribution of the observations considered by statistical models in the previous section. In addition, for both methods in network tomography and those based on information propagation models, the recovered network typically only indicates the existence of edges and does not necessarily promote specific signal characteristics on the graph. As we shall see, this is a clear difference from the GSP models that are discussed in the following section.

2.3. Graph learning: a signal representation perspective

There is clearly a growing interest in the signal processing community to analyze signals that are supported on the vertex set of weighted graphs, leading to the fast-growing field of GSP (Shuman *et al.* 2013; Sandryhaila and Moura 2013). GSP enables the processing and analysis of signals that lie on structured but irregular domains by generalizing classical signal processing concepts, tools and methods, such as time-frequency analysis and filtering, on graphs (Shuman *et al.* 2013; Sandryhaila and Moura 2013; Ortega *et al.* 2018).

Consider a weighted graph $\mathcal{G}(\mathcal{V}, \mathcal{E}, \mathbf{W})$ with the vertex set $\mathcal{V}$ of cardinality N, edge set $\mathcal{E}$ and weighted adjacency matrix $\mathbf{W}$. As mentioned in section 1.3, when the graph is undirected, the combinatorial graph Laplacian matrix $\mathbf{L}$ admits a complete set of orthonormal eigenvectors with the associated eigenvalues via the eigencomposition:

$$\mathbf{L} = \mathbf{U}\mathbf{\Lambda}\mathbf{U}^\top, \qquad [2.3]$$

where $\mathbf{U}$ is the eigenvector matrix that contains the eigenvectors as columns, and $\mathbf{\Lambda}$ is the eigenvalue matrix $\mathbf{diag}(\lambda_0, \lambda_1, \cdots, \lambda_{N-1})$ that contains the eigenvalues along the diagonal. Conventionally, the eigenvalues are sorted in an increasing order, and for a connected graph we have: $0 = \lambda_0 < \lambda_1 \leq \cdots \leq \lambda_{N-1}$. As pointed out in section 1.3, the Laplacian matrix $\mathbf{L}$ enables a generalization of the notion of frequency and Fourier transform for graph signals (Hammond *et al.* 2011). Alternatively, a graph Fourier transform may also be defined using the adjacency matrix $\mathbf{W}$, and this definition can be used in directed graphs (Sandryhaila and Moura 2013). Furthermore, both $\mathbf{L}$ and $\mathbf{W}$ can be interpreted as a general class of shift operators on graphs (Sandryhaila and Moura 2013).

The above operators are used to represent and process signals on a graph in a similar way to traditional signal processing. To see this more clearly, consider two equations of central importance in signal processing: $\mathcal{D}\mathbf{c} = \mathbf{x}$ for the synthesis view and $\mathcal{A}\mathbf{x} = \mathbf{b}$ for the analysis view. In the synthesis view, the signal $\mathbf{x}$ is represented as a linear combination of atoms that are columns of a representation matrix $\mathcal{D}$, with $\mathbf{c}$ being the coefficient vector. In the context of GSP, a graph signal is defined as a function $\mathbf{x} : \mathcal{V} \to \mathbb{R}^N$ that assigns a scalar value to each vertex of a graph $\mathcal{G}$. Therefore, the representation $\mathcal{D}$ of the graph signal $\mathbf{x}$ is realized via $\mathcal{F}(\mathcal{G})$, i.e. a function of $\mathcal{G}$. In the analysis view of GSP, given $\mathcal{G}$ and $\mathbf{x}$ and with a design for $\mathcal{F}$ (that defines $\mathcal{A}$), we study the characteristics of $\mathbf{x}$ encoded in $\mathbf{b}$. Examples include the generalization of the Fourier and wavelet transforms for graph signals (Hammond *et al.* 2011; Sandryhaila and Moura 2013), which are defined based on mathematical properties of a given graph $\mathcal{G}$, or graph dictionaries, which can be trained by taking information from both $\mathcal{G}$ and $\mathbf{x}$ into account (Zhang *et al.* 2012; Thanou *et al.* 2014).

Although most GSP approaches focus on developing techniques for analyzing signals on a predefined or known graph, there is a growing interest in addressing the problem of learning graph topologies from observed signals, especially when the topology is not readily available (i.e. not pre-defined given the application domain). This offers a new perspective to the problem of graph learning, by especially focusing on the representation of the observed signals on the learned graph. Indeed, this corresponds to a synthesis view of the signal processing model: given $\mathbf{x}$, with some designs for $\mathcal{F}$ and $\mathbf{c}$, we would like to infer $\mathcal{G}$. Therefore, a model that captures the relationship between the signal representation and the graph is of crucial importance, which, together with graph operators, such as the adjacency/Laplacian matrices or the graph shift operators (Sandryhaila and Moura 2013), contributes to specific designs for $\mathcal{F}$. Moreover, assumptions on the structure or properties of $\mathbf{c}$ also play an important role in determining the characteristics of the resulting signal representation. Therefore, graph learning frameworks that are developed from a signal representation perspective have the unique advantage of enforcing certain desirable representations of the observed signals. More specifically, compared to the statistical and physics perspectives described in the previous section, it introduces one more important ingredient that can be used as a regularizer for complicated inference problems: the frequency or spectral representation of the observations. In what follows, we will review three models for signal representation on graphs, which lead to various methodologies for inferring graph topologies from the observed signals.

2.3.1. *Models based on signal smoothness*

The first model we consider is a smoothness model, under which the signal takes similar values at neighboring vertices. Practical examples of this model could be temperature observed at different locations in a flat geographical region, or ratings on movies by individuals in a social network. The measure of smoothness of a signal $\mathbf{x}$ on the graph $\mathcal{G}$ is usually defined by the graph Laplacian regularizer (GLR) introduced in section 1.5:

$$\mathcal{Q}_{\mathbf{x}}(\mathbf{L}) = \mathbf{x}^{\top}\mathbf{L}\mathbf{x} = \frac{1}{2}\sum_{i,j} w_{ij}[\mathbf{x}(i) - \mathbf{x}(j)]^2, \qquad [2.4]$$

where w_{ij} is the ijth entry of the adjacency matrix $\mathbf{W}$ and $\mathbf{L}$ is the Laplacian matrix. Clearly, $\mathcal{Q}_{\mathbf{x}}(\mathbf{L}) = 0$ when $\mathbf{x}$ is a constant signal over the graph (i.e. a DC signal with no variation). More generally, we can see that given the same ℓ_2-norm, the smaller the value $\mathcal{Q}_{\mathbf{x}}(\mathbf{L})$, the more similar the signal values at neighboring vertices (i.e. the lower the variation of $\mathbf{x}$ is with respect to $\mathcal{G}$). One natural criterion is therefore to learn a graph (or equivalently its Laplacian matrix $\mathbf{L}$), such that the signal variation on the resulting graph, i.e. the GLR $\mathcal{Q}_{\mathbf{x}}(\mathbf{L})$, is small. As an example, for the same signal, learning a graph in Figure 2.1(a) leads to a smoother signal representation in terms of $\mathcal{Q}_{\mathbf{x}}(\mathbf{L})$, than by learning a graph in Figure 2.1(c). The criterion of minimizing $\mathcal{Q}_{\mathbf{x}}(\mathbf{L})$ or its variants with powers of $\mathbf{L}$ has been proposed in a number of existing approaches, such as the ones in Lake and Tenenbaum (2010); Daitch *et al.* (2009); Hu *et al.* (2015a).

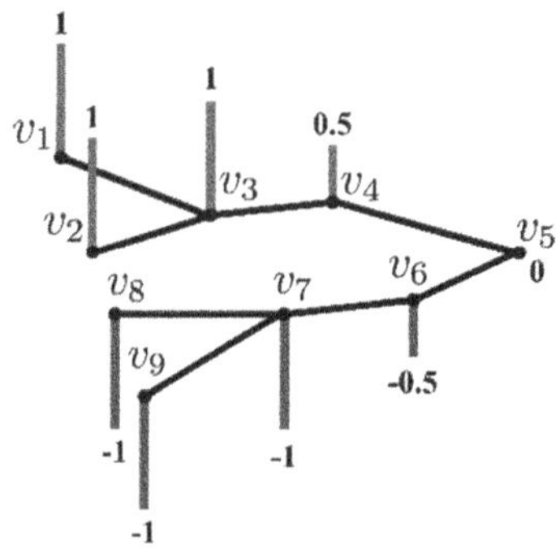

(a) A smooth signal on the graph with $\mathcal{Q}(\mathbf{L}) = 1$.

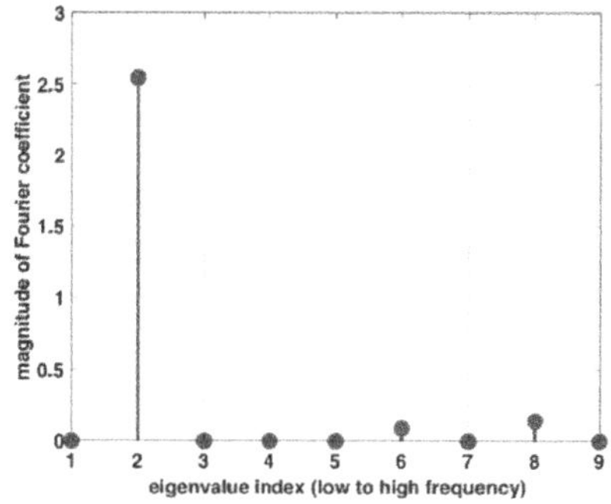

(b) The Fourier coefficients of the smooth signal in the graph spectral domain.

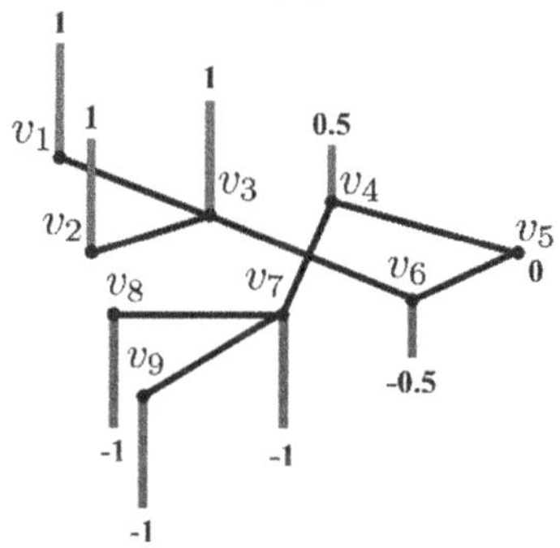

(c) A less smooth signal on the graph with $\mathcal{Q}(\mathbf{L}) = 5$.

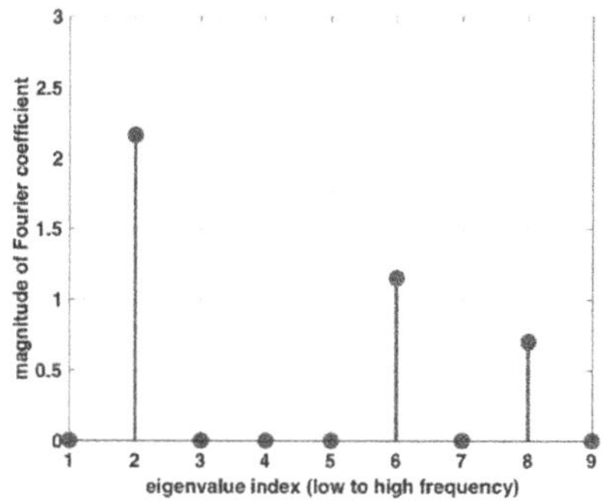

(d) The Fourier coefficients of the less smooth signal in the graph spectral domain.

Figure 2.1. *The different choices of graph lead to different representations of the same signal. The signal forms a smooth representation on the graph (a) as its values vary slowly along the edges of the graph, and it mainly consists of low-frequency components in the graph spectral domain (b). The representation of the same signal is less smooth on the graph (c), which consists of both low and high frequency components in the graph spectral domain (d). Figure from (Dong* et al. *2019) with permission. For a color version of this figure, see www.iste.co.uk/cheung/graph.zip*

A procedure to infer a graph that favors the smoothness of the graph signals, can be obtained using the synthesis model $\mathcal{F}(\mathcal{G})\mathbf{c} = \mathbf{x}$, and this is the idea behind the approach in Dong *et al.* (2016). Specifically, considering a factor analysis model with the choice of $\mathcal{F}(\mathcal{G}) = \mathbf{U}$, we have:

$$\mathbf{x} = \mathbf{U}\mathbf{c} + \epsilon, \quad [2.5]$$

where $\mathbf{U}$ is the eigenvector matrix of the Laplacian $\mathbf{L}$, and $\epsilon \sim \mathcal{N}(\mathbf{0}, \sigma_\epsilon^2\mathbf{I})$ is additive Gaussian noise. With a further assumption that $\mathbf{c}$ follows a Gaussian distribution with a precision matrix $\boldsymbol{\Lambda}$:

$$\mathbf{c} \sim \mathcal{N}(\mathbf{0}, \boldsymbol{\Lambda}^{\dagger}), \tag{2.6}$$

where $\boldsymbol{\Lambda}^{\dagger}$ is the Moore–Penrose pseudo-inverse of the eigenvalue matrix of $\mathbf{L}$, and $\mathbf{c}$ and $\boldsymbol{\epsilon}$ are statistically independent; it is shown in Dong *et al.* (2016) that the signal $\mathbf{x}$ follows a GMRF model:

$$\mathbf{x} \sim \mathcal{N}(\mathbf{0}, \mathbf{L}^{\dagger} + \sigma_{\epsilon}^{2}\mathbf{I}). \tag{2.7}$$

This leads to the formulation of a problem in which one jointly infers the graph Laplacian and the latent variable $\mathbf{c}$:

$$\min_{\mathbf{U},\boldsymbol{\Lambda},\mathbf{c}} \|\mathbf{x} - \mathbf{U}\mathbf{c}\|_2^2 + \alpha\, \mathbf{c}^{\top}\boldsymbol{\Lambda}\mathbf{c}, \tag{2.8}$$

where α is a non-negative regularization parameter related to the assumed noise level σ_{ϵ}^{2}. By making the change of variables $\mathbf{y} = \mathbf{U}\mathbf{c}$ and recalling that the matrix of Laplacian eigenvectors $\mathbf{U}$ is orthornormal, one arrives at the equivalent problem:

$$\min_{\mathbf{L},\mathbf{y}} \|\mathbf{x} - \mathbf{y}\|_2^2 + \alpha\, \mathbf{y}^{\top}\mathbf{L}\mathbf{y}, \tag{2.9}$$

in which the GLR appears. Therefore, these particular modeling choices for $\mathcal{F}$ and $\mathbf{c}$ lead to a procedure for inferring a graph, over which the observation $\mathbf{x}$ is smooth. Note that there is a one-to-one mapping between the Laplacian matrix $\mathbf{L}$ and a weighted undirected graph, so inferring $\mathbf{L}$ is equivalent to inferring $\mathcal{G}$.

By taking the matrix form of the observations and adding an ℓ_2 penalty, the authors of Dong *et al.* (2016) propose to solve the following optimization problem:

$$\begin{aligned} &\min_{\mathbf{L},\,\mathbf{Y}} \;\; ||\mathbf{X} - \mathbf{Y}||_F^2 + \alpha\, \mathrm{tr}(\mathbf{Y}^{\top}\mathbf{L}\mathbf{Y}) + \beta||\mathbf{L}||_F^2, \\ &\text{s.t.} \;\; \mathrm{tr}(\mathbf{L}) = N, \; \mathbf{L} \in \mathcal{L}, \end{aligned} \tag{2.10}$$

where $\mathrm{tr}(\cdot)$ and $||\cdot||_F$ represent the trace and Frobenius norm of a matrix, respectively, and α and β are non-negative regularization parameters. The trace constraint acts as a normalization factor that fixes the volume of the graph, and $\mathcal{L}$ is the set of valid Laplacian matrices. This constitutes the problem of finding $\mathbf{Y}$ that is close to the data observations $\mathbf{X}$, while at the same time ensuring that $\mathbf{Y}$ is smooth on the learned graph represented by its Laplacian matrix $\mathbf{L}$. The Frobenius norm of $\mathbf{L}$ is added to control the distribution of the edge weights and is inspired by the approach in Hu *et al.* (2015a). The problem is solved via alternating minimization in Dong *et al.* (2016), in which the step of solving for $\mathbf{L}$ bears similarity to the optimization in Hu *et al.* (2015a). In summary, the approach in Dong *et al.* (2016) emphasizes the characteristics of

GSP-based graph learning approaches, i.e. enforcing desirable signal representations through the learning process.

Following the idea of promoting signal smoothness, a more general graph learning approach has been proposed in Kalofolias (2016). By considering the rows of the data matrix $\mathbf{X}$ as feature vectors for the vertices, the authors first define a pairwise distance matrix $\mathbf{Z} \in \mathbb{R}^{N\times N}$, whose ijth entry is:

$$z_{ij} = ||\mathbf{X}_{i\cdot} - \mathbf{X}_{j\cdot}||_2^2, \tag{2.11}$$

where $\mathbf{X}_{i\cdot}$ and $\mathbf{X}_{j\cdot}$ represent the ith and jth row of $\mathbf{X}$, respectively. The smoothness measure of $\mathbf{X}$ can then be rewritten as:

$$\mathrm{tr}(\mathbf{X}^\top \mathbf{L}\mathbf{X}) = \frac{1}{2}\mathrm{tr}(\mathbf{W}\mathbf{Z}) = \frac{1}{2}\mathrm{tr}||\mathbf{W}\circ\mathbf{Z}||_1, \tag{2.12}$$

where $\circ$ represents the Hadamard product. The authors then propose to solve the following general problem:

$$\begin{aligned} \min_{\mathbf{W}} \quad & \frac{1}{2}\mathrm{tr}||\mathbf{W}\circ\mathbf{Z}||_1 + f(\mathbf{W}), \\ \text{s.t.} \quad & \mathbf{W} \in \mathcal{W}, \end{aligned} \tag{2.13}$$

where $f(\mathbf{W})$ is a regularization term that promotes certain structural properties of the graph, and $\mathcal{W}$ is the set of valid adjacency matrices for undirected graphs. It can easily be seen that the formulation of equation [2.10] is a special case (by dropping the data fidelity term) of equation [2.13]. The authors further propose the following form of $f(\mathbf{W})$:

$$f(\mathbf{W}) = -\alpha\mathbf{1}^\top \log(\mathbf{W}\mathbf{1}) + \frac{\beta}{2}||\mathbf{W}||_F^2, \tag{2.14}$$

where the logarithmic barrier on the node degree vector $\mathbf{W1}$ promotes the connectivity of the nodes, and the Frobenius norm helps control the sparsity of $\mathbf{W}$, by penalizing large (but not small) entries in $\mathbf{W}$. In addition to the different modeling of the structural properties of the graph, the formulation of the learning problem in terms of the adjacency matrix $\mathbf{W}$, rather than the Laplacian $\mathbf{L}$, also leads to more computationally efficient learning algorithms (Kalofolias 2016). In general, the family of graph learning problems in the form of equation [2.13] all promote signal smoothness via the GLR, hence, the representation of the graph signals is similar to that in Dong *et al.* (2016).

As we have seen, the smoothness property of the graph signal is associated with a multivariate Gaussian distribution, which also underlies the idea of classical approaches for learning graphical models, such as the graphical Lasso. Following the same design for $\mathcal{F}$ and slightly different ones for $\mathbf{\Lambda}$, compared to Dong *et al.* (2016), the authors of Egilmez *et al.* (2017) have proposed to solve a similar objective compared to the graphical Lasso, but with the constraints that the solutions correspond to different types of graph Laplacian matrices (e.g. the combinatorial or generalized Laplacian). The basic idea in the latter approach is to identify GMRF models such that the precision matrix has the form of a graph Laplacian. Their work generalizes the classical graphical Lasso formulation and the formulation proposed in Lake and Tenenbaum (2010), to precision matrices restricted to have a Laplacian form. From a probabilistic perspective, the problems of interest correspond to a maximum *a posteriori* (MAP) parameter estimation of GMRF models, whose precision matrix is a graph Laplacian. In addition, the proposed approach allows for the incorporation of prior knowledge on graph connectivity, which, if applicable, can help improve the performance of the graph inference algorithm.

It is also worth mentioning that the approaches in Dong *et al.* (2016); Kalofolias (2016); Egilmez *et al.* (2017) learn a graph topology without any explicit constraint on the number of edges in the learned graph. This information, if available, can be incorporated in the learning process. For example, the work of Chepuri *et al.* (2017) has proposed to learn a graph with a desired number of edges by selecting the ones that lead to the smallest $\mathcal{Q}_{\mathbf{x}}(\mathbf{L})$.

To summarize, in the global smoothness model, the objective of minimizing the original, or a variant of the GLR $\mathcal{Q}_{\mathbf{x}}(\mathbf{L})$, can be interpreted as having $\mathcal{F}(\mathcal{G}) = \mathbf{U}$ and $\mathbf{c}$ following a multivariate Gaussian distribution. However, different learning algorithms may differ in both the output of the algorithm and the computational complexity. For instance, the approaches in Kalofolias (2016); Chepuri *et al.* (2017) learn an adjacency matrix, while the approaches in Dong *et al.* (2016); Egilmez *et al.* (2017) learn a graph Laplacian matrix or its variants. In terms of complexity, the approaches in Dong *et al.* (2016); Kalofolias (2016); Egilmez *et al.* (2017) all solve a quadratic program (QP), with efficient implementations provided in the latter two, based on primal-dual techniques and block-coordinate descent algorithms, respectively. On the other hand, the method in Chepuri *et al.* (2017) involves a sorting algorithm that scales with the desired number of edges.

Finally, it is important to note that $\mathcal{Q}_{\mathbf{x}}(\mathbf{L})$ is a measure for *global* smoothness on $\mathcal{G}$ in the sense that a small $\mathcal{Q}_{\mathbf{x}}(\mathbf{L})$ implies a small variation of signal values along *all* of the edges in the graph, and the signal energy is mostly concentrated in the low-frequency components in the graph spectral domain. Although global smoothness is often a desirable property for signal representation, it can also be limiting in other scenarios. The second class of models that we introduce in the following section relaxes this constraint, by allowing a more flexible representation of the signal in terms of its spectral characteristics.

2.3.2. *Models based on spectral filtering of graph signals*

The second graph signal model that we consider goes beyond the global smoothness of the signal on the graph and focuses more on the general family of graph signals that are generated by applying a filtering operation to a latent (input) signal. In particular, the filtering operation may correspond to the diffusion of an input signal on the graph. Depending on the type of graph filter and the input signal, the generated signal can have different frequency characteristics (e.g. bandpass signals) and localization properties (e.g. locally smooth signals). Moreover, this family of algorithms is more appropriate than the one based on a globally smooth signal model for learning graph topologies, when the observations are the result of a diffusion process on a graph. Particularly, the graph diffusion model can be widely applied in real world scenarios to understand the distribution of heat (sources) (Chung 2007), such as the propagation of a heat wave in geographical spaces, the movement of people in buildings or vehicles in cities and the shift of people's interest towards certain subjects on social media platforms (Ma *et al.* 2008).

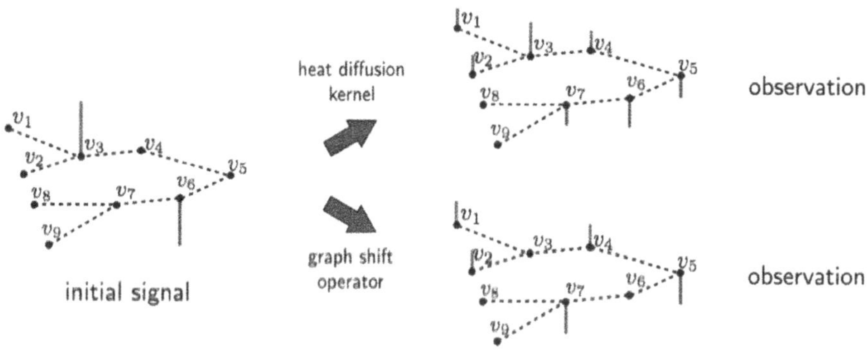

Figure 2.2. *Diffusion processes on the graph defined by a heat diffusion kernel (top right) and a graph shift operator (bottom right). Figure from (Dong* et al. *2019) with permission. For a color version of this figure, see www.iste.co.uk/cheung/graph.zip*

In this type of model, the graph filters and the input signals may be interpreted as the functions $\mathcal{F}(\mathcal{G})$ and the coefficients $\mathbf{c}$ in our synthesis model, respectively. The existing methods in literature therefore differ in the assumptions on $\mathcal{F}$, as well as the distribution of $\mathbf{c}$. In particular, $\mathcal{F}$ may be defined as an arbitrary (polynomial) function of a matrix related to the graph (Segarra *et al.* 2017a; Pasdeloup *et al.* 2018), or a well-known diffusion kernel such as the heat diffusion kernel (Thanou *et al.* 2017) (see Figure 2.2 for two examples). The assumptions on $\mathbf{c}$ can also vary, with the most prevalent ones being zero-mean Gaussian distribution and sparsity. Broadly speaking, we can distinguish the graph learning algorithms belonging to this family in two different categories. The first category models the graph signals as stationary processes on graphs, where the eigenvectors of a graph operator, such as the adjacency/Laplacian matrix or a shift operator, are estimated from the sample covariance matrix of the observations in the first step. The eigenvalues are then estimated in the second step to obtain the operator. The second category poses the graph learning problem as a dictionary learning problem with a prior on the coefficients $\mathbf{c}$. In what follows, we will give a few representative examples of both categories, which differ in terms of graph filters as well as input signal characteristics.

2.3.2.1. *Stationarity-based learning frameworks*

The main characteristic of this line of work is that, given a stationarity assumption, the eigenvectors of a graph operator are estimated by the empirical covariance matrix of the observations. In particular, the graph signal $\mathbf{x}$ can be generated from:

$$\mathbf{x} = \beta_0 \Pi_{k=1}^{\infty} (\mathbf{I} - \beta_k \mathbf{S}) \mathbf{c} = \sum_{k=0}^{\infty} \alpha_k \mathbf{S}^k \mathbf{c}, \qquad [2.15]$$

for some set of the parameters $\{\alpha\}$ and $\{\beta\}$. The latter implies that an underlying diffusion process exists in the graph operator $\mathbf{S}$, which can be the adjacency matrix, Laplacian, or a variation thereof, that produces the signal $\mathbf{x}$ from the input signal $\mathbf{c}$. By assuming a finite polynomial degree K, the generative signal model becomes:

$$\mathbf{x} = \mathcal{F}(\mathcal{G}) \mathbf{c} = \sum_{k=0}^{K} \alpha_k \mathbf{S}^k \mathbf{c}, \qquad [2.16]$$

where the connectivity matrix of $\mathcal{G}$ is captured through the graph operator $\mathbf{S}$. Usually, $\mathbf{c}$ is assumed to be a zero-mean graph signal with covariance matrix $\boldsymbol{\Sigma}_c = \mathbb{E}[\mathbf{c}\mathbf{c}^\top]$. In addition, if $\mathbf{c}$ is white and $\boldsymbol{\Sigma}_c = \mathbf{I}$, equation [2.15] is equivalent to assuming that the graph process $\mathbf{x}$ is *stationary* in $\mathbf{S}$. This assumption of stationarity is important in estimating the eigenvectors of the graph operator. Indeed, since the graph operator $\mathbf{S}$ is often a real and symmetric matrix, its eigenvectors are also eigenvectors of the covariance matrix $\boldsymbol{\Sigma}_x$. As a matter of fact:

$$\begin{aligned} \boldsymbol{\Sigma}_x = \mathbb{E}[\mathbf{x}\mathbf{x}^\top] &= \mathbb{E}\left[\sum_{k=0}^{K} \alpha_k \mathbf{S}^k \mathbf{c} \Big(\sum_{k=0}^{K} \alpha_k \mathbf{S}^k \mathbf{c}\Big)^\top\right] \\ &= \sum_{k=0}^{K} \alpha_k \mathbf{S}^k \Big(\sum_{k=0}^{K} \alpha_k \mathbf{S}^k\Big)^\top = \mathbf{U} \left(\sum_{k=0}^{K} \alpha_k \boldsymbol{\Lambda}^k\right)^2 \mathbf{U}^\top, \end{aligned} \qquad [2.17]$$

where we have used the assumption that $\boldsymbol{\Sigma}_c = \mathbf{I}$ and the eigendecomposition $\mathbf{S} = \mathbf{U}\boldsymbol{\Lambda}\mathbf{U}^\top$. Given a sufficient number of graph signals, the eigenvectors of the graph operator $\mathbf{S}$ can therefore be approximated by the eigenvectors of the empirical covariance matrix of the observations. To recover $\mathbf{S}$, the second step of the process would then be to learn its eigenvalues.

The authors in Pasdeloup *et al.* (2018) follow the aforementioned reasoning and model the diffusion process by powers of the normalized Laplacian matrix. More precisely, they propose an algorithm for characterizing and then computing a set of admissible diffusion matrices, which defines a polytope. In general, this polytope corresponds to a continuum of graphs that are all consistent with the observations. To obtain a particular solution, an additional criterion is required. Two such criteria are proposed: one that encourages the resulting graph to be sparse, and another that encourages the recovered graph to be *simple* (i.e. a graph in which no vertex has a connection to itself, hence an adjacency matrix with only zeros along the diagonal).

Similarly, in Segarra *et al.* (2017a), after obtaining the eigenvectors of a graph shift operator, the graph learning problem is equivalent to learning its eigenvalues, under the constraints that the shift operator obeys some desired properties, such as sparsity. The optimization problem of (Segarra *et al.* 2017a) can be written as:

$$\begin{aligned} &\min_{\mathbf{S},\, \boldsymbol{\Psi}} \ f(\mathbf{S}), \\ &\text{s.t.} \ \ \mathbf{S} = \mathbf{U}\boldsymbol{\Psi}\mathbf{U}^{\top}, \ \mathbf{S} \in \mathcal{S}, \end{aligned} \tag{2.18}$$

where $f(\cdot)$ is a convex function applied on $\mathbf{S}$ that imposes the desired properties of $\mathbf{S}$, e.g. sparsity via an entry-wise ℓ_1-norm, $\boldsymbol{\Psi}$ is the diagonal eigenvalue matrix of $\mathbf{S}$ and $\mathcal{S}$ is the constraint set of $\mathbf{S}$ being a valid graph operator, e.g. non-negativity of the edge weights. The stationarity assumption is further relaxed in Shafipour *et al.* (2018). However, all of these approaches are based on the assumption that the sample covariance of the observed data and the graph operator have the same set of eigenvectors. Thus, their performance depends on the accuracy of eigenvectors obtained from the sample covariance of data, which can be difficult to guarantee, especially when the number of data samples is small, relative to the number of vertices in the graph.

Given the limitation in estimating the eigenvectors of the graph operator from the sample covariance, the work of (Egilmez *et al.* 2019) has proposed a different approach. They have formulated the problem of graph learning as a graph system identification problem where, by assuming that the observed signals are the output of a system with a graph-based filter given a certain input, the goal is to learn a weighted graph (a graph Laplacian matrix) and the graph-based filter (a function of the graph Laplacian matrices). The algorithm is based on the minimization of a regularized maximum likelihood criterion, and it is valid under the assumption that the graph filters are one-to-one functions, i.e. increasing or decreasing in the space of eigenvalues, such as a heat diffusion kernel. More specifically, the system input is assumed to be multivariate white Gaussian noise (hence the stationarity assumption on the observed signals), and equation [2.17] is again used for computing an initial estimate of the eigenvectors. However, different from previous studies (Segarra *et al.* 2017a; Pasdeloup *et al.* 2018) where these eigenvectors are used directly in forming the graph operators, in Egilmez *et al.* (2019) they are used to compute the graph Laplacian: after initializing the filter parameter, the algorithm iterates until convergence between the following three steps: (a) pre-filter the sample covariance using the inverse of the graph filter; (b) estimate a graph Laplacian from the pre-filtered covariance matrix by solving a maximum likelihood optimization criterion, using an algorithm proposed in Egilmez *et al.* (2017); (c) update the filter parameter based on the current estimate of the graph Laplacian. Therefore, compared to Segarra *et al.* (2017a); Pasdeloup *et al.* (2018), this approach may lead to a more accurate inference of the graph operator (graph Laplacian in this case).

2.3.2.2. *Graph dictionary-based learning frameworks*

The methods that belong to this category are based on the notion of spectral graph dictionaries for efficient signal representation. Specifically, the authors in Thanou *et al.* (2017); Maretic *et al.* (2017) assume a different graph signal diffusion model, where the data consists of (sparse) combinations of overlapping local patterns that reside on the graph. These patterns may describe localized events or specific processes appearing at different vertices of the graph, such as traffic bottlenecks in transportation networks or rumor sources in social networks. The graph signals are then viewed as observations at different time instants of a few processes, which start at different nodes of an unknown graph and diffuse with time. Such signals can be represented as the combination of graph heat kernels or, more generally, of localized graph kernels. Both algorithms can be considered as a generalization of dictionary learning to graph signals. Dictionary learning (Rubinstein *et al.* 2010; Tosic and Frossard 2011) is an area of research in signal processing and machine learning where the signals are represented as a linear combination of simple components, i.e. atoms, in an (often) overcomplete basis. Signal decompositions with overcomplete dictionaries offer a way to efficiently approximate or process signals, such that the important characteristics are revealed by the sparse signal representation. Due to these desirable properties, dictionary learning has been extended to the representation of graph signals, and has eventually been applied to the problem of graph inference.

Next, we provide more details on one of the above-mentioned algorithms. The authors in Thanou *et al.* (2017) have focused on graph signals generated from heat diffusion processes, which are useful in identifying processes evolving near a starting seed node. An illustrative example of such a signal can be found in Figure 2.3, in which the graph Laplacian matrix is used to model the diffusion of heat throughout a graph. The concatenation of a set of heat diffusion operators at different time instances defines a graph dictionary that is further used to represent the graph signals. Hence, the graph signal model becomes:

$$\mathbf{x} = \mathcal{F}(\mathcal{G})\mathbf{c} = [e^{-\tau_1 \mathbf{L}} \; e^{-\tau_2 \mathbf{L}} \; \ldots \; e^{-\tau_S \mathbf{L}} \,]\mathbf{c} = \sum_{s=1}^{S} e^{-\tau_s \mathbf{L}} \mathbf{c}_s, \qquad [2.19]$$

which is a linear combination of different heat diffusion processes evolving on the graph. In this synthesis model, the coefficients $\mathbf{c}_s$ corresponding to a subdictionary $e^{-\tau_s \mathbf{L}}$ can be seen as a graph signal that goes through a heat diffusion process on the graph. The signal component $e^{-\tau_s \mathbf{L}}\mathbf{c}_s$ can then be interpreted as the result of this diffusion process at time τ_s. It is interesting to note that the parameter τ_s in the model carries a notion of scale. In particular, when τ_s is small, the ith column of $e^{-\tau_s \mathbf{L}}$, i.e. the atom centered at node v_i of the graph, is mainly localized in a small neighborhood of v_i. As τ_s becomes larger, it reflects information about the graph at a larger scale around v_i. Thus, the signal model can be seen as an additive model of the diffusion processes of S initial graph signals that undergo a diffusion model with different diffusion times.

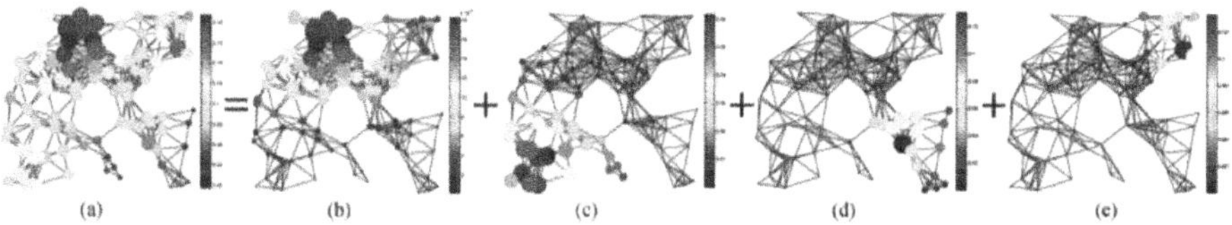

Figure 2.3. *(a) A graph signal. (b-e) Its decomposition in four localized simple components. Each component is a heat diffusion process ($e^{-\tau \mathbf{L}}$) at time τ that has started from different network nodes. The size and the color of each ball indicates the value of the signal at each vertex of the graph. Figure from Thanou* et al. *(2017) with permission. For a color version of this figure, see www.iste.co.uk/cheung/graph.zip*

An additional assumption on the above signal model is that the diffusion processes are expected to start from only a few nodes of the graph, at specific times, and spread over the entire graph over time[1]. This assumption can be formally captured by imposing a sparsity constraint on the latent variable $\mathbf{c}$. The graph learning problem can be cast as a structured dictionary learning problem, where the dictionary is defined by the unknown graph Laplacian matrix. The latter can then be estimated as a solution of the following optimization problem:

$$\min_{\mathbf{L},\, \mathbf{C},\, \tau} \quad \|\mathbf{X} - \mathbf{D}\mathbf{C}\|_F^2 + \alpha \textstyle\sum_{m=1}^{M} \|\mathbf{c}_m\|_1 + \beta \|\mathbf{L}\|_F^2,$$

1 When no locality assumptions are imposed (e.g. large τ_s) and a single diffusion kernel is used in the dictionary, the model reduces to a global smoothness model.

$$\text{s.t.} \quad \mathbf{D} = [e^{-\tau_1 \mathbf{L}} \; e^{-\tau_2 \mathbf{L}} \dots e^{-\tau_S \mathbf{L}}], \; \{\tau_s\}_{s=1}^{S} \geq 0, \qquad \text{tr}(\mathbf{L}) = N, \; \mathbf{L} \in \mathcal{L}, \tag{2.20}$$

where the constraint set $\mathcal{L}$ is a set of valid Laplacian matrices, as in equation [2.10]. Following the same reasoning, the work in Maretic *et al.* (2017) extends the heat diffusion dictionary to the more general family of polynomial graph kernels. Both approaches propose to recover the graph Laplacian matrix, by assuming that the graph signals can be sparsely represented by a dictionary that consists of graph diffusion kernels.

In summary, from the perspective of spectral filtering, and in particular network diffusion, the function $\mathcal{F}(\mathcal{G})$ is one that helps define a meaningful diffusion process on the graph via the graph Laplacian, heat diffusion kernel, or other more general graph shift operators. This directly leads to the slightly different output of the learning algorithms in various studies (Pasdeloup *et al.* 2018; Segarra *et al.* 2017a; Thanou *et al.* 2017). The choice of the coefficients $\mathbf{c}$, on the other hand, determines specific characteristics of the graph signals, such as stationarity or sparsity. In terms of computational complexity, the methods in Pasdeloup *et al.* (2018); Segarra *et al.* (2017a); Thanou *et al.* (2017) all involve the computation of eigenvectors, followed by solving a linear program (LP) and a semidefinite program (SDP), respectively.

2.3.3. *Models based on causal dependencies on graphs*

The models described in the previous two sections are mainly designed for learning undirected graphs that are also the predominant consideration in current GSP literature. Undirected graphs are associated with symmetric Laplacian matrices $\mathbf{L}$, which admit a complete set of orthonormal eigenvalues and eigenvectors that conveniently provide a notion of frequency for signals on graphs. However, it is often the case that in some application domains, learning directed graphs are more desirable, like in the cases where the directions of edges may be interpreted as causal dependencies between the variables that the vertices represent. For example, in brain analysis, even though the inference of an undirected *functional connectivity* between the regions of interest (ROIs) is certainly of interest, a directed *effective connectivity* may reveal extra information about the causal dependencies between those regions (Friston 1994; Shen *et al.* 2019). The third class of models that we discuss is therefore one that allows for the inference of such directed dependencies.

Mei and Moura (2017) have proposed a causal graph process based on the idea of sparse vector autoregressive (SVAR) estimation (Songsiri and Vandenberghe 2010; Bolstad *et al.* 2011). In their model, the signal at time step t, $\mathbf{x}[t]$, is represented as a linear combination of its observations in the past T time steps and a random noise process $\mathbf{n}[t]$:

$$\begin{aligned}\mathbf{x}[t] &= \mathbf{n}[t] + \sum_{j=1}^{\mathrm{T}} P_j(\mathbf{W})\mathbf{x}[t-j] \\ &= \mathbf{n}[t] + \sum_{j=1}^{\mathrm{T}} \sum_{k=0}^{j} a_{jk} \mathbf{W}^k \mathbf{x}[t-j],\end{aligned} \qquad [2.21]$$

where $P_j(\mathbf{W})$ is a degree j polynomial of the (possibly directed) adjacency matrix $\mathbf{W}$, with coefficients a_{jk} (see Figure 2.4 for an illustration). Clearly, this model admits the design of $\mathcal{F}(\mathcal{G}) = P_i(\mathbf{W})$ and $\mathbf{c} = \mathbf{x}[t-i]$ in forming a one time-lagged copy of the signal $\mathbf{x}[t]$. For temporal observations $\mathbf{X} = \big(\mathbf{x}[0]\ \mathbf{x}[1]\ \cdots\ \mathbf{x}[M-1]\big)$, the authors have therefore proposed to solve the following optimization problem:

$$\min_{\mathbf{W},\mathbf{a}} \frac{1}{2} \sum_{t=T}^{M-1} \left\| \mathbf{x}[t] - \sum_{j=1}^{\mathrm{T}} P_j(\mathbf{W})\mathbf{x}[t-j] \right\|_2^2 + \alpha\, ||\mathrm{vec}(\mathbf{W})||_1 + \beta\, ||\mathbf{a}||_1, \quad [2.22]$$

where $\mathrm{vec}(\mathbf{W})$ is the vectorized form of $\mathbf{W}$, $\mathbf{a} = \big(a_{10}\ a_{11}\ \cdots\ a_{jk}\ \cdots\ a_{TT}\big)$ is a vector of all the polynomial coefficients a_{jk}, and the entry-wise ℓ_1-norm is imposed on $\mathbf{W}$ and $\mathbf{a}$ for promoting sparsity. Due to non-convexity introduced by the matrix polynomials, the problem in equation [2.22] is solved in three steps, i.e. solving sequentially for $P_j(\mathbf{W})$ $\mathbf{W}$, and $\mathbf{a}$. In summary, in the SVAR model, the specific designs of $\mathcal{F}$ and $\mathbf{c}$ lead to a particular generative process of the observed signals on the learned graph. Similar ideas can also be found in the Granger causality or vector autoregressive models (VARMs) (Roebroeck *et al.* 2005; Goebel *et al.* 2003).

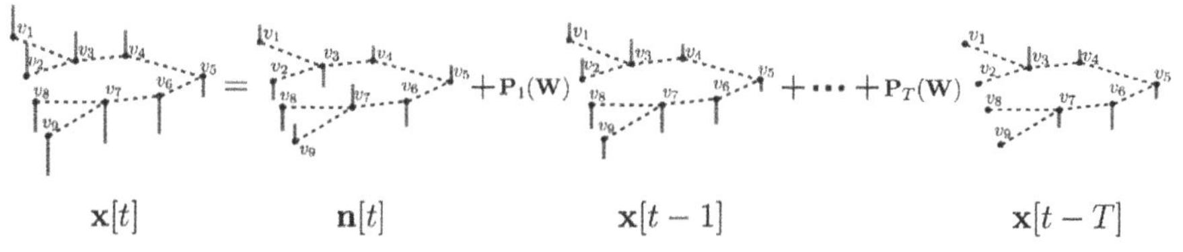

Figure 2.4. *A graph signal $\mathbf{x}$ at time step t is modeled as a linear combination of its observations in the past T time steps and a random noise process $\mathbf{n}[t]$. Figure from Dong* et al. *(2019) with permission. For a color version of this figure, see www.iste.co.uk/cheung/graph.zip*

Structural equation models (SEMs) are another popular approach for inferring directed graphs (Kaplan 2009; McLntosh and Gonzalez-Lima 1994). In the SEMs, the signal observation $\mathbf{x}$ at time step t is modeled as:

$$\mathbf{x}[t] = \mathbf{W}\mathbf{x}[t] + \mathbf{E}\mathbf{y}[t] + \mathbf{n}[t], \qquad [2.23]$$

where the first term in equation [2.23] consists of endogenous variables, which define the signal value at each variable as a linear combination of the values at its neighbors in the graph, and the second term represents exogenous variables $\mathbf{y}[t]$ with a coefficient matrix $\mathbf{E}$. The third term represents observation noise, which is similar to that in equation [2.21]. The endogenous component of the signal implies a choice of $\mathcal{F}(\mathcal{G}) = \mathbf{W}$ (which can again be directed) and $\mathbf{c} = \mathbf{x}[t]$ and, similar to the SVAR model, enforces a certain generative process of the signal on the learned graph.

As we can see, causal dependencies on the graph, either between different components of the signal or between its present and past observations, can be conveniently modeled in a straightforward manner by choosing $\mathcal{F}(\mathcal{G})$ as a polynomial of the adjacency matrix of a directed graph, and choosing the coefficients $\mathbf{c}$ as the present or past signal observations. As a result, the methods in Mei and Moura (2017); Baingana and Giannakis (2017); Shen *et al.* (2019) are all able to learn an asymmetric graph adjacency matrix, which is a potential advantage compared to methods based on the previous two models. Furthermore, the SEMs can be extended to track network topologies that evolve dynamically (Baingana and Giannakis 2017) and deal with highly correlated data (Traganitis *et al.* 2017), or combined with the SVAR model, which leads to structural vector autoregressive models (SVARMs) (Chen *et al.* 2011). Interested readers are referred to Giannakis *et al.* (2018) for a recent review of the related models. In these extensions of the classical models, the designs of $\mathcal{F}$ and $\mathbf{c}$ can be generalized accordingly to link the signal representation and the learned graph topology. Finally, as an overall comparison, the differences between the methods that are based on the three models discussed in this chapter are summarized in Table 2.1.

2.3.4. *Connections with the broader literature*

We have seen that GSP-based approaches can be unified by the viewpoint of learning graph topologies that enforce desirable representations of the signals on the learned graph. This offers a new interpretation of the traditional statistical and physically motivated models. First, as a typical example of approaches for learning graphical models, the graphical Lasso solves the optimization problem of equation [2.2], in which the trace term $\mathrm{tr}(\widehat{\mathbf{\Sigma}}\mathbf{\Theta}) = \frac{1}{M}\mathrm{tr}(\mathbf{X}^\top \mathbf{\Theta} \mathbf{X})$ is under the choice of the precision matrix $\mathbf{\Theta}$ as the graph Laplacian $\mathbf{L}$, equivalent to the average GLR $\mathcal{Q}_{\mathbf{X}}(\mathbf{L})$ for signals as columns of $\mathbf{X}$, as well as the trace term in the problem of equation [2.10]. This is the case for the approach in Lake and Tenenbaum (2010), which proposed to consider $\mathbf{\Theta} = \mathbf{L} + \frac{1}{\sigma^2}\mathbf{I}$ as a regularized Laplacian to fit into the formulation of equation [2.2]. The graphical Lasso approach can therefore be interpreted as one that promotes the global smoothness of the signals on the learned topology.

Method	Signal model	Assumption		Learning output	Edge directionality
		$\mathcal{F}(\mathcal{G})$	$\mathbf{c}$		
(Dong *et al.* 2016)	Global smoothness	Eigenvector matrix	i.i.d. Gaussian	Laplacian	Undirected
(Kalofolias 2016)	Global smoothness	Eigenvector matrix	i.i.d. Gaussian	Adjacency	Undirected
(Egilmez *et al.* 2017)	Global smoothness	Eigenvector matrix	i.i.d. Gaussian	Generalized Laplacian	Undirected
(Chepuri *et al.* 2017)	Global smoothness	Eigenvector matrix	i.i.d. Gaussian	Adjacency	Undirected
(Pasdeloup *et al.* 2018)	Spectral filtering (diffusion by adjacency)	Normalized adjacency	i.i.d. Gaussian	Normalized adjacency/Laplacian	Undirected
(Segarra *et al.* 2017a)	Spectral filtering (diffusion by graph shift operator)	Graph shift operator	i.i.d. Gaussian	Graph shift operator	Undirected
(Thanou *et al.* 2017)	Spectral filtering (heat diffusion)	Heat kernel	Sparsity	Laplacian	Undirected
(Mei and Moura 2017)	Causal dependency (SVAR)	Polynomials of adjacency	Past signals	Adjacency	Directed
(Baingana and Giannakis 2017)	Causal dependency (SEM)	Adjacency	Present signal	Time-varying adjacency	Directed
(Shen *et al.* 2019)	Causal dependency (SVARM)	Polynomials of adjacency	Past and present signals	Adjacency	Directed

Table 2.1. *Comparison between different GSP-based approaches to graph learning. Table from (Dong* et al. *2019) with permission*

Second, models based on spectral filtering and causal dependencies on graphs can generally be thought of as the ones that define the generative processes of the observed signals, in particular the diffusion processes on the graph. This is achieved either explicitly, by choosing $\mathcal{F}(\mathcal{G})$ as diffusion matrices like in section 2.3.2, or implicitly, by defining the causal processes of signal generation like in section 2.3.3. Both types of models share similar philosophies as the ones developed from a physics viewpoint in section 2.2.2, in that they all propose to infer the graph topologies by modeling signals as outcomes of physical processes on the graph, especially the diffusion and cascading processes.

It is also interesting to note that certain models can be interpreted from all three viewpoints, an example being the global smoothness model. Indeed, in addition to the statistical and GSP perspectives described above, the property of global smoothness can also be observed in a square-lattice Ising model (Cipra 1987), hence admitting a physical interpretation. Despite the connections with traditional approaches, however, GSP-based approaches offer some unique advantages compared to the classical methods. On one hand, the flexibility in designing the function $\mathcal{F}(\mathcal{G})$ allows for statistical properties of the observed signals that are not limited to a Gaussian distribution, which is the predominant choice in many statistical machine learning methods. On the other hand, this also makes it easier to consider models that go beyond a simple diffusion or cascade model. For example, by using the sparsity assumption on the coefficients $\mathbf{c}$, the method in Thanou *et al.* (2017) defines the signals as the outcomes of one or more diffusion processes originating from different parts of the graph, possibly at different time steps. Similarly, by choosing different $\mathcal{F}(\mathcal{G})$ and $\mathbf{c}$, the SVAR models (Mei and Moura 2017) and the SEMs (Baingana and Giannakis 2017) correspond to the different generative processes of the signals, one based on the static network structure and the other on temporal dynamics. These design flexibilities provide more powerful modeling of the signal representation for the graph inference process.

2.4. Applications of graph learning in image processing

Image representation and coding has been one main area of interest for GSP-based methods. Images can naturally be thought of as graph signals defined on a regular grid structure, where the nodes are the image pixels and the edge weights capture the similarity between adjacent pixels. The design of new flexible graph signal representations has opened the door to new structure-aware transform coding techniques, and eventually to more efficient image compression frameworks (Cheung *et al.* 2018). Such representation permits going beyond traditional transform coding by moving from classical fixed transforms, such as the discrete cosine transform (DCT), to graph-based transforms that are better adapted to the actual image structure.

However, the design of the graph and the corresponding transform remains one of the biggest challenges in graph-based image compression. A suitable graph for effective transform coding should lead to easily compressible signal coefficients, at the cost of a small overhead for coding the graph. Most graph-based coding techniques mainly focus on images, and they construct the graph by considering pairwise similarities among pixel intensities. A few attempts at adapting the graph topology, and consequently the graph transform, exist in literature. For example, the work in Shen *et al.* (2010) introduces a graph-based transform for image depth map coding. The authors propose to construct a graph in which each pixel in the residual block is only connected to each of its immediate neighbors (e.g. four-nearest neighbors) if the connection does not cross edges detected in the residual block. Therefore, the resulting graph-based transform helps preserve such edge information by not filtering across them. A similar idea has been adopted in Hu *et al.* (2015b); Rotondo *et al.* (2015), where both works propose to construct a list of representative graph templates from which the most suitable one is then selected for each pixel block, such that the resulting graph Fourier transform is the efficient in terms of representation cost. These approaches can thus be interpreted as defining optimal graph topology and weights for the efficient block transform coding of images, although they all rely on explicit topological prior, such as the grid structure or graph templates, hence they are not fully adapted to the image signals.

Graph learning has only recently been introduced for these imaging problems. A learning model based on signal smoothness, inspired by Dong *et al.* (2016); Kalofolias *et al.* (2017), has been further extended in order to design a graph-based coding framework that takes the coding of the signal values, as well as the cost of transmitting the graph in rate distortion terms, into account (Fracastoro *et al.* 2020). In particular, the cost of coding the image signal is minimized by promoting its smoothness on the learned topology. The transmission cost of the graph itself is further controlled by adding an additional term in the optimization problem, which penalizes the sparsity of the graph Fourier coefficients of the edge weight signal.

An illustrative example of the graph-based transform coding proposed in Fracastoro *et al.* (2020), as well as its application to image compression, is shown in Figure 2.5. Briefly, the compression algorithm consists of three important parts. First, the solution to an optimization problem that takes the rate approximation of the image signal at a patch level into account, as well as the cost of transmitting the graph, provides a graph topology (Figure 2.5(a)) that defines the optimal coding strategy. Second, the GFT coefficients of the image signal on the learned graph can be used to efficiently compress the image. As we can see in Figure 2.5(b), the decay of these coefficients (in terms of their log-magnitude) is much faster than the decay of the GFT coefficients corresponding to a regular grid graph that does not involve any learning. Third, the weights of the learned graph are treated as a new edge weight signal that lies on a dual graph, whose nodes represent the edges in the learned graph, with the signal values on the nodes being the edge weights of the learned graph. Two nodes are connected in this dual graph if and only if the two corresponding edges share one common node in the learned graph. The learned graph is then transmitted by the GFT coefficients of this edge weight signal, where the decay of these coefficients is shown in Figure 2.5(c). The obtained results confirm that the GFT coefficients of the graph weights are concentrated on the low frequencies, which indicates a highly compressible graph.

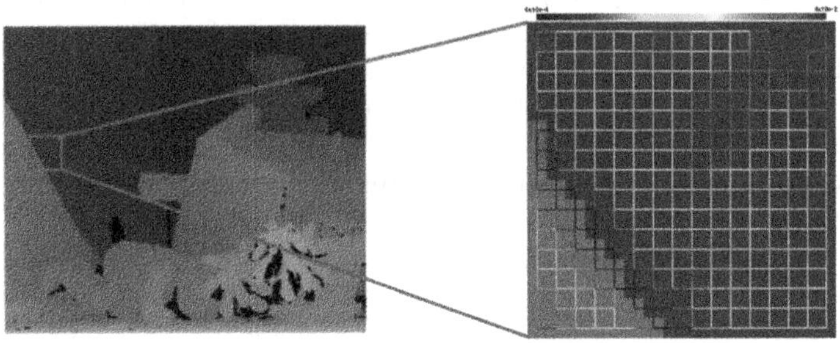

(a) The graph learned on a random patch of the image Teddy using the approach in Fracastoro *et al.* (2020).

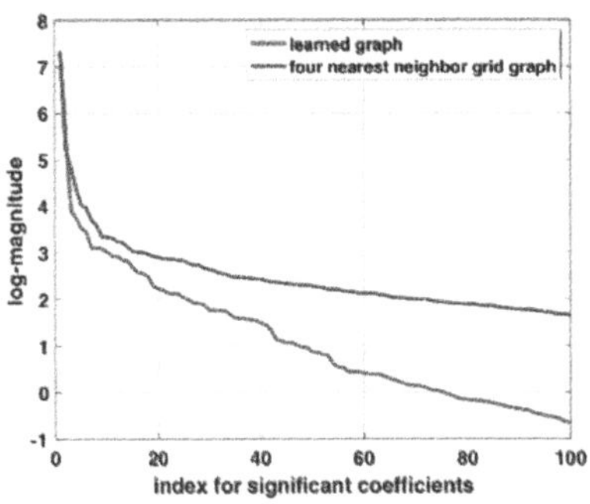

(b) Comparison between the GFT coefficients of the image signal on the learned graph and the four-nearest neighbor grid graph. The coefficients are ordered decreasingly by log-magnitude.

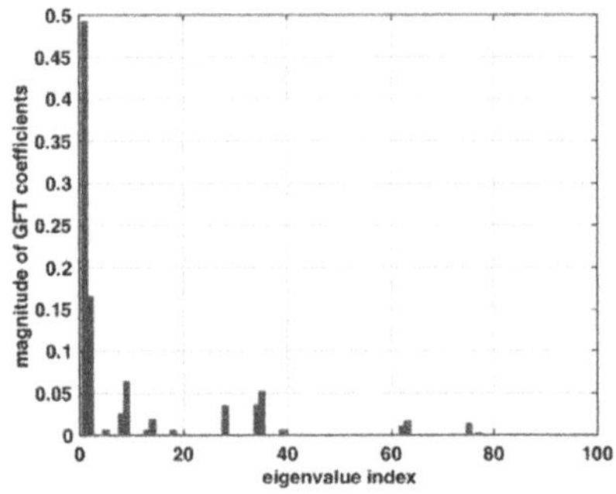

(c) The GFT coefficients of the graph weights.

Figure 2.5. *Inferring a graph for image coding. Figure from (Dong* et al. *2019) with permission. For a color version of this figure, see www.iste.co.uk/cheung/graph.zip*

Another example is the work in Lu and Ortega (2017) that introduces an efficient graph learning approach for a fast graph Fourier transform that is based on (Egilmez *et al.* 2017). The authors have considered a maximum likelihood estimation problem with additional constraints, based on a matrix factorization of the graph Laplacian matrix, such that its eigenvector matrix is a product of a block diagonal matrix and a butterfly-like matrix. The learned graph leads to a fast non-separable transform for intra-predictive residual blocks in video compression. Furthermore, the recent work in Egilmez *et al.* (2019) proposes a class of graph-based transforms for video compression, one based on the graph learning framework introduced in Egilmez *et al.* (2017) and the other based on the edge adaptive transform introduced in Shen *et al.* (2010). These efforts confirm that learning a meaningful graph can have a significant impact in graph-based image and video compression. These are only first attempts that leave much room for improvement, especially in terms of coding performance. Thus, we expect to see more research efforts in the future to fully exploit the potential of graph methods. For more discussion on the application of graph learning for image and video compression, see Chapters 4 and 5 of this book.

As a final remark, we note that graph construction methods have also been proposed in literature for graph-based regularization and learning in various imaging problems. For example, approaches such as non-local total variation (NLTV) (Gilboa and Osher 2008) and non-local total generalized variation (NLTGV) (Ranftl *et al.* 2014) can be interpreted as graph construction methods, which define weights between pixels (not restricted to immediate neighborhood) for TV-type regularization. The graph construction can also explicitly rely on the geometry of images in order to introduce geometric priors in the image processing tasks (Khasanova and Frossard 2019; Rossi and Frossard 2018).

2.5. Concluding remarks and future directions

Learning structures and graphs from data observations are an important problem in modern data analytics, and the novel signal processing approaches reviewed in this chapter have both theoretical and practical significance. On one hand, GSP provides a new theoretical framework for graph learning by utilizing signal processing tools, with a strong emphasis on the representation of the signals on the learned graph, which can be essential from a modeling viewpoint. As a result, the novel approaches developed in this field would benefit not only the inference of optimal graph topologies, but potentially also the subsequent signal processing and data analysis tasks. On the other hand, the novel signal and graph models designed from a GSP perspective may uniquely contribute to the understanding of the often complex data structure and generative processes of the observations made in real-world application domains, such as brain and social network analysis. Image processing, in particular, can benefit significantly from graph learning techniques. However, many open issues and questions exist that are worthy of further investigation. In what follows, we discuss some general directions for future work by focusing more on graph inference for image processing applications.

One important point that needs further investigation is the quality of the input images. Most of the approaches in literature have focused on the scenario where a complete set of data is observed for all the entities of interest (i.e. all vertices in the graph). However, there are often situations when observations are only partially available. For example, occlusions are common in many real-world vision applications. It is a challenge to design graph learning approaches that can handle such cases, and to study the extent to which the partial or missing pixels affect the learning performance.

Compared to the input images, it is perhaps even more important to rethink the potential outcome of the learning frameworks that should be guided by the application itself. Several important lines of thought remain largely unexplored in the current literature. First, while most of the existing work focuses on learning undirected graphs, it is certainly of interest to investigate approaches for learning directed graphs that could capture optical flow in images. The methods described in section 2.3.3, such as (Mei and Moura 2017; Baingana and Giannakis 2017; Shen *et al.* 2019), are able to learn directed graphs since they do not explicitly rely on the notion of frequency provided by the eigendecomposition of the symmetric graph adjacency or Laplacian matrices. On the other hand, it is certainly possible and desirable to extend the frequency interpretation obtained with undirected graphs to directed graphs. The proper design to be able to learn directed graphs of images by making use of recent developments in the spectral domain (e.g. (Mhaskar 2018; Girault *et al.* 2018; Sardellitti *et al.* 2017; Shafipour *et al.* 2019)) remains an interesting research question.

Second, although the current lines of work in graph learning mainly focus on the signal representation, it is also possible to put constraints directly on learned graphs by enforcing certain graph properties that go beyond the common choice of sparsity, which has been adopted explicitly in the optimization problems in many existing methods, such as the ones in Friedman *et al.* (2008); Lake and Tenenbaum (2010); Chepuri *et al.* (2017); Pasdeloup *et al.* (2018); Segarra *et al.* (2017a); Mei and Moura (2017); Baingana and Giannakis (2017). Ideally, such constraints should be imposed by the application itself. In image compression, for instance, the actual cost of coding the graph may be as equally important as the cost of coding the image signal. Such constraints should be incorporated in the graph learning framework (e.g. Fracastoro *et al.* (2020)) in order to make the learning framework more targeted to a specific application.

Last but not least, in certain applications it might not be necessary to learn the full graph topology, but some other intermediate or graph-related representation that is guided by the objective. For example, this can be an embedding of the vertices in the graph for the purpose of clustering (Dong *et al.* 2014), or a function of the graph, such as graph filters for the subsequent signal processing tasks (Segarra *et al.* 2017b). In image processing, it may be interesting to learn several graph templates instead of a single graph. Each of them could capture common patterns that are characteristic of the local regions of the image. Similar ideas have been developed in Hu *et al.* (2015b);

Rotondo *et al.* (2015), where both works propose to construct a list of representative graph templates from which the most suitable one is then selected for each pixel block, such that the resulting graph Fourier transform is efficient. However, these approaches are not fully adapted to the image signals. In these scenarios, similar to the previous point, the learning framework needs to be designed according to the end objective or application in mind.

2.6. References

Baingana, B. and Giannakis, G.B. (2017). Tracking switched dynamic network topologies from information cascades. *IEEE Transactions on Signal Processing*, 65(4), 985–997.

Banerjee, O., Ghaoui, L.E., d'Aspremont, A. (2008). Model selection through sparse maximum likelihood estimation for multivariate Gaussian or binary data. *The Journal of Machine Learning Research*, 9, 485–516.

Bolstad, A., Veen, B.D.V., Nowak, R. (2011). Causal network inference via group sparse regularization. *IEEE Transactions on Signal Processing*, 59(6), 2628–2641.

Cai, T., Liu, W., Luo, X. (2011). A constrained l1 minimization approach to sparse precision matrix estimation. *Journal of the American Statistical Association*, 106(494), 594–607.

Castro, R., Coates, M., Liang, G., Nowak, R., Yu, B. (2004). Network tomography: Recent developments. *Statistical Science*, 19(3), 499–517.

Chen, G., Glen, D.R., Saad, Z.S., Hamilton, J.P., Thomason, M.E., Gotlib, I.H., Cox, R.W. (2011). Vector autoregression, structural equation modeling, and their synthesis in neuroimaging data analysis. *Computers in Biology and Medicine*, 41(12), 1142–1155.

Chepuri, S.P., Liu, S., Leus, G., Hero, A.O. (2017). Learning sparse graphs under smoothness prior. *Proceedings of the IEEE International Conference on Acoustics, Speech and Signal Processing*, 6508–6512.

Cheung, G., Magli, E., Tanaka, Y., Ng, M.K. (2018). Graph spectral image processing. *Proceedings of the IEEE*, 106(5), 907–930.

Chung, F. (2007). The heat kernel as the pagerank of a graph. *Proceedings of the National Academy of Sciences*, 104(50), 19735–19740.

Cipra, B.A. (1987). An introduction to the Ising model. *The American Mathematical Monthly*, 94(10), 937–959.

Daitch, S., Kelner, J., Spielman, D. (2009). Fitting a graph to vector data. *Proceedings of the International Conference on Machine Learning*, 201–208.

Dempster, A.P. (1972). Covariance selection. *Biometrics*, 28(1), 157–175.

Dong, X., Frossard, P., Vandergheynst, P., Nefedov, N. (2014). Clustering on multi-layer graphs via subspace analysis on Grassmann manifolds. *IEEE Transactions on Signal Processing*, 62(4), 905–918.

Dong, X., Thanou, D., Frossard, P., Vandergheynst, P. (2016). Learning Laplacian matrix in smooth graph signal representations. *IEEE Transactions on Signal Processing*, 64(23), 6160–6173.

Dong, X., Thanou, D., Rabbat, M., Frossard, P. (2019). Learning graphs from data: A signal representation perspective. *IEEE Signal Processing Magazine*, 36(3), 44–63.

Du, N., Song, L., Smola, A.J., Yuan, M. (2012). Learning networks of heterogeneous influence. *Proceedings of Neural Information Processing Systems*, 2789–2797.

Egilmez, H.E., Pavez, E., Ortega, A. (2017). Graph learning from data under Laplacian and structural constraints. *IEEE Journal of Selected Topics in Signal Processing*, 11(6), 825–841.

Egilmez, H.E., Pavez, E., Ortega, A. (2019). Graph learning from filtered signals: Graph system and diffusion kernel identification. *IEEE Transactions on Signal and Information Processing over Networks*, 5(2), 360–374.

Egilmez, H., Chao, Y., Ortega, A. (2020). Graph-based transforms for video coding. *IEEE Transactions on Image Processing*, 29, 9330–9344.

Fortunato, S. (2010). Community detection in graphs. *Physics Reports*, 486(3–5), 75–174.

Fracastoro, G., Thanou, D., Frossard, P. (2020). Graph transform optimization with application to image compression. *IEEE Transactions on Image Processing*, 29, 419–432.

Friedman, J., Hastie, T., Tibshirani, R. (2008). Sparse inverse covariance estimation with the graphical lasso. *Biostatistics*, 9(3), 432–441.

Friston, K.J. (1994). Functional and effective connectivity in neuroimaging: A synthesis. *Human Brain Mapping*, 2(1–2), 56–78.

Giannakis, G.B., Shen, Y., Karanikolas, G.V. (2018). Topology identification and learning over graphs: Accounting for nonlinearities and dynamics. *Proceedings of the IEEE*, 106(5), 787–807.

Gilboa, G. and Osher, S. (2008). Nonlocal operators with applications to image processing. *Multiscale Modeling & Simulation*, 7(3), 1005–1028.

Girault, B., Ortega, A., Narayanan, S. (2018). Irregularity-aware graph Fourier transforms. *IEEE Transactions on Signal Processing*, 66(21), 5746–5761.

Goebel, R., Roebroeck, A., Kim, D.-S., Formisano, E. (2003). Investigating directed cortical interactions in time-resolved fMRI data using vector autoregressive modeling and Granger causality mapping. *Magnetic Resonance Imaging*, 21(10), 1251–1261.

Gomez-Rodriguez, M., Leskovec, J., Krause, A. (2010). Inferring networks of diffusion and influence. *Proceedings of the ACM SIGKDD International Conference on Knowledge Discovery and Data Mining*, Washington, DC, 1019–1028.

Gomez-Rodriguez, M., Leskovec, J., Balduzzi, D., Schölkopf, B. (2014). Uncovering the structure and temporal dynamics of information propagation. *Network Science*, (1), 26–65.

Hammond, D.K., Vandergheynst, P., Gribonval, R. (2011). Wavelets on graphs via spectral graph theory. *Applied and Computational Harmonic Analysis*, 30(2), 129–150.

Hsieh, C.-J., Dhillon, I.S., Ravikumar, P.K., Sustik, M.A. (2011). Sparse inverse covariance matrix estimation using quadratic approximation. *Proc. Neural Information Processing Systems (NIPS)*, 2330–2338.

Hu, C., Cheng, L., Sepulcre, J., Johnson, K.A., Fakhri, G.E., Lu, Y.M., Li, Q. (2015a). A spectral graph regression model for learning brain connectivity of alzheimer's disease. *PLoS ONE*, 10(5), e0128136.

Hu, W., Cheung, G., Ortega, A., Au, O.C. (2015b), Multiresolution graph Fourier transform for compression of piecewise smooth images. *IEEE Transactions on Image Processing*, 24(1), 419–433.

Kalofolias, V. (2016). How to learn a graph from smooth signals. *Proceedings of the International Conference on Artificial Intelligence and Statistics*, 51, 920–929.

Kalofolias, V., Loukas, A., Thanou, D., Frossard, P. (2017). Learning time varying graphs. *Proceedings of the IEEE International Conference on Acoustics, Speech and Signal Processing*, 2826–2830.

Kaplan, D.W. (2009). *Structural Equation Modeling: Foundations and Extensions,2nd edition*. Sage, California.

Khasanova, R. and Frossard, P. (2019). Geometry aware convolutional filters for omnidirectional images representation. *Proceedings of the International Conference on Machine Learning*, 3351–3359.

Koller, D. and Friedman, N. (2009). *Probabilistic Graphical Models: Principles and Techniques*. MIT Press, Massachusetts.

Lake, B. and Tenenbaum, J. (2010). Discovering structure by learning sparse graph. *Proceedings of the Annual Cognitive Science Conference*, 32 [Online]. Available at: https://escholarship.org/uc/item/1ww3443p.

Lu, K.-S. and Ortega, A. (2017). A graph Laplacian matrix learning method for fast implementation of graph Fourier transform. *Proceedings of the IEEE International Conference on Image Processing*, 1677–1681.

Ma, H., Yang, H., Lyu, M.R., King, I. (2008). Mining social networks using heat diffusion processes for marketing candidates selection. *17th ACM Conference on Information and Knowledge Management*, 233–242.

Maretic, H.P., Thanou, D., Frossard, P. (2017). Graph learning under sparsity priors. *Proceedings of the IEEE International Conference on Acoustics, Speech and Signal Processing*, 6523–6527.

McLntosh, A. and Gonzalez-Lima, F. (1994). Structural equation modeling and its application to network analysis in functional brain imaging. *Human Brain Mapping*, 2(1–2), 2–22.

Mei, J. and Moura, J.M.F. (2017). Signal processing on graphs: Causal modeling of unstructured data. *IEEE Transactions on Signal Processing*, 65(8), 2077–2092.

Meinshausen, N. and Bühlmann, P. (2006). High-dimensional graphs and variable selection with the Lasso. *Annals of Statistics*, 34(3), 1436–1462.

Mhaskar, H.N. (2018). A unified framework for harmonic analysis of functions on directed graphs and changing data. *Applied and Computational Harmonic Analysis*, 44(3), 611–644.

Myers, S.A. and Leskovec, J. (2010). On the convexity of latent social network inference. *Proceedings of Neural Information Processing Systems*, Vancouver, British Columbia, 1741–1749.

Ortega, A., Frossard, P., Kovačević, J., Moura, J.M.F., Vandergheynst, P. (2018). Graph signal processing: Overview, challenges and applications. *Proceedings of the IEEE*, 106(5), 808–828.

Pasdeloup, B., Gripon, V., Mercier, G., Pastor, D., Rabbat, M.G. (2018). Characterization and inference of graph diffusion processes from observations of stationary signals. *IEEE Transactions on Signal and Information Processing over Networks*, 4(3), 481–496.

Pastor-Satorras, R., Castellano, C., Mieghem, P.V., Vespignani, A. (2015). Epidemic processes in complex networks. *Reviews of Modern Physics*, 87(3), 925–979.

Poole, G. and Boullion, T. (1974). A survey on m-matrices. *SIAM Review*, 16(4), 419–427.

Ranftl, R., Bredies, K., Pock, T. (2014). Non-local total generalized variation for optical flow estimation. *Proceedings of the European Conference on Computer Vision*, 439–454.

Roebroeck, A., Formisano, E., Goebel, R. (2005). Mapping directed influence over the brain using Granger causality and fMRI. *NeuroImage*, 25(1), 230–242.

Rossi, M. and Frossard, P. (2018). Geometry-consistent light field super-resolution via graph-based regularization. *IEEE Transactions on Image Processing*, 27(9), 4207–4218.

Rotondo, I., Cheung, G., Ortega, A., Egilmez, H.E. (2015). Designing sparse graphs via structure tensor for block transform coding of images. *Proceedings of the Asia-Pacific Signal and Information Processing Association Annual Summit and Conference*, 571–574.

Rubinstein, R., Bruckstein, A.M., Elad, M. (2010). Dictionaries for sparse representation modeling. *Proceedings of the IEEE*, 98(6), 1045–1057.

Sandryhaila, A. and Moura, J.M.F. (2013). Discrete signal processing on graphs. *IEEE Transactions on Signal Processing*, 61(7), 1644–1656.

Sardellitti, S., Barbarossa, S., Lorenzo, P.D. (2017). On the graph Fourier transform for directed graphs. *IEEE Journal of Selected Topics in Signal Processing*, 11(6), 796–811.

Segarra, S., Marques, A.G., Mateos, G. and Ribeiro, A. (2017a). Network topology inference from spectral templates. *IEEE Transactions on Signal and Information Processing over Networks*, 3(3), 467–483.

Segarra, S., Mateos, G., Marques, A.G., Ribeiro, A. (2017b). Blind identification of graph filters. *IEEE Transactions on Signal Processing*, 65(5), 1146–1159.

Shafipour, R., Segarra, S., Marques, A.G., Mateos, G. (2018). Identifying the topology of undirected networks from diffused non-stationary graph signals. *IEEE Trans. Signal Process*. arXiv:1801.03862.

Shafipour, R., Khodabakhsh, A., Mateos, G., Nikolova, E. (2019). A directed graph Fourier transform with spread frequency components. *IEEE Transactions on Signal Processing*, 67(4), 946–960.

Shen, G., Kim, W.-S., Narang, S., Ortega, A., Lee, J., Wey, H. (2010). Edge-adaptive transforms for efficient depth map coding. *Picture Coding Symposium (PCS)*, 566–569.

Shen, Y., Giannakis, G.B., Baingana, B. (2019). Nonlinear structural vector autoregressive models with application to directed brain networks. *IEEE Transactions on Signal Processing*, 67(20), 5325–5339.

Shuman, D.I., Narang, S.K., Frossard, P., Ortega, A., Vandergheynst, P. (2013). The emerging field of signal processing on graphs: Extending high-dimensional data analysis to networks and other irregular domains. *IEEE Signal Processing Magazine*, 30(3), 83–98.

Songsiri, J. and Vandenberghe, L. (2010). Topology selection in graphical models of autoregressive processes. *Journal of Machine Learning Research*, 11, 2671–2705.

Thanou, D., Shuman, D.I., Frossard, P. (2014). Learning parametric dictionaries for signals on graphs. *IEEE Transactions on Signal Processing*, 62(15), 3849–3862.

Thanou, D., Dong, X., Kressner, D., Frossard, P. (2017). Learning heat diffusion graphs. *IEEE Transactions on Signal and Information Processing over Networks*, 3(3), 484–499.

Tosic, I. and Frossard, P. (2011). Dictionary learning. *IEEE Signal Processing Magazine*, 28(2), 27–38.

Traganitis, P.A., Shen, Y., Giannakis, G.B. (2017). Network topology inference via elastic net structural equation models, 146–150.

Yuan, M. and Lin, Y. (2007). Model selection and estimation in the Gaussian graphical model. *Biometrika*, 94(1), 19–35.

Zhang, C. and Florencio, D. (2013). Analyzing the optimality of predictive transform coding using graph-based models. *IEEE Signal Processing Letters*, 20(1), 106–109.

Zhang, X., Dong, X., Frossard, P. (2012). Learning of structured graph dictionaries. *Proceedings of the IEEE International Conference on Acoustics, Speech and Signal Processing*, 3373–3376.

Zhu, X. (2005). Semi-supervised learning with graphs. PhD thesis, Carnegie Mellon University, CMU-LTI-05-192.

3

Graph Neural Networks

Giulia FRACASTORO and Diego VALSESIA
Politecnico di Torino, Turin, Italy

3.1. Introduction

The recent wave of impressive results obtained in fields as varied as computer vision, natural language processing, bioinformatics and many more can be attributed to the advances in training and designing neural networks. A neural network works as a universal function approximator, so that it can use training data to learn complex input-output mappings. However, training is a non-convex optimization problem and it can be challenging to reach a satisfactory local minimum. This is why the careful design of the network, to incorporate as much prior information available for the problem at hand, is crucial to building successful models. The convolutional neural network (CNN) is a successful example of this principle and it has become the workhorse of state-of-the-art models in computer vision, audio processing and more because it explicitly exploits some underlying properties of the data. The convolutional layer leverages the three main properties of much of the natural data that is of interest, such as images: stationarity, locality and compositionality. A signal is stationary when its statistical features do not significantly change over time or space; convolution exploits this property by reusing the same filter weights over different portions of the signal, which also allows the limitation of the number of trainable parameters, preventing overfitting and vanishing gradient issues. Locality implies that most of the information about the signal can be extracted by only looking at small portions of the signal at any given time; convolution can use small filters to focus on capturing local interactions. Compositionality means that a complex model can be assembled through a hierarchical composition of small local models; CNNs stack several convolutional layers to assemble simpler local features into increasingly complex representations.

Graph Spectral Image Processing,
coordinated by Gene CHEUNG and Enrico MAGLI.

Despite their astounding successes, a few shortcomings of CNNs have recently emerged. First, CNNs can only process data defined on grids. It is well known that certain kinds of data lack a regular domain and are instead defined on the vertices of a graph. While solutions based on GSP have flourished to process this kind of signal, a data-driven universal approximator, equivalent to CNNs, is needed to tackle the most challenging problems. A second, less obvious, shortcoming of CNNs concerns traditional data types defined on grids. Some data, such as natural images, are more accurately modeled if the locality property is combined with a non-local self-similarity prior, whereby regions that are distant in the grid domain may exhibit similar features. Therefore, a more flexible domain is needed to capture this property and the graph is well suited for the task.

In light of the limitations of CNNs explained above, it is natural that a lot of research has recently focused on developing graph-convolutional neural networks (GCNNs). A GCNN uses graph-convolutional layers that can process data defined on the vertices of a graph. A graph-convolutional layer requires two inputs: a graph $\mathcal{G} = (\mathcal{V}, \mathcal{E})$ with $N = |\mathcal{V}|$ nodes and $|\mathcal{E}|$ edges, and a signal $\mathbf{X} \in \mathbb{R}^{N \times F}$, defined as an F-dimensional feature vector for each node. The output is a signal $\mathbf{Y} \in \mathbb{R}^{N \times F'}$. A graph-convolutional layer should be able to exploit the stationarity, locality and compositionality properties that made classic convolutional layers successful for grids, as well as provide efficient weight reuse to limit the number of trainable parameters. Defining a graph-convolutional layer that meets all the desiderata is a challenging task and no universally accepted solution is available at the moment. In the following, we present several alternative approaches, including the most recent advances on the topic.

This chapter will first present spectral approaches to the definition of graph-convolutional layers, drawing from literature on the GFT, and then focus on the spatial definitions of graph convolution, which have emerged as a more flexible alternative, providing superior experimental performance.

3.2. Spectral graph-convolutional layers

The first approach to generalize the convolutional layer of a CNN to signals defined on the vertices of graphs, is to leverage the graph spectral domain. According to the convolution theorem, the graph convolution of signal $\mathbf{x}_f$ with filter $\mathbf{g_f}$ is defined as:

$$\mathbf{y} = \mathbf{U}\left[(\mathbf{U}^T\mathbf{g_f}) \odot (\mathbf{U}^T\mathbf{x}_f)\right], \qquad [3.1]$$

where $\mathbf{U}$ is the GFT associated with graph $\mathcal{G}$ and $\odot$ denotes the Hadamard product. The various spectral approaches differ in how they define the filter $\mathbf{U}^T\mathbf{g_f}$. The first approach by Bruna *et al.* (2013) defines it as a diagonal matrix, and allows the full spectral graph-convolutional layer to be written as

$$\mathbf{y}_{f'} = \sum_{f=1}^{F} \mathbf{U}\Theta^{(f,f')}\mathbf{U}^T\mathbf{x}_f \qquad f = 1, \ldots, F \qquad [3.2]$$

where each $\Theta^{(f,f')} \in \mathbb{R}^{N \times N}$ for $f = 1, \ldots, F$, $f' = 1, \ldots, F'$ is a diagonal matrix representing a frequency-domain filter. Note that we consider the general case in which the input has F feature channels and the output has F' feature channels, thus $\mathbf{x}_f$ represents the graph signal corresponding to the fth input channel and $\mathbf{y}'_f$ represents the graph signal corresponding to the f'th output channel.

The spectral approach to graph convolution is desirable as it is mathematically principled, relying on the spectral domain induced by the GFT. However, several drawbacks are inherent to this method. First of all, the computational complexity is high due to the requirement of computing the GFT, and its inverse, to perform the filtering operation. While the operator itself can be precomputed offline, a multiplication with a possibly large $N \times N$ matrix is still required at every layer. A practical approximation can be to only consider the $k \ll N$ eigenvectors corresponding to the lower frequencies, at the cost of losing some representation capabilities for the higher frequencies. This definition also leads to overparameterization, by requiring $O(N)$ parameters per layer. An additional drawback of the spectral graph convolution is that it is not a localized operation in the spatial domain, i.e. the receptive field of an output node is usually the entire input graph. This is undesirable because much of the data that is of interest can be described by hierarchical compositions of local properties, which are better captured by localized operators, such as the classic convolution with small spatial filters. Finally, a fundamental limitation of the spectral approach is the restriction to a single graph. Each graph defines a different frequency domain, and the filters learned for one domain cannot be used interchangeably on another domain although some works addressed this issue by creating domain transformers (Kovnatsky *et al.* 2013; Eynard *et al.* 2015). This is an important limitation since it restricts the applicability to use cases where the graph is fixed and defined *a priori* for all train and test signals involved in the problem.

An important step toward the solution of the issues affecting the spectral approach is represented by methods, that instead of directly defining the filters as spectral multipliers, define a parametric model. In particular, the filters can be defined as polynomials of the graph Laplacian:

$$\mathbf{y}_{f'} = \sum_{f=1}^{F} p^{(f,f')}(\mathbf{L})\mathbf{x}_f \qquad f' = 1, \ldots, F' \qquad [3.3]$$

where $p^{(f,f')}(\mathbf{L}) = \sum_{k=0}^{K} \theta_k^{f,f'} \mathbf{L}^k$ is a K-degree polynomial. Note that since the Laplacian is diagonalized by the GFT basis, equation [3.3] is equivalent to

$$\mathbf{y}_{f'} = \sum_{f=1}^{F} \mathbf{U}p^{(f,f')}(\mathbf{\Lambda})\mathbf{U}^T\mathbf{x}_f \qquad f' = 1, \ldots, F' \qquad [3.4]$$

where $\mathbf{\Lambda}$ is the diagonal matrix of eigenvalues of the Laplacian, which means that the filter is constrained to be a polynomial function of the frequency. Defferrard *et al.*

(2016) propose the use of K-degree Chebyshev polynomials to have an efficient recursive implementation with a complexity of $O(KN)$ operations. Also note that this solution does not require the computation of the eigenvectors. In practice, one computes

$$\mathbf{y}_{f'} = \sum_{f=1}^{F}\sum_{k=0}^{K}\theta_k^{(f,f')}T_k(\mathbf{L})\mathbf{x}_f \qquad f = 1,\ldots,F \qquad [3.5]$$

$$T_k(\mathbf{L}) = 2\mathbf{L}T_{k-1}(\mathbf{L}) - T_{k-2}(\mathbf{L})$$

$$T_0(\mathbf{L}) = \mathbf{I}$$

$$T_1(\mathbf{L}) = \mathbf{L}$$

Another advantage of this approach is the spatial localization of the filter, which is restricted to a K-hop neighborhood, as it only involves up to the Kth power of the Laplacian. Finally, this approach requires $O(1)$ parameters per layer, solving the $O(N)$ dependence of the original spectral approach. Alternative efficient spectral approaches include using Cayley polynomials (Levie *et al.* 2018) or the Lanczos algorithm (Liao *et al.* 2019).

Kipf and Welling (2017) propose the graph convolutional network (GCN), which uses an approximation of this construction in a semi-supervised learning problem. Their approximation uses a polynomial with degree 1, which essentially amounts to a weighted spatial aggregation over a single-hop neighborhood. The GCN has successfully been applied in numerous applications and it is one of the most well-known graph convolutional neural networks.

3.3. Spatial graph-convolutional layers

A second approach to graph convolution consists of defining the convolutional operator in the spatial domain. In this case, the convolution is performed by local aggregations and the output of the convolution is a weighted average of the signal values over the neighboring nodes. Therefore, the spatial graph-convolutional layer can be represented as

$$\mathbf{y}_i = \underset{j:(i,j)\in\mathcal{E}}{\square} \mathcal{F}(\mathbf{x}_i, \mathbf{x}_j) \qquad i = 1, \ldots, N \qquad [3.6]$$

where $\square$ is an aggregation operation (e.g. sum, mean, maximum), $\mathcal{F}: \mathbb{R}^F \rightarrow \mathbb{R}^{F'}$ is a function with a set of learnable parameters and $\mathbf{x}_i, \mathbf{x}_j \in \mathbb{R}^F$ are the signal values at node i and j, respectively. Figure 3.1 shows an example of spatial aggregation. Since equation [3.6] is defined at a neighborhood level, it presents significant advantages with respect to the spectral approach. First, it is, by design, well localized in the spatial domain. Second, in the spatial approach there is no limitation to a single graph: filters learned on one graph can be generalized to other graph structures.

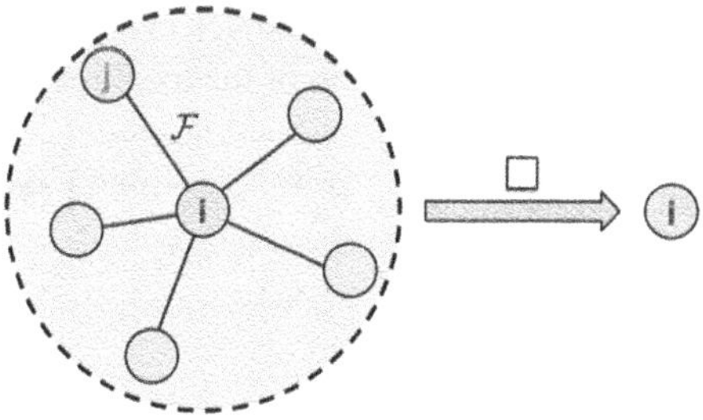

Figure 3.1. *Spatial graph-convolutional layer: an edge function $\mathcal{F}$ weighs the node feature vectors, and an aggregation function $\square$ merges them. For a color version of this figure, see www.iste.co.uk/cheung/graph.zip*

In equation [3.6], the choice of the edge function $\mathcal{F}$ is of paramount importance, because it defines the properties of the convolutional operator. For this reason, the edge function $\mathcal{F}$ has been extensively studied in literature and many definitions have been presented. A first definition of $\mathcal{F}$ has been proposed in GraphSAGE (Hamilton *et al.* 2017), where $\mathcal{F}$ and $\square$ are defined as

$$\mathcal{F}(\mathbf{x}_i, \mathbf{x}_j) = \sigma(\boldsymbol{\Theta}\mathbf{x}_j) \qquad \text{(GraphSAGE)}, \qquad [3.7]$$

$$\square = \frac{1}{d_i}\sum \qquad \text{(GraphSAGE} - \text{mean)}, \qquad [3.8]$$

$$\square = \max \qquad \text{(GraphSAGE} - \text{maxpool)}, \qquad [3.9]$$

where d_i is the degree of the ith node, σ is a non-linearity and $\boldsymbol{\Theta} \in \mathbb{R}^{F' \times F}$ is a matrix of learnable parameters. Then, in order to combine the local and global information of the data, (Wang *et al.* 2019) propose to define $\mathcal{F}$, employing the pairwise distances between the central node and its neighbors:

$$\mathcal{F}(\mathbf{x}_i, \mathbf{x}_j) = \sigma(\boldsymbol{\Theta}^{(1)}(\mathbf{x}_i - \mathbf{x}_j) + \boldsymbol{\Theta}^{(2)}\mathbf{x}_i), \qquad \text{(DGCNN)}, \qquad [3.10]$$

$$\square = \max, \qquad [3.11]$$

where $\boldsymbol{\Theta}^{(1)}, \boldsymbol{\Theta}^{(2)} \in \mathbb{R}^{F' \times F}$ are learnable parameters. In this definition, the global structure of the data, captured by the central node $\mathbf{x}_i$, is combined with the local information of the neighborhood, described by the difference $\mathbf{x}_i - \mathbf{x}_j$.

In the above definitions, the aggregation weights do not depend on the input data, but they are the same at every node. This means that the graph convolution is a non-adaptive filter, resulting in a reduced approximation capability. In order to solve this issue, several methods propose the use of weight functions that depend on the data itself. In many cases, this weight function is a predefined scalar function with some learnable parameters. For example, FeastNet (Verma *et al.* 2018) employs a softmax function over a linear transformation of the local feature vectors:

$$\mathcal{F}(\mathbf{x}_i,\mathbf{x}_j)=\sum_{m=1}^{M}\text{softmax}\left(\boldsymbol{\theta}_1^{(\mathrm{m})^\top}\mathbf{x}_\mathrm{i}+\boldsymbol{\theta}_2^{(\mathrm{m})^\top}\mathbf{x}_\mathrm{j}+\theta^{(\mathrm{m})}\right)\boldsymbol{\Theta}^{(\mathrm{m})}\mathbf{x}_\mathrm{j},\quad\text{(FeastNet)},\qquad[3.12]$$

$$\square=\frac{1}{d_i}\sum,\qquad[3.13]$$

where d_i is the degree of the ith node, M is the number of filters, $\boldsymbol{\Theta}^{(m)}\in\mathbb{R}^{F'\times F}$, $\boldsymbol{\theta}_1^{(m)},\boldsymbol{\theta}_2^{(m)}\in\mathbb{R}^F$ and $\theta^{(m)}\in\mathbb{R}$ with $m=1,...,M$ are learnable parameters. Instead, MoNet (Monti *et al.* 2017) uses parametric Gaussian kernels with learnable parameters:

$$\mathcal{F}(\mathbf{x}_i,\mathbf{x}_j)=\sum_{m=1}^{M}\exp\left(-\frac{1}{2}(u(\mathbf{x}_i,\mathbf{x}_j)-\boldsymbol{\theta}^{(m)})^\top\boldsymbol{\Omega}^{(m)^{-1}}(u(\mathbf{x}_i,\mathbf{x}_j)-\boldsymbol{\theta}^{(m)})\right)$$

$$\times\,\boldsymbol{\Theta}^{(m)}\mathbf{x}_j,\qquad\text{(MoNet)}\qquad[3.14]$$

$$\square=\sum,\qquad[3.15]$$

where $\boldsymbol{\theta}^{(m)}\in\mathbb{R}^F$, $\boldsymbol{\Omega}^{(m)}\in\mathbb{R}^{F\times F}$ and $\boldsymbol{\Theta}^{(m)}\in\mathbb{R}^{F'\times F}$ with $m=1,...,M$ are learnable parameters and $u(\mathbf{x}_i,\mathbf{x}_j)$ are local pseudo-coordinates of $\mathbf{x}_j$, with respect to $\mathbf{x}_i$. Different choices of pseudo-coordinates can be considered, we refer the reader to (Monti *et al.* 2017) for additional details. In order to reduce the number of parameters, the matrices $\boldsymbol{\Omega}^{(m)}$ are usually restricted to a diagonal form. This definition of graph convolution can thus be interpreted as a Gaussian mixture model.

By making the aggregation weights depend on the input features, both FeastNet and MoNet define adaptive filters that can be more complex than the non-adaptive filters presented in equations [3.7] and [3.10]. However, these definitions have two main limitations. First, the weight functions employed in equations [3.12] and [3.14] are scalar functions, imposing the same weight for every feature. Second, they both use a fixed predefined weight function, which may not be the optimal one. In order to overcome these issues, (Simonovsky and Komodakis 2017) propose the edge-conditioned convolution (ECC), where $\mathcal{F}$ and $\square$ are defined as

$$\mathcal{F}(\mathbf{x}_i,\mathbf{x}_j)=\begin{cases}\boldsymbol{\Theta}^{j\to i}\mathbf{x}_j, & \text{if } j\neq i\\ \boldsymbol{\Theta}\mathbf{x}_i & \text{if } j=i\end{cases}\qquad\text{(ECC)}\qquad[3.16]$$

$$\boldsymbol{\Theta}^{j\to i}=\mathrm{MLP}(\mathbf{x}_\mathrm{j}-\mathbf{x}_\mathrm{i}),\qquad[3.17]$$

$$\square=\frac{1}{d_i}\sum,\qquad[3.18]$$

where d_i is the degree of the ith node, $\mathbf{\Theta} \in \mathbb{R}^{F \times F'}$ is a matrix of learnable parameters, and $\mathbf{\Theta}^{j \to i} \in \mathbb{R}^{F \times F'}$ is the output of a fully connected network $\mathcal{H}$, which takes the node difference $\mathbf{x}_j - \mathbf{x}_i$ as input. Figure 3.2 shows a graphical example of the ECC. As seen in equation [3.16], the ECC computes a weight matrix $\mathbf{\Theta}^{j \to i}$ for each edge of the neighborhood. This is a great advantage compared to the scalar weight functions presented above, because it allows the computation of an affine transformation along every edge, making the convolutional operator more general. In addition, the function that outputs $\mathbf{\Theta}^{j \to i}$ does not have a predefined structure, but it is a general function that can be trained to be the optimal one, because of the function approximation capability of the $\mathcal{H}$ network.

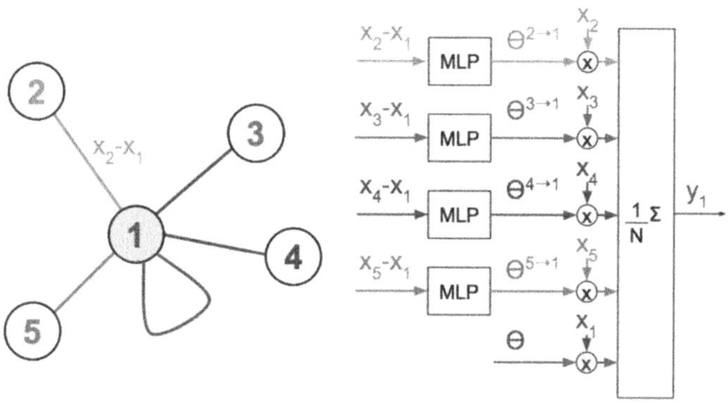

Figure 3.2. *Edge-conditioned convolution. For a color version of this figure, see www.iste.co.uk/cheung/graph.zip*

In the original definition of ECC (Simonovsky and Komodakis 2017), $\mathcal{H}$ is defined as a two-layer fully connected network. This definition may raise some relevant issues, especially when the number of features is high. The first issue is related to the risk of overparameterization. Since $\mathcal{H}$ takes $\mathbf{x}_j - \mathbf{x}_i \in \mathbb{R}^F$ as input and $\mathbf{\Theta}^{j \to i} \in \mathbb{R}^{F \times F'}$ as outputs, the number of parameters of the $\mathcal{H}$ network depends on $F^2 F'$. When F and F' are high, this number rapidly becomes excessively large, resulting in vanishing gradients or overfitting. In particular, the last layer of the $\mathcal{H}$ network is the most critical, because its output dimension has to be equal to $F \times F'$. Moreover, a second issue related to the $\mathcal{H}$ network regards memory occupation and computations. When we perform the ECC operation, we have to compute a weight matrix $\mathbf{\Theta}^{j \to i}$ for each edge j of every node, of every graph in the batch. This can easily result in very high memory requirements. For example, if we consider a K-regular graph and a batch of B graph signals with N nodes each, the memory occupation needed to store all the matrices $\mathbf{\Theta}^{j \to i}$ is equal to a tensor with dimension $B \times N \times K \times F' \times F$ and this quantity can become unmanageable when the number of feature is high. In addition, since we have to perform a matrix-vector product for each edge of the neighborhood in equation [3.16], the computational cost of the aggregation operation is $O(FF')$, becoming very expensive when F and F' are high.

To address these problems, (Valsesia *et al.* 2019) proposed a lightweight ECC. This new definition introduces some significant improvements that reduce the risk of overparameterization and increase the computational efficiency, allowing deeper networks to be built. The first improvement proposed in Valsesia *et al.* (2019) consists of imposing a structure to the parameter matrix of the last layer in the $\mathcal{H}$ network. In particular, they employ a matrix composed of multiple stacked partial circulant matrices, where only a few shifted versions of the first row are used. Figure 3.3 shows the proposed circulant approximation of the parameter matrix. This approximation helps to reduce the number of parameters of the network, since the only free parameters are in the first row of each partial circulant matrix. For example, if we impose that each partial circulant matrix has m rows, we reduce the number of parameters by a factor m.

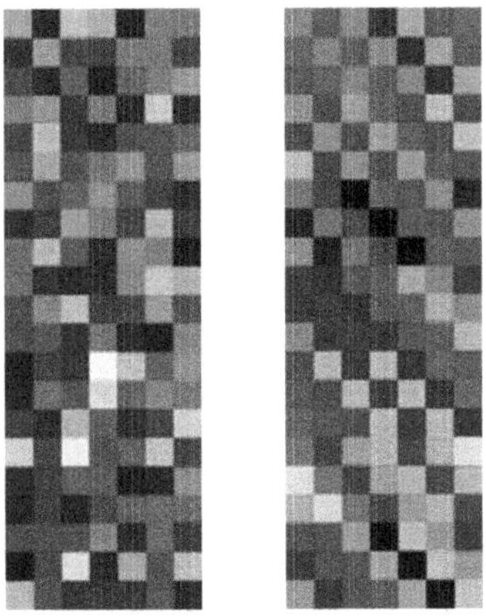

Figure 3.3. *Circulant approximation of a fully connected layer. On the left: A traditional parameter matrix without any structure. On the right: A stack of partial circulant matrices, where the weights in a row are shifted for a number of following rows before being replaced. For a color version of this figure, see www.iste.co.uk/cheung/graph.zip*

A second improvement proposed in Valsesia *et al.* (2019) is a low-rank approximation for $\mathbf{\Theta}^{j\to i}$. Inspired by the singular value decomposition of a matrix, Valsesia *et al.* (2019) propose a low-rank approximation of $\mathbf{\Theta}^{j\to i}$, where $\mathbf{\Theta}^{j\to i}$ is reduced to a sum of r outer products:

$$\mathbf{\Theta}^{j\to i} = \sum_{s=1}^{r} \kappa_s^{j\to i} \boldsymbol{\theta}_s^{j\to i,L} \boldsymbol{\theta}_s^{j\to i,R^T}, \qquad [3.19]$$

where $\boldsymbol{\theta}_s^{j\to i,L} \in \mathbb{R}^F$, $\boldsymbol{\theta}_s^{j\to i,R} \in \mathbb{R}^{F'}$, $\kappa_s^{j\to i} \in \mathbb{R}$ and $1 \leq r \leq F$. The approximation in equation [3.19] does not require that $\boldsymbol{\theta}_s^{j\to i,L}$ and $\boldsymbol{\theta}_s^{j\to i,R}$ are orthogonal, even though random initialization makes them quasi-orthogonal. Therefore, the rank of $\boldsymbol{\Theta}^{j\to i}$ can be lower than or equal to r. This approximation allows the redefinition of the last layer of the $\mathcal{H}$ network, by employing for each s, three parallel fully connected layers that output $\boldsymbol{\theta}_s^{j\to i,L}, \boldsymbol{\theta}_s^{j\to i,R}, \kappa_s^{j\to i}$ for $s = 1, \ldots, r$, instead of having a single fully connected layer that outputs the entire matrix $\boldsymbol{\Theta}^{j\to i}$. The new structure of the $\mathcal{H}$ network is shown in Figure 3.4. This approximation provides significant advantages. First, it drastically reduces the memory occupation, since it is only possible to store $\boldsymbol{\theta}_s^{j\to i,L}$, $\boldsymbol{\theta}_s^{j\to i,R}$ and $\kappa_s^{j\to i}$, instead of the entire matrix $\boldsymbol{\Theta}^{l,j\to i}$. Second, the low-rank approximation allows the computation burden of the convolution to be lowered, because $\mathcal{F}$ can be reduced as follows

$$\mathcal{F}(\mathbf{x}_i, \mathbf{x}_j) = \boldsymbol{\Theta}^{j\to i}\mathbf{x}_j = \sum_{s=1}^{r} \kappa_s^{j\to i}\boldsymbol{\theta}_s^{j\to i,L}\boldsymbol{\theta}_s^{j\to i,R^T}\mathbf{x}_j,$$

where the computation of the matrices $\boldsymbol{\Theta}^{j\to i}$ is not required and the computational cost of the decoupled operation is only $O(r(F + F'))$.

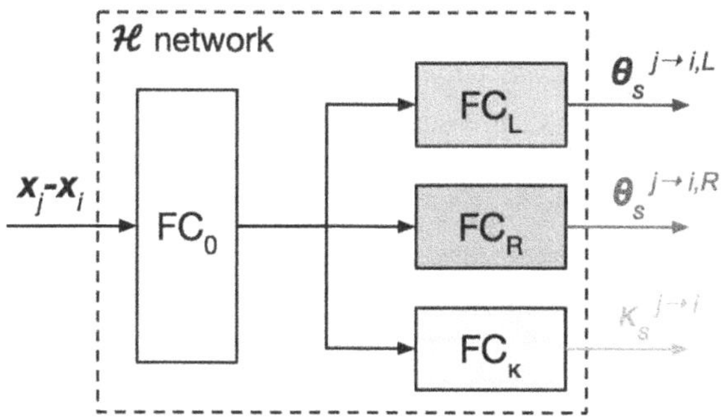

Figure 3.4. *Low-rank approximation of the last layer of the $\mathcal{H}$ network. For a color version of this figure, see www.iste.co.uk/cheung/graph.zip*

3.4. Concluding remarks

Graph neural networks have recently gained attention because of their ability to process data with an irregular domain. In this chapter, we presented an overview of the state-of-the-art methods in this field, but several challenges still need to be tackled. In particular, while spatial approaches seem to be more successful than spectral-based approaches, the research community is still striving to define a graph-convolutional layer that presents all of the desirable properties, and a definition that is universally accepted is still missing. Computational complexity is also still a major issue, especially for the adaptive approaches deriving weights as functions of features.

3.5. References

Bruna, J., Zaremba, W., Szlam, A., LeCun, Y. (2013). Spectral networks and locally connected networks on graphs. arXiv preprint arXiv:1312.6203.

Defferrard, M., Bresson, X., Vandergheynst, P. (2016). Convolutional neural networks on graphs with fast localized spectral filtering. *Advances in Neural Information Processing Systems*, 29, 3844–3852.

Eynard, D., Kovnatsky, A., Bronstein, M.M., Glashoff, K., Bronstein, A.M. (2015). Multimodal manifold analysis by simultaneous diagonalization of Laplacians. *IEEE Transactions on Pattern Analysis and Machine Intelligence*, 37(12), 2505–2517.

Hamilton, W., Ying, Z., Leskovec, J. (2017). Inductive representation learning on large graphs. *Advances in Neural Information Processing Systems*, 30, 1024–1034.

Kipf, T.N. and Welling, M. (2017). Semi-supervised classification with graph convolutional networks. *International Conference on Learning Representations (ICLR) 2017*. arXiv:1609.02907.

Kovnatsky, A., Bronstein, M.M., Bronstein, A.M., Glashoff, K., Kimmel, R. (2013). Coupled quasi-harmonic bases. *Computer Graphics Forum*, 32(issue 2, part 4), 439–448.

Levie, R., Monti, F., Bresson, X., Bronstein, M.M. (2018). Cayleynets: Graph convolutional neural networks with complex rational spectral filters. *IEEE Transactions on Signal Processing*, 67(1), 97–109.

Liao, R., Zhao, Z., Urtasun, R., Zemel, R. (2019). LanczosNet: Multi-scale deep graph convolutional networks. *International Conference on Learning Representations (ICLR) 2019*. arXiv:1901.01484.

Monti, F., Boscaini, D., Masci, J., Rodola, E., Svoboda, J., Bronstein, M.M. (2017). Geometric deep learning on graphs and manifolds using mixture model cnns. *Proceedings of the IEEE Conference on Computer Vision and Pattern Recognition*, 5115–5124.

Simonovsky, M. and Komodakis, N. (2017). Dynamic edge-conditioned filters in convolutional neural networks on graphs. *IEEE Conference on Computer Vision and Pattern Recognition (CVPR)*, 29–38.

Wang, Y., Sun, Y., Liu, Z., Sarma, S.E., Bronstein, M.M., Solomon, J.M. (2019). Dynamic graph CNN for learning on point clouds. *ACM Transactions on Graphics (TOG)*, 38(5).

Valsesia, D., Fracastoro, G., Magli, E. (2019). Deep graph-convolutional image denoising. arXiv preprint arXiv:1907.08448.

Verma, N., Boyer, E., Verbeek, J. (2018). Feastnet: Feature-steered graph convolutions for 3D shape analysis. *Proceedings of the IEEE Conference on Computer Vision and Pattern Recognition*, 2598–2606.

PART 2

Imaging Applications of Graph Signal Processing

4

Graph Spectral Image and Video Compression

Hilmi E. EGILMEZ[1], **Yung-Hsuan CHAO**[1] **and Antonio ORTEGA**[2]
[1] *Qualcomm Technologies Inc., San Diego, USA*
[2] *University of Southern California, Los Angeles, USA*

4.1. Introduction

Transforms are integral parts of many state-of-the-art compression systems and standards, and are used to provide compact spectral representations for signals to be compressed. This chapter presents methods for building graph Fourier transforms (GFTs)[1] for image and video compression. A key insight developed in this chapter is that classical transforms, such as the discrete sine/cosine transform (DST/DCT) or the Karhunen–Loeve transform (KLT), can be interpreted from a graph perspective. Thus, the ideas presented in this chapter can be viewed as extensions of these conventional techniques, where changes to some of the underlying assumptions are made. We consider two sets of techniques for designing graphs, from which the associated GFTs are derived:

– *Graph learning oriented GFT (GL-GFT)* approaches aim to find graphs that best fit a collection of image/video block data. Similar to the KLT, these methods are data driven, but unlike the KLT they learn the graph (rather than the transform) and introduce regularization parameters as part of the learning.

---

1 The reader is referred to the Introduction for a detailed discussion of GFTs. The separable and non-separable block GFTs used throughout this chapter are formally introduced in definitions 4.1 and 4.2.

Graph Spectral Image Processing,
coordinated by Gene CHEUNG and Enrico MAGLI.

– *Block-adaptive GFT (BA-GFT)* approaches are based on adaptively updating graph weights according to important features (such as texture and discontinuities) in an image/video block. A typical approach to design these transforms is to start with a graph with equal weights, such as the one corresponding to the DCT, and adapt its weights to incorporate signal characteristics.

The potential advantages of these methods over traditional techniques are theoretically analyzed and empirically demonstrated in what follows.

DEFINITION 4.1 (Separable GFT).– Let $\mathbf{U}_{\text{row}}$ and $\mathbf{U}_{\text{col}}$ be $N \times N$ GFTs associated with two line graphs with N vertices, then the separable GFT (S-GFT) of $\mathbf{X}$ is

$$\widehat{\mathbf{X}} = \mathbf{U}_{\text{col}}^{\top}\mathbf{X}\mathbf{U}_{\text{row}}, \qquad [4.1]$$

where $\mathbf{U}_{\text{row}}$ and $\mathbf{U}_{\text{col}}$ are applied to each row and each column of an $N \times N$ block signal $\mathbf{X}$, respectively.

DEFINITION 4.2 (Non-separable GFT).– Let $\mathbf{U}$ be an $N^2 \times N^2$ GFT associated with a graph with N^2 vertices, then the non-separable GFT (NS-GFT) of $N \times N$ block signal $\mathbf{X}$ is

$$\widehat{\mathbf{X}} = \text{block}(\mathbf{U}^{\top}\text{vec}(\mathbf{X})), \qquad [4.2]$$

where $\mathbf{U}$ is applied on vectorized signal $\mathbf{x} = \text{vec}(\mathbf{X})$, and the block operator restructures the signal back in block form.

Figure 4.1 illustrates separable and non-separable transformation applied on a block.

Box 4.1. *Formal definitions of separable and non-separable GFTs*

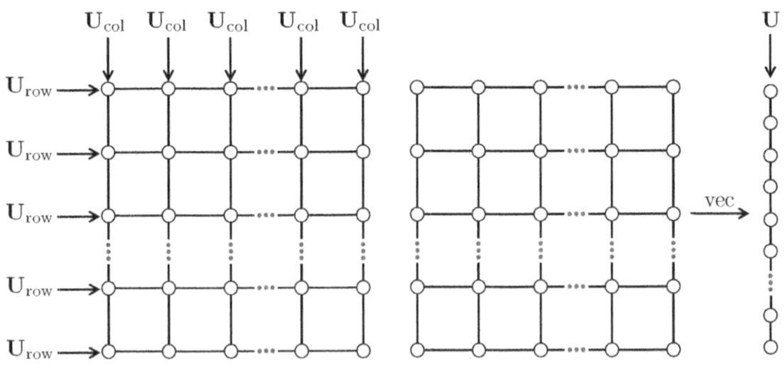

Figure 4.1. *Separable and non-separable transforms: (Left) For an $N \times N$ block, separable transforms are composed of two (possibly distinct) $N \times N$ transforms, $\mathbf{U}_{row}$ and $\mathbf{U}_{col}$, applied to rows and columns of the block. (Right) Non-separable transforms apply a $N^2 \times N^2$ linear transformation using $\mathbf{U}$*

4.1.1. *Basics of image and video compression*

Predictive transform coding is a fundamental compression technique adopted in many block-based image and video compression systems, where block signals are initially predicted from a set of available (coded) reference pixels, and then the resulting residual block signals are transformed (generally by an orthogonal transformation) to decorrelate residual pixel values for effective compression. After prediction and transformation steps, a typical image/video compression system applies quantization and entropy coding to convert transform coefficients into a stream of bits. Figure 4.2 illustrates a representative encoder–decoder architecture comprising three main steps: (i) prediction, (ii) transformation, and (iii) quantization and entropy coding. Important examples of such image and video coding architectures include *Joint Photographic Experts Group* (JPEG) (Pennebaker and Mitchell 1992), *Advanced Video Coding* (AVC) (Wiegand *et al.* 2003), *High-Efficiency Video Coding* (HEVC) (Sullivan *et al.* 2012), VP9 (Mukherjee *et al.* 2013), *AOMedia Video 1* (AV1) (Chen *et al.* 2018) and *Versatile Video Coding* (VVC) (Bross *et al.* 2020) standards.

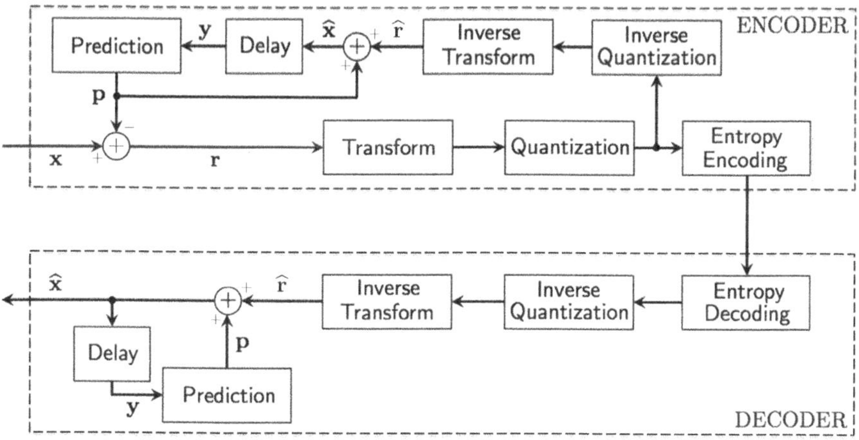

Figure 4.2. *Building blocks of a typical image/video encoder and decoder consisting of three main steps, which are (i) prediction, (ii) transformation, and (iii) quantization and entropy coding. This figure is adapted from Figure 1 in Egilmez* et al. *(2020a)*

In predictive transform coding of video, prediction is typically carried out by choosing one multiple intra- or inter-prediction mode in order to exploit spatial and temporal redundancies between block signals[2]. The transformation step allows efficient spectral domain representations for (prediction) residual signals, whose transform domain energy is typically concentrated in a few transform coefficients. Therefore, for the effective coding of images and videos it is important to develop transforms that are well adapted to the statistics of residual signals.

In video coding standards predating HEVC and VP9, a single transform, typically the type-2 discrete cosine transform (DCT-2), is applied in a separable manner to rows and columns of each residual block. The main problem of using fixed block transforms is the implicit assumption that all residual blocks share the same statistical properties. However, image/video blocks can have very diverse statistical characteristics depending on the video content and the prediction mode. The HEVC standard (Sullivan *et al.* 2012) has aimed to address this problem by allowing the use of the type-7 discrete sine transform (DST-7), in addition to the DCT-2 for small (i.e. 4×4) intra-predicted blocks. In order to further increase the diversity in transform selection, AV1 (Chen *et al.* 2018) enables combinations of DST-7 and DCT-8 for block sizes up to 16×16. Moreover, VVC employs *multiple transform selection* (MTS) among five candidates based on DCT-2, DCT-8 and DST-7, where larger block transforms up to size 32×32 are supported (Said and Zhao 2018). Finally, it has been shown that better compression can be achieved by using data-driven transform designs that adapt to statistical properties of residual blocks (Zhao *et al.* 2016b; Said *et al.* 2016; Egilmez *et al.* 2020a).

4.1.2. *Literature review*

The majority of studies on data-driven transform designs for image and video coding focus on methods based on KLT (Goyal 2001). For example, a mode-dependent transform (MDT) scheme was proposed by Ye and Karczewicz (2008), where a KLT is derived for each intra-prediction mode. More recently, in Takamura and Shimizu (2013); Arrufat *et al.* (2014), the MDT scheme is similarly implemented in the HEVC standard, where a single KLT is trained for each intra prediction mode offered in HEVC. Considerable coding gains over the MDT method can be achieved by using the rate-distortion optimized transformation (RDOT) scheme (Zou *et al.* 2013; Zhao *et al.* 2016a,b; Said *et al.* 2016; Egilmez *et al.* 2020a), which suggests designing multiple transforms for each prediction mode, so that the encoder can select a transform (from the predetermined set of transforms) by minimizing a rate-distortion (RD) cost. Since the RDOT scheme allows the flexibility of selecting a transform on a per-block basis on the encoder side, it provides better adaptation to residual blocks with different statistical characteristics, compared to the MDT scheme.

---

2 Note that image coding can be viewed as part of video compression, since videos consist of multiple frames (i.e. images), some of which are only coded via intra prediction.

Most MDT-and RDOT-based schemes in literature rely on KLTs estimated by the eigenvectors of sample covariance matrices. However, it has been shown that sample covariances may not be good estimators for the true covariances of models due to overfitting, especially when the number of samples is small (Johnstone and Lu 2009; Ravikumar *et al.* 2011), and more accurate estimates can be obtained using sparse inverse covariance estimation methods (Friedman *et al.* 2008; Ravikumar *et al.* 2011; Egilmez *et al.* 2017b; Egilmez *et al.* 2019). The approaches discussed in this chapter aim to estimate more accurate and robust models by learning graphs whose spectra (i.e. GFTs) provide effective representations for image/video signals. While graphs can be learned from covariance data, the number of graph parameters to be learned (edge and vertex weights) is much smaller than what is required to design a KLT. This alleviates the overfitting problem, so that these graph-based transforms can be viewed as regularized variants of KLTs.

4.1.3. *Outline of the chapter*

This chapter provides an overview of GL-GFT and BA-GFT construction methods for image and video compression. Section 4.2 presents the models defined based on graph Laplacian matrices with probabilistic interpretations. Section 4.3 discusses two well-established methodologies for GL-GFT and BA-GFT construction. Specifically, a graph learning framework with a maximum-likelihood (ML) criterion for GL-GFT construction, and an edge-adaptive GFT scheme for BA-GFT are discussed in detail. The coding benefits of these methods are both empirically and theoretically justified. Concluding remarks, open issues and future potential lines of work are presented in section 4.4.

4.2. Graph-based models for image and video signals

The graph spectral approaches discussed in this chapter aim to find graph Laplacian matrices, which denote the inverse covariances for the models of interest. Then, the GFTs are derived based on these graph Laplacians. Figure 4.3 illustrates this basic idea of *designing spectral transforms from graphs*, where a graph Laplacian can be learned directly from the samples of data (e.g. on a per-block basis) or from their aggregated statistics (e.g. sample covariance).

Specifically, the models considered in this chapter are defined based on the class of matrices called generalized graph Laplacians (GGLs)[3], which can be used to represent the pairwise relations between samples of image/video signals. In a probabilistic sense, these models correspond to a special class of Gaussian Markov

3 In the Introduction, generalized graph Laplacians are also called loopy graph Laplacians.

random field (GMRF), whose precision (inverse covariance) matrix is a GGL:

$$\mathbf{x} \sim \mathsf{N}(\mathbf{0}, \mathbf{L}^{-1}) : \quad \mathsf{p}(\mathbf{x}|\mathbf{L}) = \frac{1}{(2\pi)^{n/2} \det(\mathbf{L})^{-1/2}} \exp\left(-\frac{1}{2}\mathbf{x}^\top \mathbf{L}\mathbf{x}\right), \qquad [4.3]$$

where the random vector $\mathbf{x} \in \mathbb{R}^n$ has a zero-mean multivariate Gaussian distribution[4] defined by graph Laplacian $\mathbf{L}$. The entries of $\mathbf{L}$ can be interpreted in terms of conditional dependence relations among the variables in $\mathbf{x}$,

$$\mathsf{E}\left[x_i \,|\, (\mathbf{x})_{\mathcal{S}\setminus\{i\}}\right] = -\frac{1}{(\mathbf{L})_{ii}} \sum_{j \in \mathcal{S}\setminus\{i\}} (\mathbf{L})_{ij} x_j \qquad [4.4]$$

$$\mathsf{Prec}\left[x_i \,|\, (\mathbf{x})_{\mathcal{S}\setminus\{i\}}\right] = (\mathbf{L})_{ii} \qquad [4.5]$$

$$\mathsf{Corr}\left[x_i x_j \,|\, (\mathbf{x})_{\mathcal{S}\setminus\{i,j\}}\right] = -\frac{(\mathbf{L})_{ij}}{\sqrt{(\mathbf{L})_{ii}(\mathbf{L})_{jj}}} \quad i \neq j, \qquad [4.6]$$

where $\mathcal{S} = \{1, \ldots, n\}$ is the index set for $\mathbf{x} = [x_1, \ldots, x_n]^\top$. The conditional expectation in equation [4.4] gives the best minimum mean square error (MMSE) estimate of x_i using all other random variables. The relation in equation [4.5] corresponds to the *precision* of x_i, and equation [4.6] to the *partial correlation* between x_i and x_j (i.e. correlation between random variables x_i and x_j given all other variables in $\mathbf{x}$). If x_i and x_j are conditionally independent, then there is no edge between the corresponding vertices v_i and v_j in the associated graph (i.e. $(\mathbf{L})_{ij} = 0$).

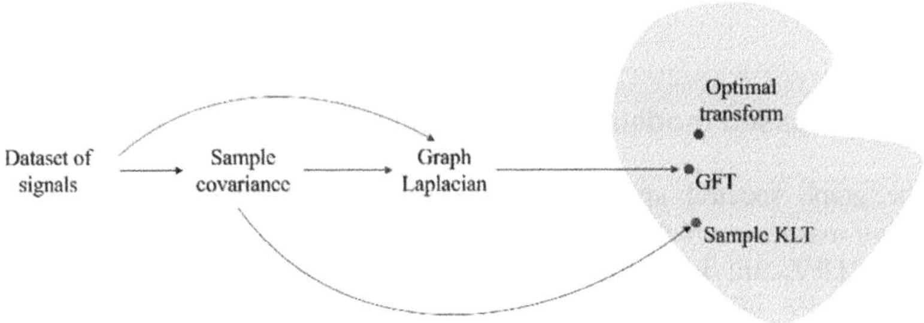

Figure 4.3. *Illustration of spectral transform design methodology. Traditional methods use a sample covariance to derive an approximation of KLT (sample KLT). This chapter focuses on approaches to find a graph Laplacian from data for designing GFTs (whose steps are represented by red arrows). For a color version of this figure, see www.iste.co.uk/cheung/graph.zip*

4 The zero-mean assumption is made to simplify the notation. The models can be trivially extended to GMRFs with non-zero mean.

In the statistical modeling of image and video signals (e.g. auto-regressive models), it is generally assumed that adjacent pixel values are positively correlated (with a correlation coefficient close to one) (Jain 1989; Tekalp 2015). This assumption is intuitively reasonable, since neighboring pixel values in image/video blocks are often similar to each other due to spatial and temporal redundancy. The GMRF model in equation [4.3] nicely captures this property, as partial correlations in equation [4.6] are always non-negative (i.e. off-diagonal elements of $\mathbf{L}$ are non-positive) by the definition of generalized graph Laplacians. In literature, such models are classified as *attractive GMRFs* (Koller and Friedman 2009; Rue and Held 2005; Egilmez *et al.* 2017b), which are well suited for effectively representing image/video signals (Egilmez *et al.* 2017b, 2016).

4.2.1. *Graph-based models for residuals of predicted signals*

This section presents GMRF models based on equation [4.3] for residual signals obtained from intra-inter-prediction. In section 4.2.1.1, a general 2D block model is introduced. section 4.2.1.2 discusses more specific 1D line models, with rigorous derivations of two separate GMRFs for intra- and inter-predicted blocks.

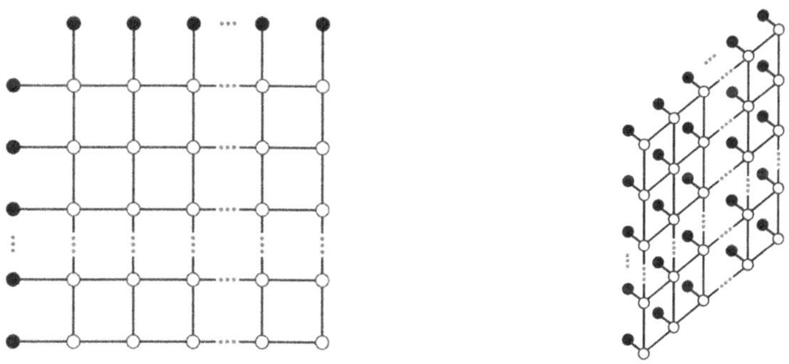

(a) GMRF for intra-predicted blocks (b) GMRF for inter-predicted blocks

Figure 4.4. *2D GMRF models for intra- and inter-predicted signals. Filled vertices correspond to reference pixels obtained (a) from neighboring blocks and (b) from other frames via motion compensation. Unfilled vertices denote the pixels to be predicted and then transform coded. This figure is adapted from Figure 3 in Egilmez* et al. *(2020a)*

4.2.1.1. *A general model for residual signals*

We introduce a general 2D GMRF model for intra-inter-predicted $N \times N$ block signals, depicted in Figure 4.4, by deriving the precision matrix of the residual signal $\mathbf{r}$, which is obtained after predicting the signal $\mathbf{x} = [x_1 \, x_2 \, \cdots \, x_n]^\top$, with $n = N^2$, from n_p reference samples in $\mathbf{y} = [y_1\ y_2\ \cdots\ y_{n_p}]^\top$ (i.e. predicting unfilled vertices from black filled vertices in Figure 4.4), where $\mathbf{x}$ and $\mathbf{y}$ are zero-mean and jointly Gaussian with respect to the following attractive GMRF:

$$\mathsf{p}([\begin{smallmatrix}\mathbf{x}\\ \mathbf{y}\end{smallmatrix}] \mid \mathbf{L}) = \frac{1}{(2\pi)^{n/2}\det(\mathbf{L})^{-1/2}}\exp\left(-\frac{1}{2}[\begin{smallmatrix}\mathbf{x}\\ \mathbf{y}\end{smallmatrix}]^\top \mathbf{L}\,[\begin{smallmatrix}\mathbf{x}\\ \mathbf{y}\end{smallmatrix}]\right). \quad [4.7]$$

The precision matrix $\mathbf{L}$ and the covariance matrix $\boldsymbol{\Sigma} = \mathbf{L}^{-1}$ (assuming that $\mathbf{L}$ is invertible[5]) can be partitioned as follows (Rue and Held 2005; Zhang *et al.* 2015):

$$\mathbf{L} = \begin{bmatrix}\mathbf{L}_{\mathbf{x}} & \mathbf{L}_{\mathbf{xy}}\\ \mathbf{L}_{\mathbf{yx}} & \mathbf{L}_{\mathbf{y}}\end{bmatrix} = \begin{bmatrix}\boldsymbol{\Sigma}_{\mathbf{x}} & \boldsymbol{\Sigma}_{\mathbf{xy}}\\ \boldsymbol{\Sigma}_{\mathbf{yx}} & \boldsymbol{\Sigma}_{\mathbf{y}}\end{bmatrix}^{-1} = \boldsymbol{\Sigma}^{-1}. \quad [4.8]$$

Irrespective of the type of prediction, the MMSE prediction of $\mathbf{x}$ from the reference samples (given $\mathbf{y}$) is

$$\mathbf{p} = \mathsf{E}[\mathbf{x}|\mathbf{y}] = \boldsymbol{\Sigma}_{\mathbf{xy}}\boldsymbol{\Sigma}_{\mathbf{y}}^{-1}\mathbf{y} = -\mathbf{L}_{\mathbf{x}}^{-1}\mathbf{L}_{\mathbf{xy}}\mathbf{y}, \quad [4.9]$$

and the resulting residual vector is $\mathbf{r} = \mathbf{x} - \mathbf{p}$ with covariance

$$\begin{aligned}\boldsymbol{\Sigma}_{\mathbf{r}} = \boldsymbol{\Sigma}_{\mathbf{x}|\mathbf{y}} &= \mathsf{E}[\mathbf{r}\mathbf{r}^\top] = \mathsf{E}[(\mathbf{x}-\mathbf{p})(\mathbf{x}-\mathbf{p})^\top]\\ &= \mathsf{E}[\mathbf{x}\mathbf{x}^\top + \mathbf{p}\mathbf{p}^\top - 2\mathbf{x}\mathbf{p}^\top]\\ &= \boldsymbol{\Sigma}_{\mathbf{x}} + \boldsymbol{\Sigma}_{\mathbf{xy}}\boldsymbol{\Sigma}_{\mathbf{y}}^{-1}\boldsymbol{\Sigma}_{\mathbf{yx}} - 2\boldsymbol{\Sigma}_{\mathbf{xy}}\boldsymbol{\Sigma}_{\mathbf{y}}^{-1}\boldsymbol{\Sigma}_{\mathbf{yx}}\\ &= \boldsymbol{\Sigma}_{\mathbf{x}} - \boldsymbol{\Sigma}_{\mathbf{xy}}\boldsymbol{\Sigma}_{\mathbf{y}}^{-1}\boldsymbol{\Sigma}_{\mathbf{yx}}.\end{aligned} \quad [4.10]$$

By the matrix inversion lemma in Woodbury (1950), the precision matrix of the residual $\mathbf{r}$ is in fact equal to $\mathbf{L}_{\mathbf{x}}$, the submatrix in equation [4.8],

$$\boldsymbol{\Sigma}_{\mathbf{r}}^{-1} = (\boldsymbol{\Sigma}_{\mathbf{x}} - \boldsymbol{\Sigma}_{\mathbf{xy}}\boldsymbol{\Sigma}_{\mathbf{y}}^{-1}\boldsymbol{\Sigma}_{\mathbf{yx}})^{-1} = \mathbf{L}_{\mathbf{x}}. \quad [4.11]$$

Since we also have $\boldsymbol{\Sigma}_{\mathbf{x}} = (\mathbf{L}_{\mathbf{x}} - \mathbf{L}_{\mathbf{xy}}\mathbf{L}_{\mathbf{y}}^{-1}\mathbf{L}_{\mathbf{yx}})^{-1}$ by Woodbury (1950), the desired precision matrix can also be written as

$$\boldsymbol{\Omega}_{\text{residual}} = \boldsymbol{\Sigma}_{\mathbf{r}}^{-1} = \mathbf{L}_{\mathbf{x}} = \boldsymbol{\Sigma}_{\mathbf{x}}^{-1} + \mathbf{L}_{\mathbf{xy}}\mathbf{L}_{\mathbf{y}}^{-1}\mathbf{L}_{\mathbf{yx}}. \quad [4.12]$$

This construction leads us to the following theorem, formally stating the conditions for a residual signal (i.e. $\mathbf{r}$) to be modeled by an attractive GMRF.

5 Note that GGLs without any vertex weights (without self-loops) correspond to combinatorial Laplacians, which are not invertible by definition. The reader is referred to (Egilmez *et al.* 2017b) for models associated with combinatorial Laplacians.

THEOREM 4.1.– Let the signals $\mathbf{x} \in \mathbb{R}^n$ and $\mathbf{y} \in \mathbb{R}^{n_p}$ be distributed based on the attractive GMRF model in equation [4.7], with precision matrix $\mathbf{L}$. If the residual signal $\mathbf{r}$ is estimated by the minimum mean square error (MMSE) prediction of $\mathbf{x}$ from $\mathbf{y}$ (i.e. $\mathbf{r} = \mathbf{x} - \mathsf{E}[\mathbf{x}|\mathbf{y}]$), then the residual signal $\mathbf{r}$ is distributed as an attractive GMRF, whose precision matrix $\boldsymbol{\Omega}_{\text{residual}}$ in equation [4.12] is a generalized graph Laplacian.

The proof follows from equations [4.7]–[4.12] where the inverse covariance of residual signal $\mathbf{r}$, $\boldsymbol{\Sigma}_{\mathbf{r}}^{-1}$, is shown to be equal to $\mathbf{L}_{\mathbf{x}}$. Since $\mathbf{L}_{\mathbf{x}}$ is a submatrix of $\mathbf{L}$ in equation [4.8] and $\mathbf{L}$ is a GGL, $\boldsymbol{\Omega}_{\text{residual}} = \mathbf{L}_{\mathbf{x}}$ is also a GGL. Hence, $\mathbf{r}$ is distributed as an attractive GMRF whose precision is $\mathbf{L}_{\mathbf{x}}$.

In what follows, we present particular cases of the above general model for 1D intra- and inter-predicted residuals. The related derivations will then be used to analyze models' relations with discrete sine and cosine transforms in section 4.2.2.

4.2.1.2. *1D line models for residual signals*

In order to model rows/columns of $N \times N$ block residual signals, we construct 1D GMRFs based on first-order autoregressive (AR) processes. Depending on the type of prediction (i.e. intra/inter), a different set of reference samples, denoted as $\mathbf{y}$, is used to predict $\mathbf{x}$, representing $n = N$ samples in a row/column of a block[6].

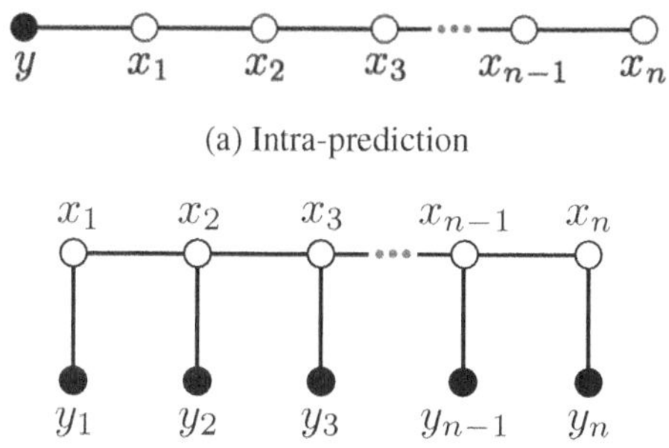

(b) Inter-prediction

Figure 4.5. *1D GMRF models for (a) intra- and (b) inter-predicted signals. Black-filled vertices represent the reference pixels and unfilled vertices denote pixels to be predicted and then transform coded. This figure is adapted from Figure 2 in Egilmez* et al. *(2020a)*

Intra-predicted signal model: For the modeling of intra-predicted signals as 1D GMRFs, illustrated in Figure 4.5(a), the following stochastic difference equations

6 For 1D models, the number of vertices (n) is equal to the number of pixels (N) in a row/column of an $N \times N$ block.

generalize existing models by allowing arbitrary ρ_i correlation parameters, which instead were fixed in Han *et al.* (2012); Hu *et al.* (2015a), for $i = 0, \dots, n-1$,

$$
\begin{aligned}
x_1 &= \rho_0(y+d) + e_1 \\
x_2 &= \rho_1 x_1 + e_2 \\
&\vdots \\
x_{n-1} &= \rho_{n-2} x_{n-2} + e_{n-1} \\
x_n &= \rho_{n-1} x_{n-1} + e_n
\end{aligned}
\qquad [4.13]
$$

where the reference sample y is used to predict n samples in $\mathbf{x} = [x_1\ x_2\ \cdots\ x_n]^\top$. The random variable $d \sim \mathsf{N}(0, \sigma_d^2)$ models the distortion due to compression in the reference sample y, and $e_i \sim \mathsf{N}(0, \sigma_e^2)$ is the noise in x_i with a fixed variance σ_e^2. The spatial correlation coefficients between samples are denoted by $\rho_0, \rho_1, \dots, \rho_{n-1}$, and we also assume that random variables d and e_i are independent incase of $i = 1, \dots, n$.

In the following, we present a rigorous derivation for the precision matrix of the residual signal ($\mathbf{r}$) obtained, based on the optimal MSE prediction from y. The reader may skip the intermediate steps and go straight to the resulting precision matrix stated in equation [4.17].

The relations in equation [4.13] can be written more compactly as $\mathbf{Q}\mathbf{x} = \mathbf{y}_1 + \mathbf{d}_1 + \mathbf{e}$, where $\mathbf{y}_1 = [(\rho_0 y)\ 0 \cdots 0]^\top$, $\mathbf{d}_1 = [(\rho_0 d)\ 0 \cdots 0]^\top$, $\mathbf{e} = [e_1\ e_2 \cdots e_n]^\top$ and

$$
\mathbf{Q} = \begin{bmatrix}
1 & 0 & \cdots & \cdots & \cdots & 0 \\
-\rho_1 & 1 & 0 & & & \vdots \\
0 & -\rho_2 & 1 & \ddots & & \vdots \\
\vdots & \ddots & \ddots & \ddots & \ddots & \vdots \\
\vdots & & \ddots & -\rho_{n-2} & 1 & 0 \\
0 & \cdots & \cdots & 0 & -\rho_{n-1} & 1
\end{bmatrix}. \qquad [4.14]
$$

Since $\mathbf{x} = \mathbf{Q}^{-1}\mathbf{y}_1 + \mathbf{Q}^{-1}(\mathbf{d}_1 + \mathbf{e})$, where $\mathbf{p} = \mathbf{Q}^{-1}\mathbf{y}_1$ is the optimal prediction for $\mathbf{x}$, the resulting residual vector is $\mathbf{r} = \mathbf{x} - \mathbf{p}$, and its covariance matrix is

$$
\boldsymbol{\Sigma}_{\mathbf{r}} = \mathbf{Q}^{-1}\mathsf{E}\left[(\mathbf{e}+\mathbf{d}_1)(\mathbf{e}+\mathbf{d}_1)^\top\right](\mathbf{Q}^{-1})^\top. \qquad [4.15]
$$

Inverting the covariance matrix gives us the precision matrix,

$$\mathbf{L}_{\text{intra}} = \boldsymbol{\Sigma}_{\mathbf{r}}{}^{-1} = \mathbf{Q}^{\top}\left(\mathsf{E}\left[\mathbf{e}\mathbf{e}^{\top} + \mathbf{d}_1\mathbf{d}_1^{\top}\right]\right)^{-1}\mathbf{Q}$$
$$= \frac{1}{\sigma_e^2}\mathbf{Q}^{\top}\begin{bmatrix} \frac{1}{1+\beta_d} & 0 & \cdots & 0 \\ 0 & 1 & & \vdots \\ \vdots & & \ddots & 0 \\ 0 & \cdots & 0 & 1 \end{bmatrix}\mathbf{Q}, \qquad [4.16]$$

which defines our 1D GMRF model for the residual signal $\mathbf{r}$. The resulting $\mathbf{L}_{\text{intra}}$ can be explicitly stated as:

$$\frac{1}{\sigma_e^2}\begin{bmatrix} \frac{1}{1+\beta_d} + \rho_1^2 & -\rho_1 & 0 & \cdots & \cdots & 0 \\ -\rho_1 & 1+\rho_2^2 & -\rho_2 & 0 & & \vdots \\ 0 & -\rho_2 & 1+\rho_3^2 & -\rho_3 & \ddots & \vdots \\ \vdots & \ddots & \ddots & \ddots & \ddots & 0 \\ \vdots & & \ddots & -\rho_{n-2} & 1+\rho_{n-1}^2 & -\rho_{n-1} \\ 0 & \cdots & \cdots & 0 & -\rho_{n-1} & 1 \end{bmatrix} \qquad [4.17]$$

where $\beta_d = \frac{(\rho_0\sigma_d)^2}{\sigma_e^2}$.

As neighboring pixels in image/video signals are highly correlated in practice, a simplified model can be obtained by letting ρ_i for $i = 0, 1, \ldots, n-1$ be equal to 1, so that the resulting $\mathbf{L}_{\text{intra}}$ reduces to a GGL in the following form:

$$\frac{1}{\sigma_e^2}\begin{bmatrix} \frac{1}{1+\beta_d} + 1 & -1 & 0 & \cdots & \cdots & 0 \\ -1 & 2 & -1 & 0 & & \vdots \\ 0 & -1 & 2 & -1 & \ddots & \vdots \\ \vdots & \ddots & \ddots & \ddots & \ddots & 0 \\ \vdots & & \ddots & -1 & 2 & -1 \\ 0 & \cdots & \cdots & 0 & -1 & 1 \end{bmatrix}, \qquad [4.18]$$

whose relationship with different types of discrete sine and cosine transforms, in terms of different choices of σ_d (in β_d) and σ_e, is discussed in section 4.2.2.

Inter-predicted signal model: In modeling inter-predicted signals, we have n reference samples, $y_1, \ldots, y_n$, used to predict n samples in $\mathbf{x} = [x_1\ x_2\ \cdots\ x_n]^{\top}$, as shown in Figure 4.5(b). Thus, we derive the following difference equations by incorporating multiple reference samples as

$$\begin{aligned} x_1 &= \widetilde{\rho}_1(y_1 + d_1) + e_1 \\ x_2 &= \rho_1 x_1 + \widetilde{\rho}_2(y_2 + d_2) + e_2 \\ &\vdots \\ x_{n-1} &= \rho_{n-2} x_{n-2} + \widetilde{\rho}_{n-1}(y_{n-1} + d_{n-1}) + e_{n-1} \\ x_n &= \rho_{n-1} x_{n-1} + \widetilde{\rho}_n(y_n + d_n) + e_n \end{aligned} \quad [4.19]$$

where $d_i \sim \mathsf{N}(0, \sigma_{d_i}^2)$ denotes the distortion due to compression in the reference sample y_i and $e_i \sim \mathsf{N}(0, \sigma_e^2)$ is the noise in sample x_i. The random variables e_i and d_i are also assumed to be independent. In addition to the spatial correlation coefficients $\rho_1, \ldots, \rho_{n-1}$, our model includes temporal correlation coefficients, $\widetilde{\rho}_1, \ldots, \widetilde{\rho}_n$.

As in the intra-model, we also provide a derivation of the precision matrix for inter-predicted residual signals, where the resulting precision matrix is stated in equation [4.22]. For the derivation, the recursive relations in equation [4.19] can also be written in vector form, that is $\mathbf{Qx} = \mathbf{y}_2 + \mathbf{d}_2 + \mathbf{e}$, where $\mathbf{Q}$ is given in equation [4.14], $\mathbf{y}_2 = [\widetilde{\rho}_1 y_1 \; \widetilde{\rho}_2 y_2 \cdots \widetilde{\rho}_n y_n]^\top$ and $\mathbf{d}_2 = [\widetilde{\rho}_1 d_1 \; \widetilde{\rho}_2 d_2 \cdots \widetilde{\rho}_n d_n]^\top$. We write $\mathbf{x} = \mathbf{Q}^{-1}\mathbf{y}_2 + \mathbf{Q}^{-1}(\mathbf{d}_2 + \mathbf{e})$, where $\mathbf{p} = \mathbf{Q}^{-1}\mathbf{y}_2$ is the optimal prediction for $\mathbf{x}$. So, the resulting residual vector is $\mathbf{r} = \mathbf{x} - \mathbf{p} = \mathbf{Q}^{-1}(\mathbf{d}_2 + \mathbf{e})$ and its covariance is

$$\mathbf{\Sigma_r} = \mathbf{Q}^{-1}\mathsf{E}\left[(\mathbf{e} + \mathbf{d}_2)(\mathbf{e} + \mathbf{d}_2)^\top\right](\mathbf{Q}^{-1})^\top. \quad [4.20]$$

By inverting the covariance matrix $\mathbf{\Sigma_r}$, we obtain the precision matrix

$$\begin{aligned} \mathbf{L}_{\text{inter}} &= \mathbf{\Sigma_r}^{-1} = \mathbf{Q}^\top \left(\mathsf{E}\left[\mathbf{ee}^\top + \mathbf{d}_2\mathbf{d}_2^\top\right]\right)^{-1}\mathbf{Q} \\ &= \frac{1}{\sigma_e^2}\mathbf{Q}^\top \begin{bmatrix} \frac{1}{1+\gamma_1} & 0 & \cdots & 0 \\ 0 & \frac{1}{1+\gamma_2} & & \vdots \\ \vdots & & \ddots & 0 \\ 0 & \cdots & 0 & \frac{1}{1+\gamma_n} \end{bmatrix} \mathbf{Q}, \end{aligned} \quad [4.21]$$

where $\gamma_i = \frac{(\widetilde{\rho}_i \sigma_{d_i})^2}{\sigma_e^2}$ for $i = 1, \ldots, n$. More explicitly, the precision $\mathbf{L}_{\text{inter}}$ can be written as:

$$
\frac{1}{\sigma_e^2}\begin{bmatrix}
\frac{1}{1+\gamma_1}+\frac{\rho_1^2}{1+\gamma_2} & \frac{-\rho_1}{1+\gamma_2} & 0 & \cdots & \cdots & 0 \\
\frac{-\rho_1}{1+\gamma_2} & \frac{1}{1+\gamma_2}+\frac{\rho_2^2}{1+\gamma_3} & \frac{-\rho_2}{1+\gamma_3} & \ddots & \ddots & \vdots \\
0 & \ddots & \ddots & \ddots & \ddots & \vdots \\
\vdots & \ddots & \ddots & \ddots & \frac{-\rho_{n-2}}{1+\gamma_{n-1}} & 0 \\
\vdots & \ddots & \ddots & \frac{-\rho_{n-2}}{1+\gamma_{n-1}} & \frac{1}{1+\gamma_{n-1}}+\frac{\rho_{n-1}^2}{1+\gamma_n} & \frac{-\rho_{n-1}}{1+\gamma_n} \\
0 & \cdots & \cdots & 0 & \frac{-\rho_{n-1}}{1+\gamma_n} & \frac{1}{1+\gamma_n}
\end{bmatrix} \qquad [4.22]
$$

The 1D GMRF models discussed above are attractive if $\rho_1, \rho_2, \ldots, \rho_{n-1}$ are nonnegative. In this case, the corresponding precision matrices in equation [4.17] and [4.22] are GGL matrices. Note that theorem 4.1 also applies to the 1D signal models presented in Egilmez *et al.* (2016), which are special cases of equation [4.7].

4.2.2. *DCT/DSTs as GFTs and their relation to 1D models*

Certain types of DCTs and DSTs, including DCT-2 and DST-7, are in fact special cases of GFTs derived from the Laplacians of specific line graphs. The relation between different types of DCTs and graph Laplacian matrices is discussed by Strang (1999), where DCT-2 is shown to be equal to the GFT uniquely obtained from graph Laplacians of the following form:

$$
\mathbf{L}_u = \begin{bmatrix}
c & -c & & & 0 \\
-c & 2c & -c & & \\
 & \ddots & \ddots & \ddots & \\
 & & -c & 2c & -c \\
0 & & & -c & c
\end{bmatrix} \quad \text{for } c > 0, \qquad [4.23]
$$

which represents uniformly weighted line graphs with no self-loops (i.e. all edge weights are equal to a positive constant and vertex weights to zero). Moreover, in Egilmez *et al.* (2016); Hu *et al.* (2015a), it has been shown that DST-7 is the GFT derived from a graph Laplacian matrix $\mathbf{L} = \mathbf{L}_u + \mathbf{V}$, where $\mathbf{V} = \text{diag}([c\ 0\ \cdots\ 0]^\top)$, including a self-loop at vertex v_1 with weight $f(v_1) = c$. Based on the results in Strang (1999); Püschel and Moura (2008), various other types of DCTs and DSTs can be characterized using graphs. Table 4.1 specifies the line graphs (with n vertices $v_1, v_2, \ldots, v_n$ having self-loops at v_1 and v_n) corresponding to different types of DCTs and DSTs, which are GFTs derived from Laplacians of the form $\bar{\mathbf{L}} = \mathbf{L}_u + \bar{\mathbf{V}}$, $\bar{\mathbf{V}} = \text{diag}([f(v_1)\ 0\ \cdots\ 0\ f(v_n)]^\top)$.

Vertex weights	$f(v_1)=0$	$f(v_1)=c$	$f(v_1)=2c$
$f(v_n)=0$	DCT-2	DST-7	DST-4
$f(v_n)=c$	DCT-8	DST-1	DST-6
$f(v_n)=2c$	DCT-4	DST-5	DST-2

Table 4.1. *DCTs/DSTs corresponding to $\bar{\mathbf{L}}$ with different vertex weights*

In predictive image/video coding, the use of DCT-2 and DST-7 in practice can be justified by analyzing the 1D intra- and interpredicted models in section 4.2.1.2. For this purpose, it is reasonable to suppose that parameters ρ_i for $i = 0, 1, \ldots, n-1$ approach to 1 (i.e. $\rho_i \to 1$), since video pixels are typically highly correlated in practice. Then, based on the precision matrices in equations [4.17] and [4.22], it can be shown that the optimal GFT leads to:

– DCT-2 if $\sigma_d \gg \sigma_e$ for the intra-predicted model (if intra-prediction is bad);

– DST-7 if $\sigma_d \ll \sigma_e$ for the intra-predicted model (if intra-prediction is good);

– DCT-2 if $\gamma_1 = \gamma_2 = \cdots = \gamma_n$ for the inter-predicted model (if inter-prediction is similar across pixels).

4.2.3. *Interpretation of graph weights for predictive transform coding*

By following theorem 4.1, the distribution of residual signals, denoted as $\mathbf{r} = [r_1 \cdots r_n]^\top$, can be modeled by the following GMRF, whose precision matrix is a GGL matrix $\mathbf{L}$ (i.e. $\mathbf{L} = \mathbf{\Omega}_{\text{residual}}$),

$$\mathsf{p}(\mathbf{r}|\mathbf{L}) = \frac{1}{(2\pi)^{n/2}\det(\mathbf{L})^{-1/2}}\exp\left(-\frac{1}{2}\mathbf{r}^\top\mathbf{L}\mathbf{r}\right), \qquad [4.24]$$

where the quadratic term in the exponent can be decomposed in terms of graph weights (i.e. $\mathbf{V}$ and $\mathbf{W}$) as

$$\mathbf{r}^\top\mathbf{L}\mathbf{r} = \sum_{i=1}^{n}(\mathbf{V})_{ii}\, r_i^2 + \sum_{(i,j)\in\mathcal{I}} (\mathbf{W})_{ij}\,(r_i - r_j)^2 \qquad [4.25]$$

such that $(\mathbf{W})_{ij} = -(\mathbf{L})_{ij}$, $(\mathbf{V})_{ii} = \sum_{j=1}^{n}(\mathbf{L})_{ij}$, and $\mathcal{I} = \{(i,j)\,|\,(v_i, v_j) \in \mathcal{E}\}$ is the set of index pairs of all vertices associated with the edge set $\mathcal{E}$.

From equations [4.24] and [4.25], it is clear that the distribution of the residual signal $\mathbf{r}$ depends on edge weights ($\mathbf{W}$) and vertex weights ($\mathbf{V}$) where:

– a model with larger (respectively, smaller) edge weights (e.g. $(\mathbf{W})_{ij}$) increases the probability of having a smaller (respectively, larger) squared difference between corresponding residual pixel values (e.g. r_i and r_j);

– a model with larger (respectively, smaller) vertex weights (e.g., $(\mathbf{V})_{ii}$) increases the probability of pixel values (e.g. r_i) with smaller (respectively, larger) magnitude.

In practice, a more complete characterization of the edge and vertex weights ($\mathbf{W}$ and $\mathbf{V}$) can be made by estimating $\mathbf{L}$ from data, which depends on inherent signal statistics and the type of prediction used for predictive coding. The following section presents two different strategies for finding graphs for the efficient compression of residual signals via graph spectral methods.

4.3. Graph spectral methods for compression

4.3.1. *GL-GFT design*

4.3.1.1. *Generalized graph Laplacian estimation*

As discussed in section 4.2.1 and justified in theorem 4.1, the residual signal $\mathbf{r} \in \mathbb{R}^n$ can be modeled as an attractive GMRF, $\mathbf{r} \sim \mathsf{N}(\mathbf{0}, \boldsymbol{\Sigma} = \mathbf{L}^{-1})$, whose precision matrix is a GGL denoted by $\mathbf{L}$. Assuming that we have k residual signals, $\mathbf{r}_1, \dots, \mathbf{r}_k$, sampled from $\mathsf{N}(\mathbf{0}, \boldsymbol{\Sigma} = \mathbf{L}^{-1})$, the likelihood of a candidate $\mathbf{L}$ is

$$\prod_{i=1}^{k} \mathsf{p}(\mathbf{r}_i|\mathbf{L}) = (2\pi)^{-\frac{kn}{2}} \det(\mathbf{L})^{\frac{k}{2}} \prod_{i=1}^{k} \exp\left(-\frac{1}{2}\mathbf{r}_i{}^\top \mathbf{L}\mathbf{r}_i\right). \qquad [4.26]$$

The maximization of the likelihood in equation [4.26] can be equivalently formulated as minimizing the negative log-likelihood, that is

$$\begin{aligned} \widehat{\mathbf{L}}_{\mathrm{ML}} &= \arg\min_{\mathbf{L}} \left\{ \frac{1}{2} \sum_{i=1}^{k} \mathrm{Tr}\left(\mathbf{r}_i{}^\top \mathbf{L}\mathbf{r}_i\right) - \frac{k}{2} \mathrm{logdet}(\mathbf{L}) \right\} \\ &= \arg\min_{\mathbf{L}} \left\{ \mathrm{Tr}\left(\mathbf{L}\mathbf{S}\right) - \mathrm{logdet}(\mathbf{L}) \right\} \end{aligned} \qquad [4.27]$$

where $\mathbf{S} = \frac{1}{k}\sum_{i=1}^{k} \mathbf{r}_i \mathbf{r}_i{}^\top$ is the sample covariance, and $\widehat{\mathbf{L}}_{\mathrm{ML}}$ is the ML estimate of $\mathbf{L}$. To find the best GGL from a set of residual signals $\{\mathbf{r}_1, \dots, \mathbf{r}_k\}$ in an ML sense, the following GGL estimation problem with connectivity constraints can be solved:

$$\begin{aligned} \underset{\mathbf{L} \succeq 0}{\text{minimize}} \quad & \mathrm{Tr}\left(\mathbf{L}\mathbf{S}\right) - \mathrm{logdet}(\mathbf{L}) \\ \text{subject to} \quad & (\mathbf{L})_{ij} \leq 0 \ \text{ if } (\mathbf{A})_{ij} = 1 \\ & (\mathbf{L})_{ij} = 0 \ \text{ if } (\mathbf{A})_{ij} = 0 \end{aligned} \qquad [4.28]$$

where $\mathbf{S}$ is the sample covariance of residual signals, and $\mathbf{A}$ denotes the connectivity matrix representing the graph structure (set of graph edges). In order to optimally solve equation [4.28], the *GGL estimation algorithm*, introduced in Egilmez *et al.* (2017b,a), can be used. Alternatively, the methods based on different smoothness criteria discussed in the Introduction can be used to learn graphs from data. For example, Fracastoro *et al.* (2016, 2020) propose a block-adaptive scheme for learning graphs by minimizing a regularized Laplacian quadratic term, which is used as a proxy for the actual rate-distortion cost. Moreover, Pavez *et al.* (2015) present a comparison of various instances of different graph learning problems for non-separable image modeling, and empirically validate the use of GFTs derived by GGLs for image compression. A later work by Egilmez *et al.* (2020a) justifies the use of GL-GFT designs for video coding, both theoretically and empirically, where GFTs are specifically constructed by solving equation [4.28]. Egilmez *et al.* (2020b) further demonstrate the benefit of such GFT design over the VVC reference software, VTM (Bross *et al.* 2020), in terms of coding gains.

4.3.1.2. *GL-GFT construction*

To design separable and non-separable GFTs, instances of the optimization problem for different $\mathbf{S}$ and $\mathbf{A}$ in equation [4.28], denoted as $\mathsf{GGL}(\mathbf{S}, \mathbf{A})$, are solved. Then, the optimized GGL matrices are used to derive GFTs. A representative step-by-step description of GFT construction is presented in what follows.

Graph-based separable transforms (S-GFT): For the S-GFT design, we solve two instances of equation [4.28] to optimize two separate line graphs, used to derive $\mathbf{U}_{\text{row}}$ and $\mathbf{U}_{\text{col}}$ in equation [4.1]. Since we wish to design a separable transform, each line graph can be optimized independently. Thus, our basic goal is finding the best line graph pair based on sample covariance matrices $\mathbf{S}_{\text{row}}$ and $\mathbf{S}_{\text{col}}$, created from rows and columns of residual block signals. For $N \times N$ residual blocks, the proposed S-GFT construction has the following steps:

1) create the connectivity matrix $\mathbf{A}_{\text{line}}$ representing a line graph structure with $n = N$ vertices, as in Figure 4.5;

2) obtain two $N \times N$ sample covariances, $\mathbf{S}_{\text{row}}$ and $\mathbf{S}_{\text{col}}$, from rows and columns of size N, respectively, obtained from residual blocks in the dataset;

3) solve $\mathsf{GGL}(\mathbf{S}_{\text{row}}, \mathbf{A}_{\text{line}})$ and $\mathsf{GGL}(\mathbf{S}_{\text{col}}, \mathbf{A}_{\text{line}})$, by using the GGL estimation algorithm (Egilmez *et al.* 2017b) to learn Laplacians $\mathbf{L}_{\text{row}}$ and $\mathbf{L}_{\text{col}}$ representing line graphs, respectively;

4) perform eigendecomposition on $\mathbf{L}_{\text{row}}$ and $\mathbf{L}_{\text{col}}$ to obtain GFTs, $\mathbf{U}_{\text{row}}$ and $\mathbf{U}_{\text{col}}$, which define the S-GFT, as in equation [4.1].

Graph-based nonseparable transforms (NS-GFT): Similarly, for $N \times N$ residual block signals, we propose the following steps to design an NS-GFT:

1) create the connectivity matrix $\mathbf{A}$ based on a desired graph structure. For example, $\mathbf{A}$ can represent a grid graph with $n = N^2$ vertices, as in Figure 4.4;

2) obtain $N^2 \times N^2$ sample covariance $\mathbf{S}$ using residual block signals in the dataset (after vectorizing the block signals);

3) solve the problem $\text{GGL}(\mathbf{S}, \mathbf{A})$ by using the GGL estimation algorithm (Egilmez *et al.* 2017b) to estimate a Laplacian $\mathbf{L}$;

4) perform eigendecomposition on $\mathbf{L}$ to obtain the $N^2 \times N^2$ NS-GFT, $\mathbf{U}$, defined in equation [4.2].

4.3.1.3. *Theoretical justifications for graph learning from data*

It has been shown by Gersho and Gray (1991); Goyal (2001); Mallat (2008) that KLT is optimal for the transform coding of jointly Gaussian sources in terms of mean-square error (MSE) criterion under high-bitrate assumptions. Since GMRF models lead to jointly Gaussian distributions, the corresponding KLTs are optimal, in theory. However, in practice, a KLT is obtained by the eigendecomposition of the associated sample covariance, which has to be estimated from a training dataset, where the number of data samples may not be sufficient to accurately recover the parameters. As a result, the sample covariance may lead to a poor estimation of the actual model parameters (Johnstone and Lu 2009; Ravikumar *et al.* 2011). To improve estimation accuracy and alleviate overfitting, it is often useful to reduce the number of model parameters by introducing model constraints and regularization. From the statistical learning theory perspective (Vapnik 1999; von Luxburg and Schölkopf 2011), the advantage of GL-GFT over KLT is that KLT requires learning the $O(n^2)$ model parameters, while GL-GFT only needs $O(n)$, given the connectivity constraints in equation [4.28]. Therefore, the graph learning approach discussed above provides better *generalization* in learning the signal model by taking the variance-bias tradeoff into account. This advantage can also be justified based on the following error bounds characterized in Vershynin (2012); Ravikumar *et al.* (2011). Assuming that k residual blocks are used for calculating the sample covariance $\mathbf{S}$, under a general set of assumptions, the error bound for estimating $\mathbf{\Sigma}$ with $\mathbf{S}$, derived in Vershynin (2012), is

$$||\mathbf{\Sigma} - \mathbf{S}||_F = O\left(\sqrt{\frac{n^2 \log(n)}{k}}\right), \qquad [4.29]$$

while estimating the precision matrix $\mathbf{\Omega}$ by using the graph learning approach leads to the following bound shown in Ravikumar *et al.* (2011):

$$||\mathbf{\Omega} - \mathbf{L}||_F = O\left(\sqrt{\frac{n \log(n)}{k}}\right), \qquad [4.30]$$

where $\mathbf{L}$ denotes the estimated GGL. Thus, in terms of the worst-case errors, graph learning provides a better model estimation, compared to the estimation based on the sample covariance.

Section 4.3.3 will empirically justify the advantage of GL-GFT against KLT.

4.3.2. *EA-GFT design*

4.3.2.1. *EA-GFT construction*

Typically, to construct an EA-GFT for a residual block, an image edge detector is applied to identify salient discontinuities between pixel values. Then, the edge weights of a predefined graph are modified according to the locations of detected image edges, and the resulting graph is used to derive the associated GFT. Figure 4.6 demonstrates an example of EA-GFT construction (with basis patches shown in Figure 4.6(d), where a uniformly weighted grid graph, with all of its edge weights equal to a fixed constant w_c (Figure 4.6(a)), is first created. Then, the detected image edges on a given residual block (Figure 4.6(b)) are used to determine the co-located edges in the graph, and the corresponding weights are reduced as $w_e = w_c/s_{\text{edge}}$ (Figure 4.6(c)), where $s_{\text{edge}} \geq 1$ is a parameter modeling the sharpness of image edges (i.e. the level of differences between pixel values in the presence of an image edge). Thus, a larger s_{edge} leads to smaller weights on edges connecting pixels (vertices) with an image edge in between.

Shen *et al.* (2010) originally proposed EA-GFTs specifically for depth-map compression, and Hu *et al.* (2015b) extended EA-GFTs for piecewise-smooth image compression. In Egilmez *et al.* (2015); Chao *et al.* (2016); Egilmez *et al.* (2020a), the authors demonstrated the benefit of EA-GFTs in coding intra- and inter-predicted video blocks.

From the compression perspective, the EA-GFT construction can also be viewed as a classification procedure, so that each residual block (e.g. in Figure 4.6(b)) is assigned to a class of signals associated with an attractive GMRF, whose corresponding graph (i.e. GGL) is determined by s_{edge} and image edge detection, based on a threshold T_{edge} (e.g. in Figure 4.6(c)).

According to the simulation results on residual data in Egilmez *et al.* (2020a), coding gains are observed for a sufficiently large s_{edge}. The following section provides a theoretical bound for s_{edge} where coding gains can be achieved.

4.3.2.2. *Theoretical justifications for EA-GFT*

For the sake of simplicity, the analysis is based on 1D GMRF models consisting of a single image edge, whose location l is uniformly distributed as

$$\mathsf{P}(l = j) = \begin{cases} \frac{1}{N-1} & \text{for } j = 1, \ldots, N-1 \\ 0 & \text{otherwise} \end{cases} \qquad [4.31]$$

where N is the number of pixels (i.e. vertices) on the line graph depicted in Figure 4.7. This construction leads to a Gaussian mixture distribution based on $M = N - 1$ attractive GMRFs,

$$\mathsf{p}(\mathbf{x}) = \sum_{j=1}^{M} \mathsf{P}(l = j)\, \mathsf{N}(\mathbf{0}, \mathbf{\Sigma}_j) \qquad [4.32]$$

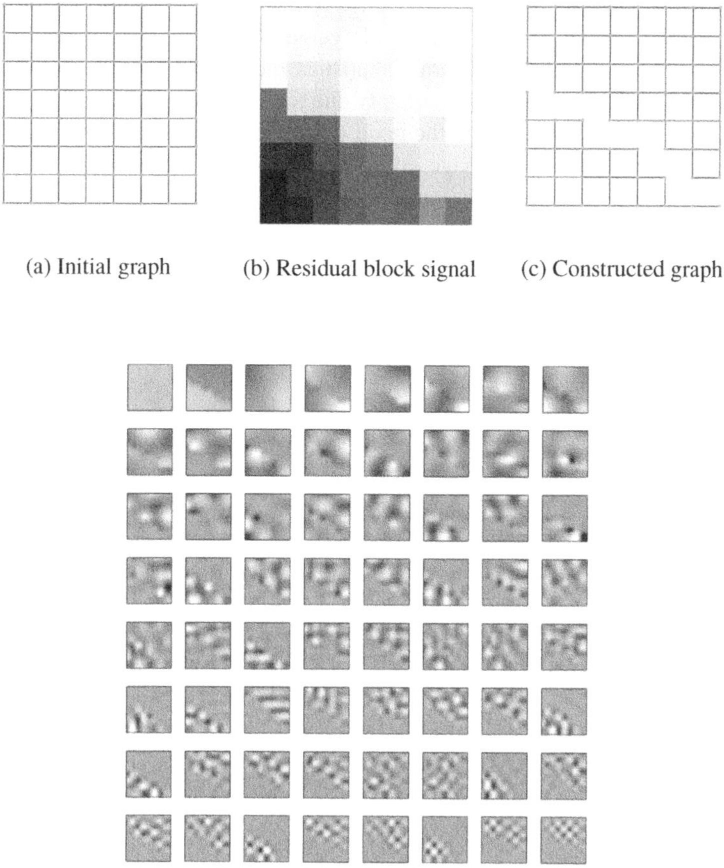

(a) Initial graph (b) Residual block signal (c) Constructed graph

(d) Basis patches

Figure 4.6. *An illustration of graph construction for a given 8×8 residual block signal, where $w_c = 1$ and $w_e = w_c/s_{edge} = 0.1$ where $s_{edge} = 10$. The resulting basis patches are also depicted. For a color version of this figure, see www.iste.co.uk/cheung/graph.zip*

with $\mathbf{\Sigma}_j$ denoting the covariance of the jth attractive GMRF, whose corresponding graph has an image edge between pixels v_j and v_{j+1}, as illustrated in Figure 4.7. Note that the graph weights are all equal to w_c. Since $\mathbf{x}$ follows a Gaussian mixture distribution, the KLT obtained from the covariance of $\mathbf{x}$ (which implicitly performs a second-order approximation of the distribution) is suboptimal in MSE sense (Effros *et al.* 2004). With many possible image edge locations and different orientations, the corresponding distribution may contain a large number of mixtures (large M), which makes learning a model from average statistics inefficient. On the other hand, the proposed EA-GFT removes the uncertainty due to the random variable l, by detecting the location of the image edge in the pixel (vertex) domain, and then constructing a GFT based on the detected image edge. Yet, EA-GFT requires the allocation of additional bits to represent the image edge (side) information, while KLT only allocates bits for coding transform coefficients.

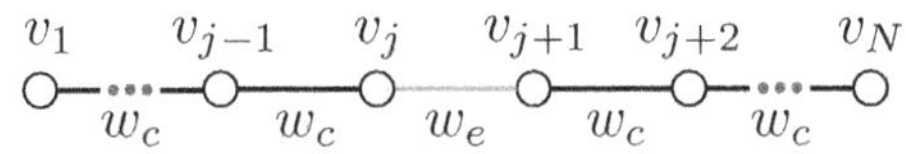

Figure 4.7. *A 1D graph-based model with an image edge at location $l = j$. All black colored edges have weights equal to w_c, and the gray edge between vertices v_j and v_{j+1} is weighted as $w_e = w_c/s_{edge}$. This figure is adapted from Figure 5 in Egilmez* et al. *(2020a)*

In order to demonstrate the rate-distortion tradeoff between KLT and EA-GFT, we use classical rate-distortion theory results with high-bitrate assumptions (Gersho and Gray 1991; Goyal 2001; Mallat 2008), in which the distortion (D) can be written as a function of bitrate (R),

$$D(\bar{R}) = \frac{N}{12} 2^{2\bar{H}_d} 2^{-2\bar{R}} \qquad [4.33]$$

with

$$\bar{R} = \frac{R}{N} \quad \text{and} \quad \bar{H}_d = \frac{1}{N} \sum_{i=1}^{N} H_d((\mathbf{c})_i) \qquad [4.34]$$

where R denotes the total bitrate allocated to code transform coefficients in $\mathbf{c} = \mathbf{U}^\top \mathbf{x}$, and $H_d((\mathbf{c})_i)$ is the differential entropy of transform coefficient $(\mathbf{c})_i$. For EA-GFT, R is allocated to code both transform coefficients ($R^{\text{coeff}}_{\text{EA-GFT}}$) and side information ($R^{\text{edge}}$), so we have

$$R = R^{\text{coeff}}_{\text{EA-GFT}} + R^{\text{edge}} = R^{\text{coeff}}_{\text{EA-GFT}} + \log_2(M) \qquad [4.35]$$

while for KLT, the bits are only allocated to code transform coefficients ($R_{\text{KLT}}^{\text{coeff}}$), so that $R = R_{\text{KLT}}^{\text{coeff}}$. Figure 4.8 shows the coding gain of EA-GFT over KLT for different sharpness parameters (i.e. s_{edge}) in terms of the following metric, called coding gain,

$$\text{cg}(D_{\text{EA-GFT}}, D_{\text{KLT}}) = 10\log_{10}\left(\frac{D_{\text{EA-GFT}}}{D_{\text{KLT}}}\right) \qquad [4.36]$$

where $D_{\text{EA-GFT}}$ and D_{KLT} denote distortion levels measured at a high-bitrate regime for EA-GFT and KLT, respectively. EA-GFT provides better compression for negative cg values in Figure 4.8, which appear when the sharpness of edges s_{edge} is large (e.g. $s_{\text{edge}} > 10$).

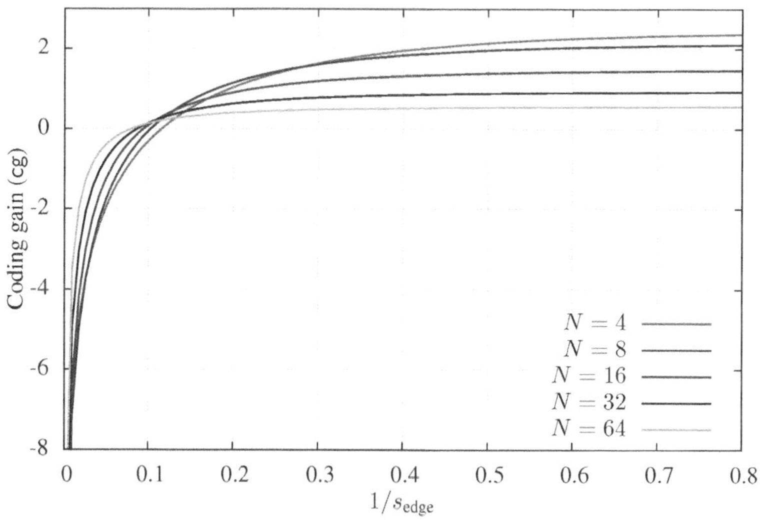

Figure 4.8. *Coding gain (cg) versus s_{edge} for block sizes with $N = 4, 8, 16, 32, 64$. EA-GFT provides better coding gain (i.e. cg is negative) when s_{edge} is larger than 10 across different block sizes. This figure is adapted from Figure 6 in Egilmez* et al. *(2020a). For a color version of this figure, see www.iste.co.uk/cheung/graph.zip*

Note that the distortion function in equation [4.33] is derived based on high-bitrate assumptions. To characterize rate-distortion tradeoff for different bitrates, we use the reverse water-filling technique (Cover and Thomas 1999; Gersho and Gray 1991), by varying the parameter θ in equation [4.38] to obtain rate and distortion measures as follows:

$$R(D) = \sum_{i=1}^{N} \frac{1}{2}\log_2\left(\frac{\lambda_i}{D_i}\right) \qquad [4.37]$$

where λ_i is the ith eigenvalue of the signal covariance, and

$$D_i = \begin{cases} \lambda_i & \text{if } \lambda_i \geq \theta \\ \theta & \text{if } \theta < \lambda_i \end{cases} \qquad [4.38]$$

so that $D = \sum_{i=1}^{N} D_i$.

Figure 4.9 illustrates the coding gain formulated in equation [4.36] achieved at different bitrates, where each curve corresponds to a different s_{edge} parameter. Similar to Figure 4.8, EA-GFT leads to a better compression if the sharpness of edges, s_{edge}, is large (e.g. $s_{\text{edge}} > 10$ for $R/N > 0.6$)[7]. At low-bitrates (e.g. $R/N < 0.6$), EA-GFT can perform worse than KLT for $s_{\text{edge}} = 20, 40$, yet EA-GFT outperforms as the bitrate increases.

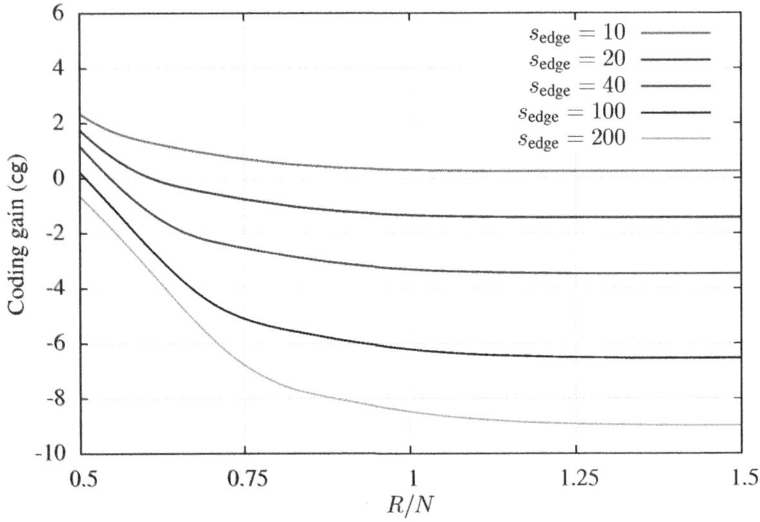

Figure 4.9. *Coding gain* (cg) *versus bits per pixel* (R/N) *for different edge sharpness parameters* $s_{edge} = 10, 20, 40, 100, 200$. *EA-GFT provides better coding gain (i.e.* cg *is negative) if* s_{edge} *is larger than* 10 *for different block sizes. This figure is adapted from Figure 7 in Egilmez* et al. *(2020a). For a color version of this figure, see www.iste.co.uk/cheung/graph.zip*

7 In practice, $R/N > 0.6$ is typically achieved at quantization parameters (Sullivan *et al.* 2012) smaller than 32 in video coding.

4.3.3. *Empirical evaluation of GL-GFT and EA-GFT*

4.3.3.1. *Experimental setup*

For the experiments, two residual block datasets (one for training and the other for testing) are generated. The residual blocks are collected by using HEVC reference software (HM version 14) (Sullivan *et al.* 2012). For the training dataset, residual blocks are obtained by encoding five video sequences, *City*, *Crew*, *Harbour*, *Soccer* and *Parkrun*, and for the test dataset, we use five different video sequences, *BasketballDrill*, *BQMall*, *Mobcal*, *Shields* and *Cactus*. In both datasets, residual blocks are classified based on the side information provided by the HEVC encoder. Specifically, intra-predicted blocks are classified based on 35 intra-prediction modes offered in HEVC. Similarly, inter-predicted blocks are classified into seven different classes using prediction unit (PU) partitions, such that two square PU partitions are grouped as one class and the other six rectangular PU partitions determine other classes. Hence, we have $35+7=42$ classes in total. For each class and block size, the optimal S-GFT, NS-GFT and KLT are designed using the residual blocks in training dataset, while EA-GFTs are constructed based on the detected image edges.

To evaluate the performance of transforms, the mode-dependent transform (MDT) and the rate-distortion optimized transform (RDOT) schemes are adopted. In a nutshell, the MDT scheme assigns a single transform trained for each mode and each block size. In the RDOT scheme, the best transform is selected from a set of transforms $\mathcal{T}$ by minimizing the rate-distortion cost $J(\lambda_{\text{rd}}) = D + \lambda_{\text{rd}} R$ (Ortega and Ramchandran 1998). In the simulations, different transform sets are chosen for each mode (i.e. class) and block size. Specifically, the RDOT scheme selects either the DCT or the transform designed for each mode and block size pair, so that the encoder has two transform choices for each block. Note that, this requires the encoder to send one extra bit to identify the RD optimized transform on the decoder side. For EA-GFTs, the necessary graph (i.e. image edge) information is also sent by using the arithmetic edge codec (Daribo *et al.* 2014). After the transformation of a block, the resulting transform coefficients are uniformly quantized, and then entropy coded using arithmetic coding (Said 2004). The compression performance is measured in terms of Bjontegaard-delta rate (BD-rate) (Bjontegaard 2001).

4.3.3.2. *Compression results*

Table 4.2 presents the overall coding gains achieved by using KLTs, S-GFT and NS-GFT with the RDOT scheme for intra- and inter-predicted blocks. According to the results, GFT outperforms KLT (obtained from sample covariances), irrespective of the prediction type and coding scheme. Figure 4.10 further demonstrates the advantage of the proposed approach over KLT when fewer number of training samples are available[8], where the performance difference between NS-GFT and KLT is increased as the number of available training samples are reduced. Specifically, the BD-rate gap between NS-GFT and KLT increases from 0.7 to 1.5% when 20% of the training data is used instead of the complete data. This validates the observation in section 4.3.1.3 that the graph learning method leads to a more robust transform and provides a better generalization than KLT. Table 4.2 also shows that NS-GFT performs substantially better than S-GFT for coding intra-predicted blocks, while for inter-blocks, S-GFT performs slightly better than NS-GFT. This is because inter-predicted residuals tend to have a separable structure, yet intra-residuals have more directional structures, which are better captured by using non-separable transforms.

Transform	Intra-coding	Inter-coding
KLT	−6.02%	−3.28%
S-GFT	−4.61%	**−3.89%**
NS-GFT	**−6.70%**	−3.68%

Table 4.2. *Comparison of KLT, S-GFT and NS-GFT with RDOT scheme in terms of BD-rate (% bitrate reduction) with respect to the DCT. Smaller (negative) BD-rates mean better compression*

Table 4.3 compares the RDOT coding performance of KLTs, S-GFTs and NS-GFTs on residual blocks with different prediction modes. In the RDOT scheme, the transform sets are $\mathcal{T}_{\text{KLT}} = \{\text{DCT}, \text{KLT}\}$, $\mathcal{T}_{\text{S-GFT}} = \{\text{DCT}, \text{S-GFT}\}$ and $\mathcal{T}_{\text{NS-GFT}} = \{\text{DCT}, \text{NS-GFT}\}$, which consist of DCT and a trained transform for each mode and block size. The results show that NS-GFT consistently outperforms KLT for all prediction modes. Similar to Table 4.2, S-GFT provides slightly better compression compared to KLT and S-GFT. Also for angular modes (e.g. diagonal mode) in intra-predicted coding, NS-GFT significantly outperforms S-GFT as expected.

Table 4.4 demonstrates the RDOT coding performance of EA-GFTs for different modes. As shown in the table, the contribution of EA-GFT within the transform set $\mathcal{T}_{\text{GL-GFT+EA-GFT}} = \{\text{DCT}, \text{GL-GFT}, \text{EA-GFT}\}$ is limited to 0.3% for intra-predicted coding, while it is approximately 0.8% for inter-coding. On the other hand, if the transform set is selected as $\mathcal{T}_{\text{EA-GFT}} = \{\text{DCT}, \text{EA-GFT}\}$, the contribution of EA-GFT provides considerable coding gains, which are approximately 0.5% for intra-predicted and 1% for inter-predicted coding.

8 This experiment is conducted by randomly sampling 20%, 40%, 60% and 80% of the data from the original dataset and is repeated 20 times to estimate average BD-rates. The BD-rates at 100% correspond to the results in Table 4.2, where the complete training dataset is used.

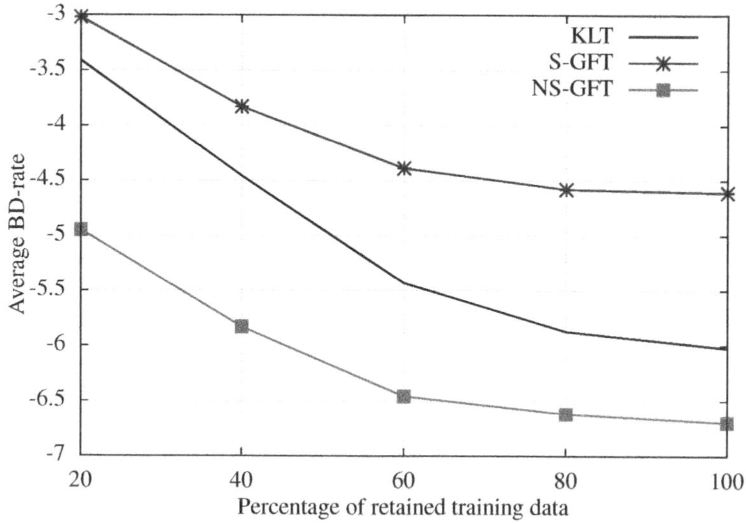

Figure 4.10. *BD-rates achieved for coding intra-predicted blocks with the RDOT scheme based on KLT, S-GFT and NS-GFT, which are trained on datasets with fewer number of samples. This figure is adapted from Figure 14 in Egilmez* et al. *(2020a). For a color version of this figure, see www.iste.co.uk/cheung/graph.zip*

Transform set	Intra-prediction Diagonal	Intra-prediction Horizontal	Intra-prediction All modes	Inter-prediction Square	Inter-prediction Rectangular	Inter-prediction All modes
$\mathcal{T}_{\text{KLT}}$	−7.68	−6.14	−6.02	−3.47	−2.93	−3.35
$\mathcal{T}_{\text{S-GFT}}$	−3.32	−6.45	−4.61	−3.95	−3.25	−3.89
$\mathcal{T}_{\text{NS-GFT}}$	−8.74	−6.53	−6.70	−3.84	−3.18	−3.68

Table 4.3. *Comparison of KLT, S-GFT and NS-GFT for coding of different prediction modes in terms of BD-rate with respect to the DCT. Smaller (negative) BD-rates mean better compression*

In a nutshell[9], the experimental results demonstrate that GL-GFTs can provide considerable coding gains with respect to standard transform coding schemes using DCT. In comparison with the traditional KLT construction, GL-GFTs are more robust and provide better generalization. Although coding gains obtained by including EA-GFTs in addition to GL-GFTs in the RDOT scheme are limited, only using EA-GFTs provides considerable coding gains over DCT.

9 The reader is referred to Egilmez *et al.* (2020a) for a more detailed discussion, with extended versions of the experimental results presented in Tables 4.2, 4.3 and 4.4.

4.4. Conclusion and potential future work

In this chapter, we presented an overview of graph spectral methods for image and video compression. Particularly, two well-established approaches based on graph learning (GL-GFT) and edge-based block adaptation (EA-GFT) are discussed in detail. We also give theoretical justifications for these two strategies and show that well-known transforms such as DCTs and DSTs are special cases of GFTs, and graphs can be used to design generalized separable transforms. The experimental results demonstrated that GL-GFTs can provide considerable coding gains with respect to standard transform coding schemes using DCT. In comparison with the KLTs obtained from sample covariances, GL-GFTs are more robust and provide better generalization. Although coding gains obtained by including EA-GFTs in addition to GL-GFTs in the RDOT scheme are limited, using EA-GFTs only provides considerable coding gains over DCT.

Transform set	Intra-prediction			Inter-prediction		
	Diagonal	Horizontal	All modes	Square	Rectangular	All modes
$\mathcal{T}_{\text{GL-GFT}}$	-8.74	-6.53	-6.70	-3.95	-3.25	-3.89
$\mathcal{T}_{\text{EA-GFT}}$	-0.69	-0.66	-0.54	-1.01	-0.73	-0.93
$\mathcal{T}_{\text{GL-GFT+EA-GFT}}$	-9.07	-6.89	-7.01	-4.80	-3.65	-4.73

Table 4.4. *The contribution of GL-GFT and EA-GFT in terms of BD-rate with respect to the DCT. Smaller (negative) BD-rates mean better compression*

Potential future directions on graph spectral methods for image and video compression can be summarized as follows:

– Graph learning approaches considering different smooth signal models (such as diffusion-based models) can be used to capture more diverse residual characteristics and provide better GFT designs. For example, the graph learning algorithms in Dong *et al.* (2016); Egilmez *et al.* (2019) can be accommodated for graph Laplacian estimation from smooth block signals on graphs.

– EA-GFT based on more sophisticated edge models may be investigated. Early findings by Chao *et al.* (2016) show that edges with smoother transitions, called ramp edges, provide a better coding efficiency for intra-predicted residuals than the step edges discussed in this chapter. On the other hand, step edges provide a better performance than ramp edges in coding inter-predicted blocks.

– Block-adaptive strategies beyond edge-detection based methods have not been investigated in detail. Other salient features such as smoothness and textures can be accommodated in block-adaptive schemes. For example, the work by Fracastoro *et al.* (2020) proposes a scheme for image compression with GFTs, where a graph is constructed for each block signal by minimizing a regularized Laplacian quadratic term used as the proxy for rate-distortion cost.

– Efficient signaling schemes for sending graph information (both structure and weights) can be useful to accommodate adaptive coding techniques, where graphs needed to be sent as side information in order to construct inverse GFTs for decoding.

– In addition to block-based image/video compression systems, GFTs can also be accommodated to support segmentation-oriented coding techniques, such as the one introduced in Fracastoro *et al.* (2015), deriving GFTs on a per-segment basis so that each image segment has its own optimized graph.

– Graph wavelet transforms (Hammond *et al.* 2011) and vertex-domain approaches, such as lifting transforms on graphs (Martinez-Enriquez *et al.* 2011), may provide further coding efficiency with their localized transformation capabilities as alternatives to GFTs.

4.5. References

Arrufat, A., Philippe, P., Deforges, O. (2014). Non-separable mode dependent transforms for intra coding in HEVC. *IEEE Visual Communications and Image Processing Conference*, 61–64.

Bjontegaard, G. (2001). Calculation of average PSNR differences between RD-curves. *ITU-T Q.6/SG16 VCEG-M33 Contribution*, 48.

Bross, B., Chen, J., Liu, S., Wang, Y.-K. (2020). Versatile video coding (draft 10), Output document JVET-S2001, Joint Video Exploration Team (JVET) of ITU-T SG16 WP3 and ISO/IEC JTC 1/SC29/WG11.

Chao, Y.H., Egilmez, H.E., Ortega, A., Yea, S., Lee, B. (2016). Edge adaptive graph-based transforms: Comparison of step/ramp edge models for video compression. *IEEE International Conference on Image Processing (ICIP)*, 1539–1543.

Chen, Y., Mukherjee, D., Han, J., Grange, A., Xu, Y., Liu, Z., Parker, S., Chen, C., Su, H., Joshi, U., Chiang, C., Wang, Y., Wilkins, P., Bankoski, J., Trudeau, L., Egge, N., Valin, J., Davies, T., Midtskogen, S., Norkin, A., de Rivaz, P. (2018). An overview of core coding tools in the AV1 video codec. *Picture Coding Symposium (PCS)*, 41–45.

Cover, T. and Thomas, J. (1999). *Elements of Information Theory*. John Wiley & Sons, New York.

Daribo, I., Florencio, D., Cheung, G. (2014). Arbitrarily shaped motion prediction for depth video compression using arithmetic edge coding, *IEEE Transactions on Image Processing*, 23(11), 4696–4708.

Dong, X., Thanou, D., Frossard, P., Vandergheynst, P. (2016). Learning Laplacian matrix in smooth graph signal representations. *IEEE Transactions on Signal Processing*, 64(23), 6160–6173.

Effros, M., Feng, H., Zeger, K. (2004). Suboptimality of the Karhunen-Loeve transform for transform coding. *IEEE Transactions on Information Theory*, 50(8), 1605–1619.

Egilmez, H.E., Said, A., Chao, Y.-H., Ortega, A. (2015). Graph-based transforms for inter predicted video coding. *IEEE International Conference on Image Processing (ICIP)*, 3992–3996.

Egilmez, H.E., Chao, Y.H., Ortega, A., Lee, B., Yea, S. (2016). GBST: Separable transforms based on line graphs for predictive video coding. *IEEE International Conference on Image Processing (ICIP)*, 2375–2379.

Egilmez, H.E., Pavez, E., Ortega, A. (2017a). GLL: Graph Laplacian learning package, version 1.0. [Online]. Available at: https://github.com/STAC-USC/Graph_Learning.

Egilmez, H.E., Pavez, E., Ortega, A. (2017b). Graph learning from data under Laplacian and structural constraints. *IEEE Journal of Selected Topics in Signal Processing*, 11(6), 825–841.

Egilmez, H.E., Pavez, E., Ortega, A. (2019). Graph learning from filtered signals: Graph system and diffusion kernel identification. *IEEE Transactions on Signal and Information Processing over Networks*, 5(2), 360–374.

Egilmez, H., Chao, Y., Ortega, A. (2020a). Graph-based transforms for video coding. *IEEE Transactions on Image Processing*, 29, 9330–9344.

Egilmez, H.E., Teke, O., Said, A., Seregin, V., Karczewicz, M. (2020b). Parametric graph-based separable transforms for video coding. *CoRR* [Online]. Available at: https://arxiv.org/abs/1911.06981.

Fracastoro, G., Verdoja, F., Grangetto, M., Magli, E. (2015). Superpixel-driven graph transform for image compression. *IEEE International Conference on Image Processing (ICIP)*, 2631–2635.

Fracastoro, G., Thanou, D., Frossard, P. (2016). Graph transform learning for image compression. *Picture Coding Symposium (PCS)*, 1–5.

Fracastoro, G., Thanou, D., Frossard, P. (2020). Graph transform optimization with application to image compression. *IEEE Transactions on Image Processing*, 29, 419–432.

Friedman, J., Hastie, T., Tibshirani, R. (2008). Sparse inverse covariance estimation with the graphical lasso. *Biostatistics*, 9(3), 432–441.

Gersho, A. and Gray, R.M. (1991). *Vector Quantization and Signal Compression*. Kluwer Academic Publishers, Massachusetts.

Goyal, V.K. (2001). Theoretical foundations of transform coding. *IEEE Signal Processing Magazine*, 18(5), 9–21.

Han, J., Saxena, A., Melkote, V., Rose, K. (2012). Jointly optimized spatial prediction and block transform for video and image coding. *IEEE Trans. Image Process*, 21(4), 1874–1884.

Hammond, D.K., Vandergheynst, P., Gribonval, R. (2011). Wavelets on graphs via spectral graph theory. *Applied and Computational Harmonic Analysis*, 30(2), 129–150.

Hu, W., Cheung, G., Ortega, A. (2015a). Intra-prediction and generalized graph Fourier transform for image coding. *IEEE Signal Processing Letters*, 22(11), 1913–1917.

Hu, W., Cheung, G., Ortega, A., Au, O.C. (2015b). Multiresolution graph Fourier transform for compression of piecewise smooth images. *IEEE Transactions on Image Processing*, 24(1), 419–433.

Jain, A.K. (1989). *Fundamentals of Digital Image Processing*. Prentice-Hall, Inc., New Jersey.

Johnstone, I.M. and Lu, A.Y. (2009). On consistency and sparsity for principal components analysis in high dimensions. *Journal of the American Statistical Association*, 104(486), 682–693.

Koller, D. and Friedman, N. (2009). *Probabilistic Graphical Models: Principles and Techniques*. MIT Press, Massachusetts.

von Luxburg, U. and Scholkopf, B. (2011). Statistical learning theory: Models, concepts, and results. *Handbook of the History of Logic,* Volume 10. Elsevier North Holland, Amsterdam.

Mallat, S. (2008). *A Wavelet Tour of Signal Processing: The Sparse Way*, 3rd edition. Academic Press, Massachusetts.

Martinez-Enriquez, E., Diaz-de Maria, E., Ortega, A. (2011). Video encoder based on lifting transforms on graphs. *18th IEEE International Conference on Image Processing (ICIP)*, 3509–3512.

Mukherjee, D., Bankoski, J., Bultje, R.S., Grange, A., Han, J., Koleszar, J., Wilkins, P., Xu, Y. (2013). The latest open-source video codec VP9 – An overview and preliminary results. *Proc. 30th Picture Coding Symp.*, California.

Ortega, A. and Ramchandran, K. (1998). *Rate-distortion Methods for Image and Video Compression. IEEE Signal Process. Mag.*, 15(6), 23–50.

Pavez, E., Egilmez, H.E., Wang, Y., Ortega, A. (2015). GTT: Graph template transforms with applications to image coding. *Picture Coding Symposium (PCS)*, 199–203.

Pennebaker, W.B. and Mitchell, J.L. (1992). *JPEG Still Image Data Compression Standard*, 1st edition. Kluwer Academic Publishers, Massachusetts.

Püschel, M. and Moura, J.M.F. (2008). Algebraic signal processing theory: 1-D space. *IEEE Transactions on Signal Processing*, 56(8), 3586–3599.

Ravikumar, P., Wainwright, M., Yu, B., Raskutti, G. (2011). High dimensional covariance estimation by minimizing l1-penalized log-determinant divergence. *Electronic Journal of Statistics (EJS)*, 5, 935–980.

Rue, H. and Held, L. (2005). *Gaussian Markov Random Fields: Theory and Applications. Monographs on Statistics and Applied Probability*. Chapman & Hall, London.

Said, A. (2004). Introduction to arithmetic coding: Theory and practice. Technical Report HPL-2004-76, HP Labs.

Said, A. and Zhao, X. (2018). CE6: Summary report on transforms and transform signalling. Input document JVET-K0026. Joint Video Exploration Team (JVET) of ITU-T SG16 WP3 and ISO/IEC JTC 1/SC29/WG11, Ljubljana, Slovenia.

Said, A., Zhao, X., Karczewicz, M., Egilmez, H.E., Seregin, V., Chen, J. (2016). Highly efficient non-separable transforms for next generation video coding. *Picture Coding Symposium (PCS 2016)*, 1–1.

Shen, G., Kim, W.-S., Narang, S., Ortega, A., Lee, J., Wey, H. (2010). Edge-adaptive transforms for efficient depth map coding. *Picture Coding Symposium (PCS)*, 566–569.

Strang, G. (1999). The discrete cosine transform. *SIAM Rev.*, 41(1), 135–147.

Sullivan, G.J., Ohm, J.-R., Han, W.-J., Wiegand, T. (2012). Overview of the High Efficiency Video Coding (HEVC) standard. *IEEE Trans. Circuits Syst. Video Technol.*, 22(12), 1649–1668.

Takamura, S. and Shimizu, A. (2013). On intra coding using mode dependent 2D-KLT. *Proc. 30th Picture Coding Symp.*, California, 137–140.

Tekalp, A.M. (2015). *Digital Video Processing*, 2nd edition. Prentice Hall Press, New Jersey.

Vapnik, V.N. (1999). An overview of statistical learning theory. *IEEE Transactions on Neural Networks*, 10(5), 988–999.

Vershynin, R. (2012). How close is the sample covariance matrix to the actual covariance matrix? *Journal of Theoretical Probability*, 25(3), 655–686.

Wiegand, T., Sullivan, G.J., Bjontegaard, G., Luthra, A. (2003). Overview of the H.264/AVC video coding standard. *IEEE Transactions on Circuits and Systems for Video Technology*, 13(7), 560–576.

Woodbury, M.A. (1950). Inverting modified matrices. *Statistical Research Group Memorandum Reports*, Princeton University, New Jersey.

Ye, Y. and Karczewicz, M. (2008). Improved H.264 intra coding based on bi-directional intra prediction, directional transform, and adaptive coefficient scanning. *IEEE International Conference on Image Processing (ICIP)*, 2116–2119.

Zhang, C., Florencio, D., Chou, P. (2015). Graph signal processing – A probabilistic framework. Technical Report MSR-TR-2015-31, Microsoft Research [Online]. Available at: http://research.microsoft.com/apps/pubs/default.aspx?id=243326.

Zhao, X., Chen, J., Karczewicz, M., Zhang, L., Li, X., Chien, W. J. (2016a). Enhanced multiple transform for video coding. *Data Compression Conference (DCC)*, Snowbird, Utah.

Zhao, X., Chen, J., Said, A., Seregin, V., Egilmez, H.E., Karczewicz, M. (2016b). NSST: Non-separable secondary transforms for next generation video coding. *Picture Coding Symposium (PCS 2016)*, 1–1.

Zou, F., Au, O., Pang, C., Dai, J., Zhang, X., Fang, L. (2013). Rate-distortion optimized transforms based on the lloyd-type algorithm for intra block coding. *IEEE Journal of Selected Topics in Signal Processing*, 7(6), 1072–1083.

5

Graph Spectral 3D Image Compression

Thomas MAUGEY[1], Mira RIZKALLAH[2], Navid MAHMOUDIAN BIDGOLI[1], Aline ROUMY[1] and Christine GUILLEMOT[1]

[1] *Inria Rennes – Bretagne Atlantique, Rennes, France*

[2] *École Centrale Nantes, France*

In recent years, new types of image modalities have emerged. Most of them were developed in the context of immersive experience, such as virtual/augmented reality and 6 degree-of-freedom (6-DoF) visualization. These images tend to represent the three-dimensional (3D) world and are, therefore, referred to as 3D images. This implies that the dimension of the images captured with such devices is huge, and requires efficient compression algorithms. Another specificity of these new images is that their shapes differ from the traditional two-dimensional (2D) array of pixels, which, in turn, implies that most of the existing compression algorithms are not directly adapted. Generally, these algorithms are used after some mappings onto 2D planes, which may lead to sub-optimality. As explained in the previous chapters, graph signal processing techniques are able to handle irregular topology, and thus become a natural tool for compressing these new images efficiently.

This chapter is organized into three main parts: section 5.1 presents an introduction to the different 3D image modalities considered in the chapter; section 5.2 presents the overall graph-based coding scheme of a 3D image and section 5.3 details the different graph construction techniques.

5.1. Introduction to 3D images

5.1.1. *3D image definition*

3D images are classically defined in opposition to 2D images. Indeed, by "2D images", we usually denote the images acquired with a traditional pinhole camera under a perspective projection model, i.e. an everyday camera. By "3D", we usually consider everything that is not "2D", hence including many types of capturing devices. The word "3D" is general and does not always mean that the image is defined in an $\mathbb{R}^3$ topology. However, it always reflects the fact that extra information (implicit or explicit) exists and can be used to assist the image processing task. In particular, in the context of graph signal processing, this extra information helps to define the graph $\mathcal{G}$. This information corresponds to the scene *geometry*. In this section, we review the most common types of 3D images from the most explicit to the most implicit geometry.

5.1.2. *Point clouds and meshes*

A point cloud refers to a set of data points in space (Figure 5.1). Each point is defined by a 3D coordinate and a color value. Point clouds are usually captured with one or several laser devices (such as lidar or time-of-flight cameras) coupled with a classical image camera or cameras (more details will be given in Chapter 7). They are used when remote visualization is desired (medical imaging, virtual tours, virtual reality, etc.). For realistic rendering, a high number of points are used (several millions), increasing the need for efficient compression algorithms. Point clouds are sometimes converted to polygon or triangular meshes, in which a mesh topology is added that describes the connectivity between 3D points. Both point clouds and 3D meshes can be static or dynamic. In case of dynamic content (sequences of frames), the number of points and the topology may vary between frames. The geometry of a point cloud is called *explicit*, as it directly corresponds to the set of 3D coordinates. The compression of this *geometry* data is based on decorrelation's principle to eliminate the statistical redundancy.

The compression of 3D mesh *geometry* has been widely studied. A survey of such methods can be found in Maglo *et al.* (2015) and Portaneri *et al.* (2019). In general, static mesh compression approaches are divided into three categories: single-rate algorithms try to build a compact representation of an input mesh. In progressive algorithms, as in Mendes *et al.* (2015), the input mesh is iteratively decimated until a base mesh is generated. This provides successive levels of details for the input mesh, in which a coarse version of the mesh can be quickly displayed to the user and this coarse mesh is progressively refined as more data are decoded. Random-accessible algorithms (Maglo *et al.* 2013) enable the decompression of only the required parts of the mesh in order to avoid the need to load and decompress the full model.

In contrast with 3D meshes, the lack of connectivity information in point clouds is the main difficulty to overcome. In this case, various types of tree-based representations are considered to process geometry information, including octrees (Meagher 1982), binary trees (Zhu *et al.* 2017) and kd-trees (Devillers and Gandoin 2000). Different approaches are used to compress the geometry, including methods to decompose the mesh into levels of details (Golla and Klein 2015; Kathariya *et al.* 2018), clustering methods (Zhang *et al.* 2018) and transform-based techniques (Thanou *et al.* 2016). A survey regarding the compression of point clouds is provided in Cao *et al.* (2019). The approaches used to decorrelate the point cloud data are classified into three main families: (1) 1D traversal compression techniques that provide 1D prediction using tree-based connectivity induced by the native geometric distances between the points in the cloud (Kathariya *et al.* 2018). (2) 2D projection-based methods map the 3D point cloud into 2D images/videos and then use existing image/video coding techniques to compress the data (Golla and Klein 2015). (3) 3D decorrelation techniques directly use the 3D correlation (Thanou *et al.* 2016).

Figure 5.1. *Example of a 3D point cloud:* Stanford Bunny. *For a color version of this figure, see www.iste.co.uk/cheung/graph.zip*

5.1.3. *Omnidirectional images*

Omnidirectional images, also called 360° images, describe the light information coming from *every direction* and converging on a point (the camera center). They can also be seen as spherical images, in which pixels are lying on a unitary sphere and their position on the sphere corresponds to a light ray direction. Here, the geometry is still explicit since it corresponds to the position on the sphere. It is, however, of only two dimensions (longitude and latitude), only being able to position the color point on a line (rather than at an exact 3D position for a point cloud).

Before the popularization of graph-based image processing, omnidirectional images were generally mapped to 2D images to enable traditional image compression (Figure 5.2). Different mappings exist: equirectangular (Snyder 1997), cube map (Kuzyakov and Pio 2016), rhombic dodecahedron (Fu *et al.* 2009) and Dyadic (Yu *et al.* 2015). Each of them presents certain drawbacks, such as non-uniform pixels' distribution on the sphere and connectivity loss between the different mapping surfaces. Graph-based processing techniques enable us to handle these projection limitations by embedding the non-uniform sampling in the weights of the graph or by simply processing the uniformly sampled data lying on a non-Cartesian grid.

Figure 5.2. *Example of omnidirectional image:* Sopha *(Hawary* et al. *2020). For a color version of this figure, see www.iste.co.uk/cheung/graph.zip*

5.1.4. *Light field images*

Light fields represent light rays emitted by every point in a scene and along different orientations (Ng 2006). It is described by the plenoptic function $L(x, y, z, \theta, \phi, \lambda, t)$, where x, y and z are the 3D coordinates, θ and ϕ are the direction angles, λ is the light frequency and t is the time instant. A light field camera, also known as a plenoptic camera, is able to capture, for a given position, (i) the light color and (ii) the light direction, contrary to regular cameras, which only capture the light color. In other words, light field cameras provide a finer sampling of the plenoptic function. This has been made possible, for example, by placing a 2D array of microlenses in front of the photo-receptor. The image captured by the 2D sensor (called a lenslet-based plenoptic image) is, in fact, seen as a 4D table of pixels, where two dimensions correspond to the pixel position, and two correspond to the ray angle (Figure 5.3). The scene geometry information is, this time, *implicit*, and can be retrieved and estimated from the plenoptic image content.

Estimating this geometry always relies on the *motion parallax effect*: when switching from one view to another, an object "moves" (i.e. it is not at the same position in the images), and the way it moves is directly linked to its depth (i.e. its distance from the cameras). More precisely, the closer the object, the more it moves between the different images. The whole point of geometry estimation is to estimate this motion, called *disparity*, in order to retrieve the scene *depth*.

Having very narrow baselines (distance between the views), lenslet-based plenoptic images could not be efficiently used in stereo matching techniques, as they usually involve interpolation with blurriness due to sub-pixel shifts in the spatial domain. Therefore, research has been devoted to finding different constraints and cues for estimating the depth. One way is to compute cross-correlation between microlens images to estimate the disparity map (Georgiev *et al.* 2011). Other approaches rely on structure tensors to estimate vertical and horizontal slopes in epipolar images (Wanner and Goldluecke 2012). Alleviating some ambiguities and difficulties encountered due to occlusions and large displacements, researchers have combined different cues like defocus and correspondence, in Tao *et al.* (2013), with occlusion handling (Wang *et al.* 2016). More recently, learning-based methods have been proposed (Sun *et al.* 2016; Kalantari *et al.* 2016; Shi *et al.* 2019) and show significant improvement when a sufficient amount of data is available for learning. Once a disparity map or a depth map is estimated, this geometry information needs to be compressed and transmitted. If a dense disparity map is needed, then it can be compressed using traditional image coding methods or the graph-based image compression techniques discussed in Chapter 4. If only a sparse set of disparity or depth values are needed, they can be coded using arithmetic coding techniques.

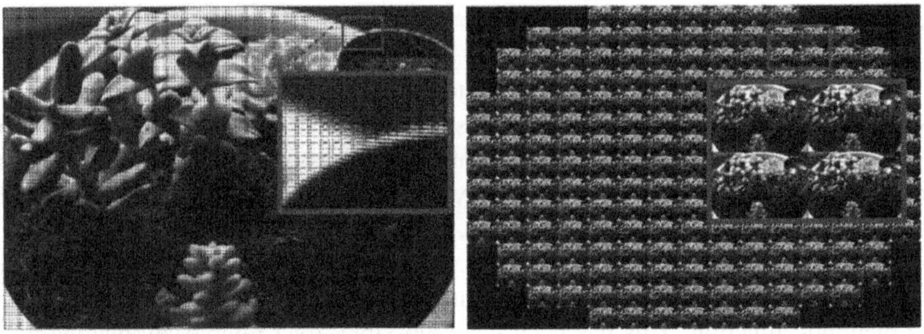

(a) Raw lenslet image (b) Sub-aperture images

Figure 5.3. *Example of light field image:* Succulents *(Jiang* et al. *2017). For a color version of this figure, see www.iste.co.uk/cheung/graph.zip*

5.1.5. *Stereo/multi-view images*

Multi-view imaging refers to a synchronous capture of a scene taken from different angles. A particular example is the stereo capture, where two points of view are captured, mimicking the human visual system, usually for rendering a 3D impression at the user's side. Since the captures are usually done with perspective cameras, no geometry information is available. However, exploiting the information from multiple points of view enables us to retrieve it, again, because of the motion parallax effect. Some other properties of the perspective projection, such as the so-called epipolar geometry, can be used to relate disparity and depth. Therefore, from multiple synchronous captures of a scene, the depth, and, thus, the 3D points position, can be retrieved. This is why multi-view imaging is considered as "3D", the geometry being *implicit* and deduced from the pixel redundancies across views.

Geometry estimation and compression is close in spirit to the ones for light field images. However, with larger baselines (distance between the views), stereo matching techniques (Barnard and Fischler 1982; Scharstein and Szeliski 2002; Hartley and Zisserman 2003) are mostly used to estimate the geometry, i.e. disparity. Those approaches can be broadly classified into two categories: the intensity-based matching technique and the feature-based matching technique. Learning-based approaches for deep stereo matching and depth estimation have also been proposed (e.g. StereoNet in Khamis *et al.* (2018) and Zbontar and LeCun (2016)).

5.2. Graph-based 3D image coding: overview

We recall that a typical 3D image is composed of two entities: (i) the color $\mathbf{x}$ (also called the texture) and (ii) the geometry γ. The compression of such data has been studied in the context of standardization (e.g. (Schwarz *et al.* 2018) for point clouds; (Ye *et al.* 2019) for 360° videos; (Wien *et al.* 2019) for multi-view; and (Schelkens *et al.* 2019) for light fields), already yielding efficient gain. However, the non-Euclidean topology inherent to these image modalities has not been taken into account properly, or only indirectly (because of mapping, for example) leading to coding sub-optimality. In this chapter, we describe how graph-based coding techniques have been able to compress color signals, exploiting the full benefit of the geometry data. We assume that the geometry information is coded *separately*, using specific coding tools (which have been reviewed in section 5.1). The global graph-based 3D image coding scheme is depicted in Figure 5.4. The color signal $\mathbf{x}$ is encoded using a *graph-based coder* that is detailed in the following text. The graph-based coding relies on a graph $\mathcal{G}$ that is constructed from the geometry information and, sometimes, from so-called *auxiliary information* $\boldsymbol{\theta}$ that is extracted from the color $\mathbf{x}$ and transmitted to the decoder.

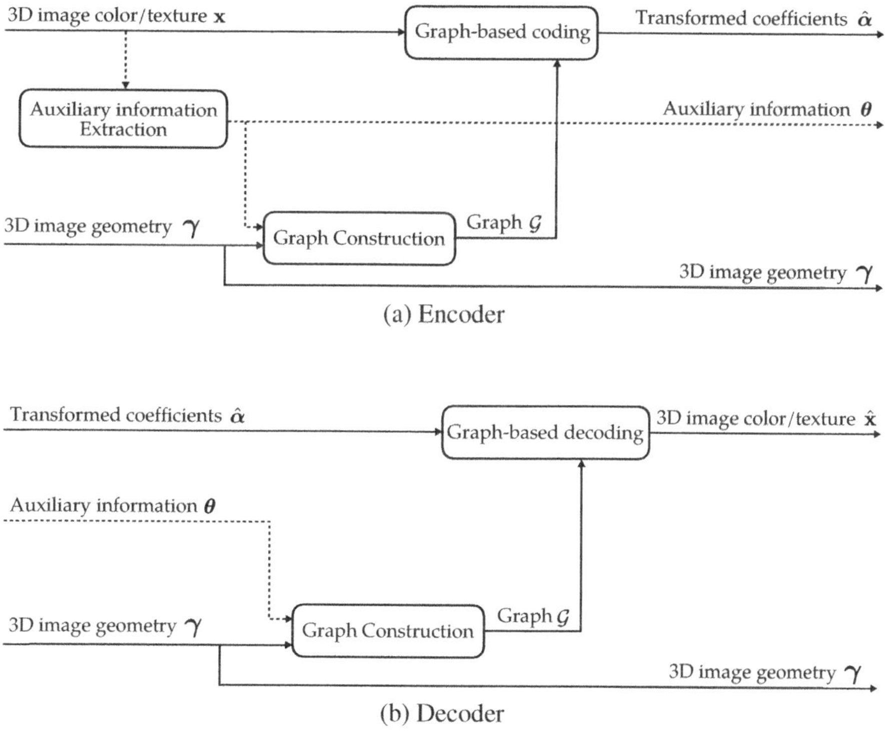

Figure 5.4. *General scheme of graph spectral 3D image compression*

Graph-based coding follows the pipeline summarized in Figure 5.5 and detailed in the following text.

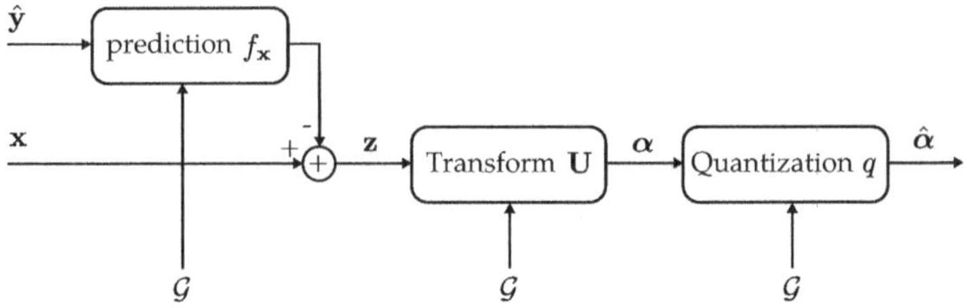

Figure 5.5. *Graph-based encoder's principles, where* $\mathbf{x}$ *is the signal to encode,* $\hat{\mathbf{y}}$ *is the previously decoded data,* $\mathbf{z}$ *is the residue,* α *the transformed coefficients and* $\hat{\alpha}$ *their quantized version*

Prediction: the prediction step is equivalent to trying to find the best approximation of a signal $\mathbf{x}$ with the knowledge acquired from previously decoded data $\hat{\mathbf{y}}$ (e.g. its neighbors), to which it might be correlated. There are several ways of having previously decoded data available. The most straightforward is when the data to encode is a dynamic 3D image[1]. In that case, the signal to encode $\mathbf{x}$ corresponds to the 3D data at time t and the already decoded data correspond to the 3D data at time $t' < t$. Another way of benefiting from already decoded data is when the 3D image is partitioned into smaller entities, each of them being encoded separately in a given order. In that case, $\mathbf{x}$ is the color signal in one of this subset, and $\hat{\mathbf{y}}$ is the color signal in one of the previously decoded subset. The way the input data are split into pieces is directly linked to the graph construction, and will be reviewed in the next section. Whatever the way $\hat{\mathbf{y}}$ is available, it is used to predict $\mathbf{x}$, and the prediction is denoted by $f_{\mathbf{x}}(\hat{\mathbf{y}})$. A residue is then computed:

$$\mathbf{z} = \mathbf{x} - f_{\mathbf{x}}(\hat{\mathbf{y}}).$$

It is expected that the residue is a signal of a smaller energy than the input color $\mathbf{x}$, hence requiring a lower bitrate to be encoded.

Transform: the transform operation is nothing else than an orthonormal basis change that enables us to (i) decorrelate the signal and (ii) compact the energy in a small number of coefficients. Because of points (i) and (ii), coding the signal in the transformed domain is more efficient than in the spatial domain. The signal to transform is the residue $\mathbf{z}$. If no prediction is done, then $\mathbf{z} = \mathbf{x}$.

1 Dynamic 3D images are intentionally not called 3D videos here, since the term video is too much attached to a 2D+t format.

i) *Decorrelate the signal*: the signal $\mathbf{z}$ can be seen as a vector of N random variables Z_i that are correlated with each other. For the sake of clarity, let us assume for the moment that one has to code this vector losslessly. The minimum achievable transmission rate is the joint entropy $H(Z_1, \ldots, Z_N)$ (in the sense of Shannon entropy (Cover and Thomas 1999)). However, this rate is difficult to achieve, in practice, because the joint probability distribution of $(Z_1, \ldots, Z_N)$ is not straightforwardly obtainable. Let us now assume that the signal has been transformed to another random vector set $(Z'_1, \ldots, Z'_N)$ because of an orthogonal basis change. The achievable rate remains the same, i.e. $H(Z'_1, \ldots, Z'_N) = H(Z_1, \ldots, Z_N)$. However, if the variable decorrelation is achieved, we have that $H(Z'_1, \ldots, Z'_N) = H(Z'_1) + \ldots + H(Z'_N)$, which means that each variable can be coded separately without any rate increase. In that case, the joint probability distribution is not needed, and the optimal rate becomes easily achievable in practice, just requiring each of the Z'_i distributions.

ii) *Compact the energy*: another interest of the transform is to compact the energy in a small number of coefficients. Since the transform is orthonormal, the amount of signal energy is the same in both signal and transform domains according to Parseval's theorem. While, in the original domain, the energy in the signal tends to be distributed relatively evenly over the nodes, with a compact representation (i.e. compact transform), the frequency coefficients do not contain the same amount of energy, as illustrated in Figure 5.6. In other words, some coefficients are more representative of the signal and have to be coded with a greater precision. On the other hand, some coefficients are less representative and can be coded coarsely or can even be withdrawn without a great impact on the decoded signal quality, but with a significant bitrate decrease. This is controlled by the quantization, as we will explain in the following text. It is important that the energy is contained in the low frequencies (e.g. power-law decay, as in the general case) or otherwise known law, which may lead to a good model (e.g. zerotree (Shapiro 1993)). The exact indices may not be exactly known in advance but are still coded efficiently using SPIHT (Wheeler and Pearlman 2000) or similar.

A widely used transform in the graph signal processing community is based on the Laplacian operator, as explained in section I.3 of the Introduction. For a graph $\mathcal{G} = (\mathcal{V}, \mathcal{E}, \mathbf{W})$, the Laplacian is defined as $\mathbf{L} = \mathbf{D} - \mathbf{W}$, whose eigenvectors are in $\mathbf{U}$ and eigenvalues are in the diagonal of $\mathbf{\Lambda}$. The graph Fourier transform is defined by $\mathbf{U}$, which means that transforming the signal $\mathbf{z}$ is done as follows:

$$\boldsymbol{\alpha} = \mathbf{U}^\top \mathbf{z}.$$

The vector $\boldsymbol{\alpha}$ is the transformed coefficients or the spectrum of signal $\mathbf{z}$. As recalled previously, a transform is efficient for compression if it (i) decorrelates and (ii) compacts the signal energy. Let us analyze why and when the Laplacian can be considered as a good transform for compression.

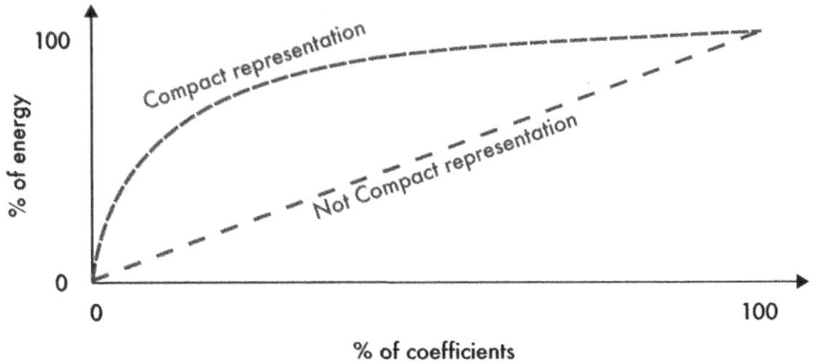

Figure 5.6. *Illustration of the difference between a compact and a not compact representation. For a color version of this figure, see www.iste.co.uk/cheung/graph.zip*

A first remark is that, based on the optimality of the Karhunen–Loeve transform (KLT) (Shalev-Shwartz and Ben-David 2014), a sufficient condition for the graph-Fourier transform to be able to decorrelate the signal $\mathbf{z}$ is that its covariance matrix looks like the Laplacian $\mathbf{L}$. In other words, the weights w_{ij} that constitute the offline elements of $\mathbf{L}$ should reflect the amount of correlation between the source Z_i and the source Z_j (still considering that the signal $\mathbf{z}$ is a realization of a vector of a random variable $(Z_1, \ldots, Z_N)$). A second remark is that the Laplacian operator, as is explained in section I.3, is strongly related to the notion of variation on the graph. In other words, the eigenvectors in $\mathbf{U}$ are ranked by their level of variation, measured by the eigenvalues λ_i, i.e. diagonal elements of $\mathbf{\Lambda}$. The graph transform is thus able to compact the signal energy in some coefficients, i.e. some eigenvectors, if the signal $\mathbf{z}$ exhibits a similar behavior to that of a few eigenvectors over the graph $\mathcal{G}$. Based on the two remarks, we are able to state that if the signal $\mathbf{z}$ is smooth on the graph, then the graph-Fourier transform is efficient for compression. By smooth, we mean that the signal $\mathbf{z}$ does not vary too much along the graph edges, or, more exactly, should vary according to the weights (i.e. large weights imply low variation). Keeping this idea in mind, a usual way to measure the smoothness of a signal is computing what we call a total variation of the signal on the graph (also called Laplacian quadratic form) as follows:

$$\mathcal{S}_{\mathbf{L}}(\mathbf{z}) = \mathbf{z}^\top \mathbf{L}\mathbf{z} = \sum_{i,j} w_{ij}(z_i - z_j)^2 = \sum_l \lambda_l \alpha_l^2. \qquad [5.1]$$

where the set of λ_l are the eigenvalues of the Laplacian matrix $\mathbf{L}$. The smaller the total variation on the graph, the more the energy of the transformed signal is concentrated in the smallest eigenvectors. The role of the graph construction is to build a graph such that the signal $\mathbf{z}$ is smooth on it. As such, the energy will be concentrated in a few coefficients corresponding to low frequencies, and only those that are the most representative need to be transmitted to the decoder side. This is controlled by the quantization step.

Quantization: as explained before, the output of the signal transform stage is a set of transform coefficients. They are, most of the time, decorrelated. A large proportion of the total signal energy is also contained in a handful of coefficients. From a compression perspective, the transform coefficients cannot be sent as they are. The number of bits needed to represent float or double values (with high precision) can easily increase greatly, without necessarily improving the signal reconstruction quality. Therefore, restrictions on the number of bits used to represent those coefficients is necessary. This is usually done by a scaling and rounding procedure, what we call a scalar uniform quantization:

$$\alpha_q = Q \times \text{round}(\frac{\alpha}{Q}). \qquad [5.2]$$

Depending on the quantization step size Q, the transform coefficients are rounded to the nearest multiple of Q. With such a procedure, the precision of the transform coefficients α_q sent can be varied and the reconstruction quality (i.e. the distortion of the signal reconstructed) is impacted.

There are different ways to define Q. The easiest way is to choose one fixed quantization step for all coefficients (de Queiroz and Chou 2017). However, since the energy is concentrated in the first coefficients, some researchers tend to group coefficients with respect to their energy (equivalent to grouping random variables with a similar distribution), and thus define different Q values per group of coefficients (Rizkallah *et al.* 2018). More sophisticated ways also exist, where quantization step sizes are optimized in a way that the reconstruction quality is maximized for a specific rate, or the rate is minimized for a certain reconstruction quality of the signal, as done in state-of-the-art video coders (such as HEVC).

In all compression schemes, the quantized coefficients are further compressed in a lossless manner because of entropy coding. This enables us to exploit the probability distribution to decrease the number of bits needed to code the quantized values. For the moment, no specific entropy coders have been redesigned for graph spectral image coders, and regular entropy coders are generally used, for example, arithmetic coding in de Queiroz and Chou (2017), CABAC in Marpe *et al.* (2003) and Rizkallah *et al.* (2019) and SPIHT in Said and Pearlman (1996) and Maugey *et al.* (2014).

5.3. Graph construction

In light of graph-based coding's principles, defined in the previous section, the graph construction plays a great role in making the coding efficient. In a nutshell, the graph should be defined such that the signal $\mathbf{x}$ to code is smooth on the graph. The description of how the graph is constructed in the different coders in paractice is described in the next section. Here, we detail the three challenges, namely, the topology, the partitioning and the weights.

Topology design: contrary to 2D images that benefit from a natural underlying 2D grid, the topology of 3D data is not straightforwardly defined and has to be carefully constructed, depending on the data type. The topology design consists of finding the edges $\mathcal{E}$ from a given set of nodes $\mathcal{V}$. Based on the previous discussions, an edge should connect two nodes if their attached signal values are correlated. We will see, in the next section, that this topology can be deduced either from the scene geometry or from the signal itself.

Local support partitioning: for various reasons, one may want to split the initial graph into subgraphs. The first motivation is that defining a transform on the whole graph is computationally intractable when N is too big, which is very common for 3D data. Secondly, the 3D signal might have some heterogeneous correlations over the whole graph. Therefore, having subgraphs enables us to localize the signal analysis, hence having more efficient compression, exactly like traditional image coders, such as JPEG, which perform the transforms locally (block of size 8×8 for both complexity and compressibility reasons). Another reason is that having subgraphs enables us to code them one after each other and thus make the prediction construction possible (see Figure 5.5).

Weights adjustment: this simply consists of estimating the matrix $\mathbf{W}$. As mentioned before, a good weight w_{ij} should depict the correlation between the random variable Z_i and Z_j.

As depicted in Figure 5.4, the graph should be constructed identically at the encoder and the decoder. Therefore, the graph construction cannot be fully driven by the input signal $\mathbf{x}$, since it is not available at the decoder. However, the geometry γ is available at the encoder and decoder and can already be used to construct the graph. As will be detailed in section 5.3, the main hypothesis behind all existing methods is that nodes that are close in space, i.e. small $||\gamma_i - \gamma_j||_2^2$, are correlated. However, this hypothesis is not always true, and some information about the signal $\mathbf{x}$ is needed to refine the graph construction. This is called the auxiliary information $\boldsymbol{\theta}$, as explained in the next section.

In the next section, we consider that no auxiliary information is sent, i.e. $R(\boldsymbol{\theta}) = 0$, while, in section 5.3.2, we explain how some $\boldsymbol{\theta}$ can be sent to increase the coding performance.

5.3.1. *Geometry-based approaches*

Here, we detail the methods where the graph is deduced *only* from the geometry. In these cases, no auxiliary information derived from the color is sent to the decoder. In order to define a graph on which the signal $\mathbf{x}$ is smooth, the following hypothesis is formulated:

HYPOTHESIS.– *The correlation between pixels decreases with the distance between their corresponding points in the 3D space:*

$$\forall(i,j),\ \ w_{ij} = \phi(||\gamma_i - \gamma_j||_2^2) \quad \textit{with } \phi \textit{ monotonically decreasing.} \qquad [5.3]$$

This hypothesis is justified by the fact that two points that are close in space are more likely to belong to the same object and therefore to have similar color. On the contrary, pixels that are far away in space do not belong to the same part of the scene, and their colors can thus be seen as independent. Despite its simplicity, this hypothesis already leads to an efficient compression performance.

Far/near model for multi-view and light field images: *The idea is to refine the Cartesian topology with new meaningful edges and weights, embedding a far/near model.*

Contrary to other 3D image modalities, multi-view and light field images are very redundant. Particularly, pixels in different views usually correspond to the same 3D point in the scene. Their color should be similar if not equal. Some works (Maugey *et al.* 2014, 2015; Su *et al.* 2017) have proposed to represent these pixels by only one node in a graph, i.e. one node per 3D point. This directly removes inter-view redundancy and only inter-pixel correlation remains. Even though it leads to interesting compression ratios, the distortion at the decoder in some cases can become high. Indeed, the Lambertian assumption that states that a 3D point's color does not depend on the viewpoint from where it is observed is usually wrong. Secondly, the image sampling grid is not exactly aligned between the views, and the different pixels do not exactly correspond to the same 3D point.

Another idea is to detect these redundant pixels and link them within a graph $\mathcal{G}$. These edges are represented in blue in Figure 5.7. At the same time, the 2D image grids in each image (red in Figure 5.7) are kept. It means that a node is connected to its four neighbors in the same view, and to its corresponding pixels in other views. While the inter-view (blue) edges connect pixels that are very likely to be correlated, the intra-view (red) edges in each 2D image grid are not always meaningful, since two neighboring pixels could correspond to two different objects. This is the reason why the weights corresponding to red edges can be refined based on the geometry information, i.e. disparity map. A far/near model can be adopted (Su *et al.* 2017):

$$w_{ij} = \begin{cases} 1 \text{ if } ||\gamma_i - \gamma_j||_2^2 < \varepsilon \\ a \quad \text{otherwise} \end{cases},$$

with a being an arbitrary small value.

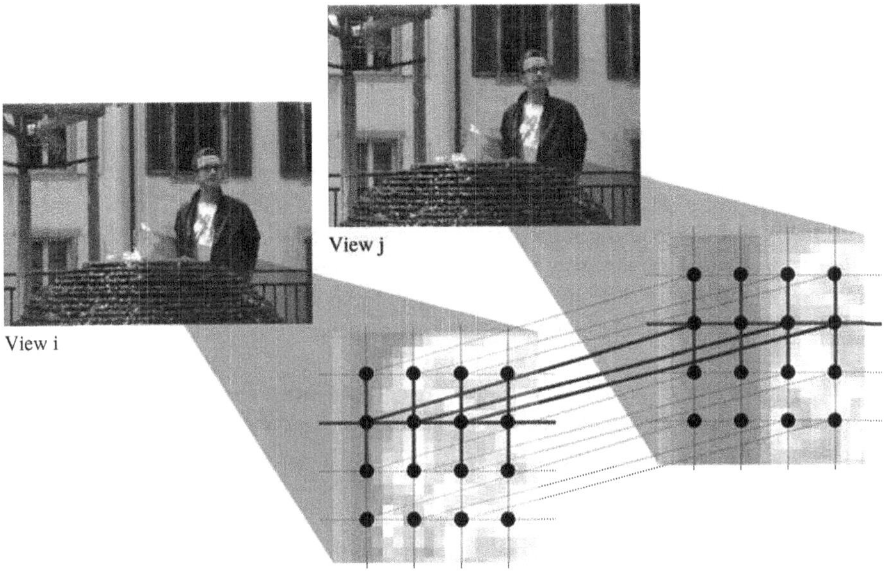

Figure 5.7. *Graph topology for light fields and multi-view images. For a color version of this figure, see www.iste.co.uk/cheung/graph.zip*

Spatial neighborhood based topology for light fields and point clouds: *the idea is to entirely rebuild the graph topology based on spatial neighborhood information and smoothly adapted weights.*

Instead of adjusting an existing grid, the graph could be reconstructed from scratch, based on the geometry information. Regarding the topology, two approaches are conceivable. In the first approach, a node could be linked to any other node in a given neighborhood (as illustrated in Figure 5.8). More formally:

$$e_{ij} \in \mathcal{E} \quad \text{if } v_i \in \mathcal{N}(v_j),$$

where $\mathcal{N}(v_j)$ stands for the neighborhood of vertex v_j. The neighborhood can simply be a ball of a given radius around vertex v_j, i.e.

$$\mathcal{N}(v_j) = \{v_i \mid ||\gamma_i - \gamma_j||_2^2 < \varepsilon\}.$$

The neighborhood can also be defined based on octrees (Schnabel and Klein 2006). The vertices v_j are usually placed at the center of cubes that pave the 3D space. A given cube is adjacent to 26 other cubes in space. These 26 cubes are taken as the neighborhood in Zhang *et al.* (2014) and de Queiroz and Chou (2017).

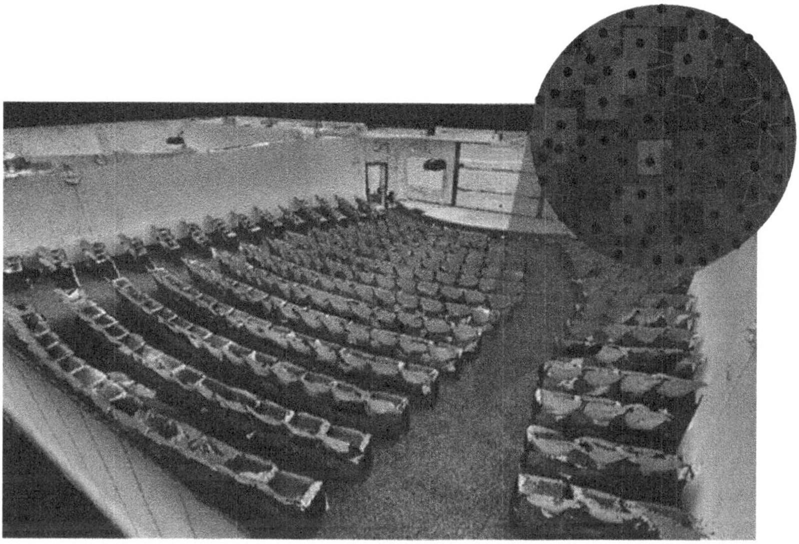

Figure 5.8. *Graph topology for point clouds. For a color version of this figure, see www.iste.co.uk/cheung/graph.zip*

In a second approach, a node can be linked to the k-nearest neighbors (Cohen *et al.* 2016). In other words, a node is linked to the k nodes that have the smallest distance $||\gamma_i - \gamma_j||_2^2$.

With both topologies, the edges do not always link nodes with the same distance. In order to fit with the hypothesis in [5.3], several continuous functions can be considered:

$$w_{ij} = \frac{1}{||\gamma_i - \gamma_j||_2^2}$$

(inverse-distance model), as in Zhang *et al.* (2014); de Queiroz and Chou (2017) or

$$w_{ij} = \exp\left(-\frac{||\gamma_i - \gamma_j||_2^2}{2\sigma^2}\right)$$

(exponential model), as in Cohen *et al.* (2016).

A similar approach can be adopted for pre-demosaic light-field compression (Chao *et al.* 2017), where red, green and blue channels are encoded separately. The pixels corresponding to each channel become irregularly spread in the image after calibration. A graph with the k-nearest neighbors is constructed and a decreasing exponential weighting function is adopted. In that case, the distance is evaluated in 2D.

Geodesic distance for 360° images: *the idea is to take into account the sphere geometry into the topology and weights.*

Before specifying the topology, the most important problem of spherical data representation is to define the position of the pixels. Several samplings of the sphere exist (Chen *et al.* 2018), each of them presenting advantages and drawbacks. In this chapter, we will only focus on two of them: equirectangular and uniform sampling. Everything that is said hereafter is compatible with any other sphere mapping.

Equirectangular sampling consists of uniformly sampling the longitude and the latitude of a sphere (as commonly done for representing the earth). The resulting pixels can then be mapped easily onto a 2D image, which makes it compatible with 2D processing tools. For this reason, it has been widely adopted. However, the pixels are not uniformly sampled on the sphere. Indeed, pixel distribution is denser at the poles than at the equator. In this case, the topology is simply derived from the 2D grid (Figure 5.9, left) but the weights can be adjusted such that this heterogeneous distribution is taken into account.

Uniform sampling consists of spreading N pixels uniformly over the sphere. Even though this problem is mathematically unsolved, pseudo-optimal solutions exist, such as the HEALPIX sampling, introduced in Gorski *et al.* (2005). In that case, the edges are built such that each pixel is linked to its k-neighbors ($k = 8$ in Figure 5.9, right). Also in this case, the weights can be defined to take into account the distance between nodes, as, for example (Mahmoudian Bidgoli *et al.* 2019):

$$w_{ij} = \exp\left(-\frac{d^2_{\text{geo}}(\gamma_i, \gamma_j)}{2\sigma^2}\right),$$

where d_{geo} stands for the geodesic distance, i.e. the shortest distance *on the sphere* between two points. The geodesic distance, or *great-circle* distance, differs from the Cartesian distance between two points and is given by:

$$d_{\text{geo}} = 2\arcsin\frac{||\gamma_i - \gamma_j||_2}{2},$$

where γ_i refers to the 3D position of the pixels lying on a unitary sphere.

Figure 5.9. *Graph topology for omnidirectional images based on two different sampling approaches. For a color version of this figure, see www.iste.co.uk/cheung/graph.zip*

5.3.2. *Joint geometry and color-based approaches*

The color pixels correlation may not always be linked to the distance in the 3D space, as is assumed in equation [5.3]. More exactly, the aforementioned hypothesis may be true in the general case, but there might be exceptions. For the latter, the color information $\mathbf{x}$ can be used to estimate these exceptions, and they are signaled with auxiliary information. Since this auxiliary information is transmitted to the decoder, it has to be as light as possible. The auxiliary information is generated because of the following rate-distortion criterion:

$$\min D(\mathbf{x}, \hat{\mathbf{x}}) + \lambda(R(\hat{\boldsymbol{\alpha}}) + R(\boldsymbol{\theta})), \qquad [5.4]$$

where $D(\mathbf{x}, \hat{\mathbf{x}})$ measures the distortion between the input color signal and the compressed one, $R(\hat{\boldsymbol{\alpha}})$ measures the rate needed to code the quantized transformed coefficients and $R(\boldsymbol{\theta})$ measures the rate needed to code the auxiliary information.

5.3.2.1. *Segmenting the graphs*

Here, we detail the "super-pixel"-like methods, which consist of dividing the high-dimensional data into coherent subgraphs. This segmentation map is deduced from the signal and has to be transmitted as auxiliary information to the decoder.

The idea of using a super-pixel-driven graph transform was first proposed for traditional 2D image compression in Fracastoro *et al.* (2015), where super-pixels are used as coding blocks. Super-pixels can be defined as perceptually meaningful atomic regions, grouping neighboring pixels with the same characteristics (like pixel intensity). They have the ability to adhere well to object borders. Examples of algorithms used to generate these kind of regions are Turbopixel (Levinshtein *et al.* 2009), VCells (Wang and Wang 2012) and the widely used and very fast SLIC algorithm (Achanta *et al.* 2012). In Fracastoro *et al.* (2015), the SLIC algorithm has been used to generate the super-pixels and a graph transform is then applied within each super-pixel before sending the coefficients, along with the segmentation map, to the decoder side.

Extending the notion of super-pixels to high-dimensional data, for multi-view and light field images, the concept of super-rays was introduced in Hog *et al.* (2017) to group light rays coming from the same 3D object, i.e. to group pixels with similar color values and spatially close in the 3D space. The method performs a k-means clustering of all light rays based on color and distance in the 3D space. To deal with dis-occlusions, the cost function can be modified, adding a term related to the depth information. With inconsistent super-rays between views, the signaling cost of such a global light field segmentation is high. That is why in Rizkallah *et al.* (2019), a super-pixel segmentation was only applied on one view, relying on both geometry and color values, and then projected, using the geometry of the scene (the depth information in hand) on the other views. By doing so, only one segmentation map and a depth (i.e. disparity) value per super-pixel need to be transmitted to be able to construct the subgraphs on the decoder side.

With the same spirit, in the case of omnidirectional data, a graph partitioning method has been proposed in Rizkallah *et al.* (2018), to divide the huge graph connecting pixels in the 360 space into coherent subgraphs. They introduce an optimized splitting algorithm, in order to achieve an effective trade-off between the distortion of the reconstructed signals on the subgraphs, the smoothness of the signal on each subgraph (i.e. the rate of the transform coefficients) and the cost of signaling the graph partitioning description. The problem has been formulated with the minimization of the following objective function:

$$\begin{aligned} \min_{\tilde{\mathcal{G}}=\{\mathcal{G}_i\}} \quad & \mathcal{D}(\tilde{\mathcal{G}}) + \gamma \mathcal{R}_C(\tilde{\mathcal{G}}) + \beta \mathcal{R}_B(\tilde{\mathcal{G}}) \\ \text{subject to} \quad & N(\mathcal{G}_i) < N_{max}, \ \forall i \end{aligned} \qquad [5.5]$$

where $\tilde{\mathcal{G}} = \{\mathcal{G}_i\}$ is the *global graph* based on the geodesic distances, in which some edges are removed. $\mathcal{D}(\tilde{\mathcal{G}})$ is the distortion between the original image and the reconstructed one, $\mathcal{R}_C(\tilde{\mathcal{G}})$ is the rate cost of the transform coefficients and $\mathcal{R}_B(\tilde{\mathcal{G}})$ is the rate cost of the boundaries for the graph partitioning description. The number of nodes in each subgraph is denoted by $N(\mathcal{G}_i)$.

Under high bitrate assumption and using the notion of signal smoothness on the subgraphs (Hu *et al.* 2015), the objective function can be simplified into:

$$\begin{aligned}&\min_{\tilde{\mathcal{G}}=\{\mathcal{G}_i\}} \sum_{i=1}^{M} \mathbf{x}_i^{\top} \mathbf{L}_i \mathbf{x}_i + \alpha \frac{1}{2} \sum_{i=1}^{M} \sum_{j \in \mathcal{N}_i} \mathcal{C}_B(ij), \\ &\text{s.t. } N(\mathcal{G}_i) < N_{max}, \ \forall i\end{aligned} \quad [5.6]$$

where $\mathbf{x}_i$ and $\mathbf{L}_i$ are the signal and the Laplacian in the ith subgraph, respectively. $\mathcal{N}_i$ stands for the neighborhood of the ith subgraph. $\mathcal{C}_B(ij)$ denotes the cost of the boundary between two subgraphs $\mathcal{G}_i$ and $\mathcal{G}_j$. An example of optimization result is shown in Figure 5.10. For more details on the problem simplification and optimization algorithm, the reader can refer to Rizkallah *et al.* (2018).

Figure 5.10. *Example of a rate-distortion optimized graph partition in an omnidirectional image. Left: in the equirectangular domain; right: in the sphere domain. For a color version of this figure, see www.iste.co.uk/cheung/graph.zip*

5.3.2.2. *Learning the graph*

In this section, we will introduce the methods that learn the weights $\mathbf{w} \in \mathbb{R}^M$ (where M is number of edges in the graph) from the signal and transmit them as auxiliary information. Therefore, the cost of transmitting the graph must be reflected in the rate-distortion criterion, as mentioned in equation [5.4].

A graph transform learning method for image coding is proposed in Fracastoro *et al.* (2016), that designs a graph by solving the rate-distortion optimization problem, while taking into account the cost of sending the graph in the optimization. More precisely, the image is split into non-overlapping blocks, and the pixels inside blocks are modeled as nodes of a graph, such that each pixel has a fixed 4-connected topology with its neighbors. The pixel intensities are treated as a signal living on the graph. Assuming a uniform scalar quantizer, the distortion depends only on the quantization step and is independent of the chosen graph. The rate-distortion optimization can be approximated by

$$\begin{aligned} &\min_{\mathbf{w}\in\mathbb{R}^M} \mathbf{x}^\top \boldsymbol{L}(\mathbf{w})\mathbf{x} + \alpha||\boldsymbol{\Psi}^\top \mathbf{w}||_1 - \beta\mathbf{1}^\top \log(\mathbf{w}) \\ &\quad \text{s. t.} \quad \forall m < M, \quad w_m \leq 1, \end{aligned} \qquad [5.7]$$

where $\boldsymbol{\Psi}$ is the eigenvector matrix of the dual graph Laplacian. α and β are regularization parameters. The first term relates the measure of signal smoothness with the sparsity of the learned graph transform coefficients to minimize the cost of coding the image signal. The cost of the graph description is introduced in the second term by treating the edge weights as another graph signal that lies on the dual graph. The logarithmic and inequality terms are added to penalize low weight values (to keep the graph connected) and to guarantee that all weights are in $(0, 1]$. Therefore the optimization problem simultaneously minimizes the cost of graph description and the cost of coding the image signal.

Later, a graph learning approach is proposed in Viola *et al.* (2018) for light field image compression, in which the graph captures similarities among neighboring views. In this case, each view is modeled as a graph node, and the edge weights relative to each pair of views are learned from the data. The learned graph weights are then losslessly compressed. At the same time, a subset of views is selected and directly encoded, using a lossy compression scheme. The remaining views are recovered using the graph, by solving an optimization problem enforcing smoothness on the representative graph among the views. In Liao *et al.* (2018), the graph structure is learned from the previous coded data to compress 3D volumetric data. Assuming the data to be modeled with a Gaussian Markov random field, they perform structure-constrained graph learning, using already coded samples to learn a sparse inverse covariance matrix. This matrix is then interpreted as a Laplacian matrix to compute a graph Fourier transform. The interested reader is referred to (Zhang *et al.* 2015) and Chapter 2 for more information about graph learning.

5.3.3. *Separable transforms*

Despite the partitioning of the huge graphs connecting pixels in high-dimensional data, the complexity of the basis function computation remains high. This is from where the need to define low complexity separable transforms that can reduce the signal correlation across all its dimensions arises.

In a general framework, the diagonalization of a Laplacian $\mathbf{L}$ can be accelerated if it can be approximated by the kronecker product of D Laplacian matrices $\mathbf{L}_d$ of smaller dimensions, i.e.

$$\mathbf{L} \approx \mathbf{L}_1 \otimes \ldots \otimes \mathbf{L}_d \otimes \ldots \otimes \mathbf{L}_D. \quad [5.8]$$

This means factorizing the graph along the signal dimensions. In this case, a basis is computed for each of the dimensions, and the transforms are successively applied for each of them. The separate diagonalization, performed on each dimension, is far less complex than the one of the full Laplacian $\mathbf{L}$, since the $\mathbf{L}_d$ are typically of a much lower size. This factorization leads to the computation of separable transforms. Separable transforms have been widely adopted for Cartesian grids (Mallat 2008; Khayam 2003). Recently, their extension to more general graph representations has emerged; for example, in Egilmez *et al.* (2016) and Pavez *et al.* (2017), the 2D graph, used to code the prediction residues in an image coder, is separated into 1D row-wise and column-wise graphs, with weights computed along the rows and columns of the image (already explained in detail in Chapter 4). In the case of light field compression, the authors in Rizkallah *et al.* (2019) separate the spatial and angular dimensions of a light field. However, defining the transform's separability notion in this framework is not straightforward because of two main reasons.

First, depending on the data correlation structures, the covariance matrix may not be exactly equivalent to the Kronecker product of elementary matrices corresponding to the different signal dimensions. In that case, an approximation of the true signal covariance matrix must be computed (Genton 2007; Aston *et al.* 2017; Loukas and Perraudin 2019).

Secondly, the topology of the graph may be such that it cannot be easily split along different dimensions in general, e.g. when the graph is slightly evolving along one dimension (time for video, angular views for light fields). In that case, the separable transforms resulting from the separation into subgraphs may lead to basis incoherence, since even a small change of topology can lead to significantly different basis functions. Eigenvectors computed independently on different shapes (i.e. corresponding to different Laplacians) can only be expected to be reasonably consistent when the shapes are approximately isometric. Whenever this assumption is violated, the basis functions do not behave consistently, and the signals defined on those Laplacians will be projected onto incompatible basis functions, and therefore we cannot guarantee any correlation to be preserved after performing the first spatial graph transform. In order to overcome this issue, we can find coupled eigenvectors, as explained in Kovnatsky *et al.* (2013); Aflalo *et al.* (2015); Kalofolias *et al.* (2017) and Rizkallah *et al.* (2019).

More precisely, suppose we have two different Laplacians $\mathbf{L}_o$ and $\mathbf{L}_i$, if we fix the eigenvectors of the Laplacian $\mathbf{L}_o$, then the coupled approximate eigenbasis of $\mathbf{L}_i$ is found by minimizing the following objective function:

$$\hat{\mathbf{U}}_i^* = \min_{\hat{\mathbf{U}}_i} off(\hat{\mathbf{U}}_i^\top \mathbf{L}_i \hat{\mathbf{U}}_i) + \lambda \text{ matching}(\mathbf{U}_o, \mathbf{U}_i), \quad [5.9]$$

$$\text{s.t. } \hat{\mathbf{U}}_i^\top \hat{\mathbf{U}}_i = \mathbf{I} \quad [5.10]$$

where we seek to minimize the weighted sum of two terms subject to the orthonormality constraint of the computed basis functions $\hat{\mathbf{U}}_i$. The first term is a diagonalization term that aims at minimizing the energy residing on off-diagonal entries $(off(\mathbf{M}) = \sum_{i \neq j} m_{ij})$. The second term aims at enforcing coherence between the two graph transforms. An example of the output of the optimization process is shown in Figure 5.11.

5.4. Concluding remarks

We have seen that graph signal processing techniques are efficient tools to compress 3D images. Indeed, in 3D imaging, the geometry in hand can directly be used to define a graph on which the color information relies. This graph is already efficient prior to compressing the pixels. Sometimes, this graph, simply derived from geometry, is not enough to capture the correlation between pixels accurately enough. We have seen that light auxiliary information could be extracted from the color to assist the graph construction, in order to enhance the compression performance.

Even though graph spectral 3D image coding is promising, some problems still occur. For example, the complexity of the graph can easily greatly increase with the data dimension. The graph spectral compression algorithms presented here all require the diagonalization of the Laplacian matrix, which can become intractable in case of large dimensions. Another aspect that requires more in-depth study is the temporal aspect. Indeed, the change of graph topology over time makes the spectral analysis complex.

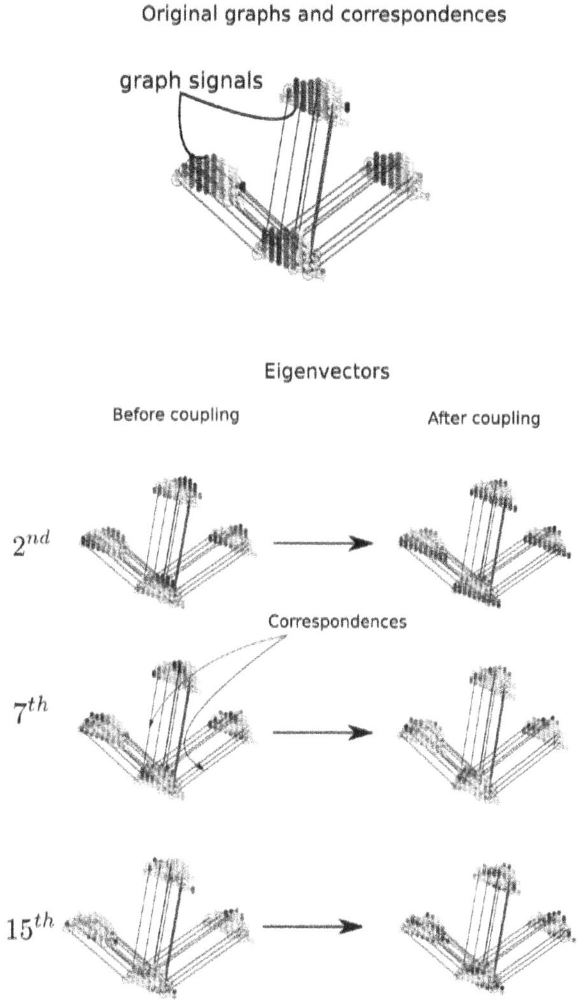

Figure 5.11. *Illustration of the output of the optimization process for a super-ray in four views of a light field. The first row corresponds to a super-ray across four views of the light field. The signal on the vertices corresponds to the color values lying on super-pixels corresponding to the same super-ray and the blue lines denote the correspondences based on the geometrical information in hand. The second to fourth rows are illustrations of basis functions before and after optimization. The signals on the vertices are the eigenvectors values. For a color version of this figure, see www.iste.co.uk/cheung/graph.zip*

5.5. References

Achanta, R., Shaji, A., Smith, K., Lucchi, A., Fua, P., Süsstrunk, S. (2012). SLIC superpixels compared to state-of-the-art superpixel methods. *IEEE Transactions on Pattern Analysis and Machine Intelligence*, 34(11), 2274–2282.

Aflalo, Y., Brezis, H., Kimmel, R. (2015). On the optimality of shape and data representation in the spectral domain. *SIAM Journal on Imaging Sciences*, 8(2), 1141–1160.

Aston, J.A., Pigoli, D., Tavakoli, S. (2017). Tests for separability in nonparametric covariance operators of random surfaces. *The Annals of Statistics*, 45(4), 1431–1461.

Barnard, S.T. and Fischler, M.A. (1982). Computational stereo. Technical report, SRI International, Artificial Intelligence Center, Menlo Park.

Cao, C., Preda, M., Zaharia, T. (2019). 3D point cloud compression: A survey. *The 24th International Conference on 3D Web Technology, Web3D'19*, New York, USA, 1–9 [Online]. Available at: https://doi.org/10.1145/3329714.3338130.

Chao, Y., Cheung, G., Ortega, A. (2017). Pre-demosaic light field image compression using graph lifting transform. *IEEE International Conference on Image Processing (ICIP)*, 3240–3244.

Chen, Z., Li, Y., Zhang, Y. (2018). Recent advances in omnidirectional video coding for virtual reality: Projection and evaluation. *Signal Processing*, 146, 66–78.

Cohen, R.A., Tian, D., Vetro, A. (2016). Attribute compression for sparse point clouds using graph transforms. *IEEE International Conference on Image Processing (ICIP)*, 1374–1378.

Cover, T. and Thomas, J. (1999). *Elements of Information Theory*. John Wiley & Sons, New York.

Devillers, O. and Gandoin, P.-M. (2000). Geometric compression for interactive transmission. *IEEE Visualization 2000*, 319–326.

Egilmez, H.E., Chao, Y.H., Ortega, A., Lee, B., Yea, S. (2016). GBST: Separable transforms based on line graphs for predictive video coding. *IEEE International Conference on Image Processing (ICIP)*, 2375–2379.

Fracastoro, G., Verdoja, F., Grangetto, M., Magli, E. (2015). Superpixel-driven graph transform for image compression. *IEEE International Conference on Image Processing (ICIP)*, 2631–2635.

Fracastoro, G., Thanou, D., Frossard, P. (2016). Graph transform learning for image compression. *IEEE Picture Coding Symposium (PCS)*, 1–5.

Fu, C.-W., Wan, L., Wong, T.-T., Leung, C.-S. (2009). The rhombic dodecahedron map: An efficient scheme for encoding panoramic video. *IEEE Transactions on Multimedia*, 11(4), 634–644.

Genton, M.G. (2007). Separable approximations of space-time covariance matrices. *Environmetrics: The Official Journal of the International Environmetrics Society*, 18(7), 681–695.

Georgiev, T., Chunev, G., Lumsdaine, A. (2011). Superresolution with the focused plenoptic camera. *Computational Imaging IX*, San Francisco, USA, 7873, 78730X.

Golla, T. and Klein, R. (2015). Real-time point cloud compression. *IEEE/RSJ International Conference on Intelligent Robots and Systems (IROS)*, 5087–5092.

Gorski, K.M., Hivon, E., Banday, A.J., Wandelt, B.D., Hansen, F.K., Reinecke, M., Bartelmann, M. (2005). Healpix: A framework for high-resolution discretization and fast analysis of data distributed on the sphere. *The Astrophysical Journal*, 622(2), 759.

Hartley, R. and Zisserman, A. (2003). *Multiple View Geometry in Computer Vision*. Cambridge University Press, Cambridge.

Hawary, F., Maugey, T., Guillemot, C. (2020). Sphere mapping for feature extraction from 360-degree fish-eye captures. *IEEE International Workshop on Multimedia Signal Processing*, 1–6.

Hog, M., Sabater, N., Guillemot, C. (2017). Super-rays for efficient light field processing. *IEEE Journal of Selected Topics in Signal Processing*, 11(7), 1187–1199.

Hu, W., Cheung, G., Ortega, A., Au, O.C. (2015). Multiresolution graph Fourier transform for compression of piecewise smooth images. *IEEE Transactions on Image Processing*, 24(1), 419–433.

Jiang, X., Le Pendu, M., Farrugia, R.A., Guillemot, C. (2017). Light field compression with homography-based low-rank approximation. *IEEE Journal of Selected Topics in Signal Processing*, 11(7), 1132–1145.

Kalantari, N.K., Wang, T.-C., Ramamoorthi, R. (2016). Learning-based view synthesis for light field cameras. *ACM Transactions on Graphics (TOG)*, 35(6), 193.

Kalofolias, V., Loukas, A., Thanou, D., Frossard, P. (2017). Learning time varying graphs. *IEEE International Conference on Acoustics, Speech and Signal Processing (ICASSP)*, 2826–2830.

Kathariya, B., Li, L., Li, Z., Alvarez, J., Chen, J. (2018). Scalable point cloud geometry coding with binary tree embedded quadtree. *IEEE International Conference on Multimedia and Expo (ICME)*, 1–6.

Khamis, S., Fanello, S., Rhemann, C., Kowdle, A., Valentin, J., Izadi, S. (2018). StereoNet: Guided hierarchical refinement for real-time edge-aware depth prediction. *Proceedings of the European Conference on Computer Vision (ECCV)*, Munich, Germany, 573–590.

Khayam, S.A. (2003). The discrete cosine transform (DCT): Theory and application. *Michigan State University*, 114, 1–31.

Kovnatsky, A., Bronstein, M.M., Bronstein, A.M., Glashoff, K., Kimmel, R. (2013). Coupled quasi-harmonic bases. *Computer Graphics Forum*, 32(2), 439–448.

Kuzyakov, E. and Pio, D. (2016). Next-generation video encoding techniques for 360 video and VR. *Facebook* [Online]. Available at: https://engineering.fb.com/2016/01/21/virtual-reality/next-generation-video-encoding-techniques-for-360-video-and-vr/ [Accessed 5 August 2020].

Levinshtein, A., Stere, A., Kutulakos, K.N., Fleet, D.J., Dickinson, S.J., Siddiqi, K. (2009). Turbopixels: Fast superpixels using geometric flows. *IEEE Transactions on Pattern Analysis and Machine Intelligence*, 31(12), 2290–2297.

Liao, W., Cheung, G., Muramatsu, S., Yasuda, H., Hayasaka, K. (2018). Graph learning fast transform coding of 3D river data. *Asia-Pacific Signal and Information Processing Association Annual Summit and Conference (APSIPA ASC)*, Honolulu, USA, 1313–1317.

Loukas, A. and Perraudin, N. (2019). Stationary time-vertex signal processing. *EURASIP Journal on Advances in Signal Processing*, 2019(1), 36 [Online]. Available at: https://doi.org/10.1186/s13634-019-0631-7.

Maglo, A., Grimstead, I., Hudelot, C. (2013). SMI 2013: POMAR: Compression of progressive oriented meshes accessible randomly. *Computers & Graphics*, 37(6), 743–752 [Online]. Available at: http://www.sciencedirect.com/science/article/pii/S0097849313000794.

Maglo, A., Lavoué, G., Dupont, F., Hudelot, C. (2015). 3D mesh compression: Survey, comparisons, and emerging trends. *ACM Computing Surveys*, 47(3), 1–41 [Online]. Available at: https://doi.org/10.1145/2693443.

Mahmoudian Bidgoli, N., Maugey, T., Roumy, A. (2019). Intra-coding of 360-degree images on the sphere. *IEEE Picture Coding Symposium (PCS)*, 1–5.

Mallat, S. (2008). *A Wavelet Tour of Signal Processing: The Sparse Way*, 3rd edition. Academic Press, Cambridge.

Marpe, D., Schwarz, H., Wiegand, T. (2003). Context-based adaptive binary arithmetic coding in the H.264/AVC video compression standard. *IEEE Transactions on Circuits and Systems for Video Technology*, 13(7), 620–636.

Maugey, T., Chao, Y., Gadde, A., Ortega, A., Frossard, P. (2014). Luminance coding in graph-based representation of multiview image. *IEEE International Conference on Image Processing (ICIP)*, 130–134.

Maugey, T., Ortega, A., Frossard, P. (2015). Graph-based representation for multiview image geometry. *IEEE Transactions on Image Processing*, 24, 1573–1586.

Meagher, D. (1982). Geometric modeling using octree encoding. *Computer Graphics and Image Processing*, 19(2), 129–147.

Mendes, C.M., Apaza-Agüero, K., Silva, L., Bellon, O.R.P. (2015). Data-driven progressive compression of colored 3D mesh. *IEEE International Conference on Image Processing (ICIP)*, 2490–2494.

Ng, R. (2006). Light field photography. PhD thesis, Stanford University.

Pavez, E., Ortega, A., Mukherjee, D. (2017). Learning separable transforms by inverse covariance estimation. *IEEE International Conference on Image Processing (ICIP)*, 285–289.

Portaneri, C., Alliez, P., Hemmer, M., Birklein, L., Schoemer, E. (2019). Cost-driven framework for progressive compression of textured meshes. *Proceedings of the 10th ACM Multimedia Systems Conference*, Amherst, USA, 175–188.

de Queiroz, R.L. and Chou, P.A. (2017). Transform coding for point clouds using a Gaussian process model. *IEEE Transactions on Image Processing*, 26(7), 3507–3517.

Rizkallah, M., De Simone, F., Maugey, T., Guillemot, C., Frossard, P. (2018). Rate distortion optimized graph partitioning for omnidirectional image coding. *26th European Signal Processing Conference (EUSIPCO)*, Rome, Italy, 897–901.

Rizkallah, M., Su, X., Maugey, T., Guillemot, C. (2019). Geometry-aware graph transforms for light field compact representation. *IEEE Transactions on Image Processing*, 1–15 [Online]. Available at: https://hal.archives-ouvertes.fr/hal-02199839.

Said, A. and Pearlman, W.A. (1996). A new, fast, and efficient image codec based on set partitioning in hierarchical trees. *IEEE Transactions on Circuits and Systems for Video Technology*, 6(3), 243–250.

Scharstein, D. and Szeliski, R. (2002). A taxonomy and evaluation of dense two-frame stereo correspondence algorithms. *International Journal of Computer Vision*, 47(1–3), 7–42.

Schelkens, P., Astola, P., Da Silva, E.A., Pagliari, C., Perra, C., Tabus, I., Watanabe, O. (2019). JPEG Pleno light field coding technologies. *SPIE Optics + Photonics 2019, Applications of Digital Image Processing XLII*, 11.

Schnabel, R. and Klein, R. (2006). Octree-based point-cloud compression. *Symposium on Point Based Graphics*, 6, 111–120.

Schwarz, S., Preda, M., Baroncini, V., Budagavi, M., Cesar, P., Chou, P.A., Cohen, R.A., Krivokua, M., Lasserre, S., Li, Z. (2018). Emerging MPEG standards for point cloud compression. *IEEE Journal on Emerging and Selected Topics in Circuits and Systems*, 9(1), 133–148.

Shalev-Shwartz, S. and Ben-David, S. (2014). *Understanding Machine Learning: From Theory to Algorithms*. Cambridge University Press, New York.

Shapiro, J.M. (1993). Embedded image coding using zerotrees of wavelet coefficients. *IEEE Transactions on Signal Processing*, 41(12), 3445–3462.

Shi, J., Jiang, X., Guillemot, C. (2019). A framework for learning depth from a flexible subset of dense and sparse light field views. *IEEE Transactions on Image Processing*, 28(12), 5867–5880.

Snyder, J.P. (1997). *Flattening the Earth: Two Thousand Years of Map Projections*. University of Chicago Press, Chicago.

Su, X., Maugey, T., Guillemot, C. (2017). Rate-distortion optimized graph-based representation for multiview images with complex camera configurations. *IEEE Transactions on Image Processing*, 26(6), 2644–2655.

Sun, X., Xu, Z., Meng, N., Lam, E.Y., So, H.K.-H. (2016). Data-driven light field depth estimation using deep convolutional neural networks. *IEEE International Joint Conference on Neural Networks (IJCNN)*, 367–374.

Tao, M.W., Hadap, S., Malik, J., Ramamoorthi, R. (2013). Depth from combining defocus and correspondence using light-field cameras. *Proceedings of the IEEE International Conference on Computer Vision*, 673–680.

Thanou, D., Chou, P.A., Frossard, P. (2016). Graph-based compression of dynamic 3D point cloud sequences. *IEEE Transactions on Image Processing*, 25(4), 1765–1778.

Viola, I., Petric Maretic, H., Frossard, P., Ebrahimi, T. (2018). A graph learning approach for light field image compression. *SPIE Optical Engineering + Applications 2018, Applications of Digital Image Processing XLI*, SPIE 12 [Online]. Available at: http://infoscience.epfl.ch/record/256601.

Wang, J. and Wang, X. (2012). VCells: Simple and efficient superpixels using edge-weighted centroidal Voronoi tessellations. *IEEE Transactions on Pattern Analysis and Machine Intelligence*, 34(6), 1241–1247.

Wang, T.-C., Efros, A.A., Ramamoorthi, R. (2016). Depth estimation with occlusion modeling using light-field cameras. *IEEE Transactions on Pattern Analysis and Machine Intelligence*, 38(11), 2170–2181.

Wanner, S. and Goldluecke, B. (2012). Globally consistent depth labeling of 4D light fields. *IEEE Conference on Computer Vision and Pattern Recognition*, 41–48.

Wheeler, F.W. and Pearlman, W.A. (2000). SPIHT image compression without lists. *IEEE International Conference on Acoustics, Speech, and Signal Processing ICASSP*, 2047–2050.

Wien, M., Boyce, J.M., Stockhammer, T., Peng, W.-H. (2019). Standardization status of immersive video coding. *IEEE Journal on Emerging and Selected Topics in Circuits and Systems*, 9(1), 5–17.

Ye, Y., Boyce, J.M., Hanhart, P. (2019). Omnidirectional 360° video coding technology in responses to the joint call for proposals on video compression with capability beyond HEVC. *IEEE Transactions on Circuits and Systems for Video Technology*, 30(5), 1241–1252.

Yu, M., Lakshman, H., Girod, B. (2015). A framework to evaluate omnidirectional video coding schemes. *IEEE International Symposium on Mixed and Augmented Reality*, 31–36.

Zbontar, J. and LeCun, Y. (2016). Stereo matching by training a convolutional neural network to compare image patches. *The Journal of Machine Learning Research*, 17(1), 2287–2318.

Zhang, C., Florencio, D., Loop, C. (2014). Point cloud attribute compression with graph transform. *IEEE International Conference on Image Processing (ICIP)*, 2066–2070.

Zhang, C., Florencio, D., Chou, P. (2015). Graph signal processing – A probabilistic framework. Technical report, Microsoft Research [Online]. Available at: http://research.microsoft.com/apps/pubs/default.aspx?id=243326.

Zhang, K., Zhu, W., Xu, Y. (2018). Hierarchical segmentation based point cloud attribute compression. *IEEE International Conference on Acoustics, Speech and Signal Processing (ICASSP)*, 3131–3135.

Zhu, W., Xu, Y., Li, L., Li, Z. (2017). Lossless point cloud geometry compression via binary tree partition and intra prediction. *IEEE 19th International Workshop on Multimedia Signal Processing (MMSP)*, 1–6.

6

Graph Spectral Image Restoration

Jiahao PANG[1] and Jin ZENG[2]
[1] *InterDigital Inc., New York, USA*
[2] *SenseTime Research, Shenzhen, China*

6.1. Introduction

Digital images are vulnerable to various degradations during acquisition, editing, compression, transmission and so on. For instance, capturing a night view with a smartphone camera will likely result in a noisy image (Yuan *et al.* 2007), while storing digital images in the widely used Joint Photographic Experts Group (JPEG) format inevitably leads to compression artifacts (Xiong *et al.* 1997). However, the quality of digital images is vital, not only to the esthetic aspect of human perception, but also to the subsequent consumption in machine tasks (e.g. recognition and segmentation). Consequently, for decades, a huge body of research has been devoted to digital image restoration; yet, it remains a crucial topic in signal processing and computer vision (Katsaggelos 2012). Representative image restoration problems include image denoising, image deblurring, image inpainting and image super-resolution (Gunturk and Li 2012), as illustrated in Figure 6.1. In this chapter, we focus on resolving this category of problems with graph spectral signal processing.

6.1.1. *A simple image degradation model*

We first introduce a sufficiently general but simple image degradation model (Aubert and Vese 1997; Milanfar 2013; Dong *et al.* 2012). We denote the original (uncorrupted) image in its vectorized form by $\mathbf{x} \in \mathbb{R}^n$ and its corrupted version by $\mathbf{y} \in \mathbb{R}^m$, and their relationship can be depicted by

$$\mathbf{y} = \mathbf{H}\mathbf{x} + \mathbf{n}, \qquad [6.1]$$

where $\mathbf{H}$ is an m-by-n matrix ($m \leq n$) that may (or may not) be known depending on the problem settings, and $\mathbf{n} \in \mathbb{R}^m$ is an additive noise term with zero mean, following a certain distribution. Our target is to recover the original image $\mathbf{x}$, given the corrupted observation $\mathbf{y}$.[1] Equipped with the model equation [6.1], different restoration problems can be formulated with different configurations of the quantities $\mathbf{H}$ and $\mathbf{n}$ as follows:

1) When $\mathbf{H}$ is an identity matrix ($m = n$ in this case), the recovery of $\mathbf{x}$ is an *image denoising* problem – the most basic task in image restoration (Milanfar 2013). A lot of works on image denoising assume the entries of $\mathbf{n}$ are independent and identically distributed (i.i.d.) Gaussian random variables (Dabov *et al.* 2007; Buades *et al.* 2005b; Rudin *et al.* 1992; Elad and Aharon 2006). With this simple setup, $\mathbf{n}$ is also referred to as additive white Gaussian noise (AWGN). However, in practice, noise distribution can be very complex (Zhang *et al.* 2017; Zeng *et al.* 2019), e.g. it can be signal-dependent and/or device-dependent, and it can be a mixture of several elementary distributions. Moreover, the quantization artifact introduced by JPEG compression is often viewed as additive noise (Xiong *et al.* 1997; Kwon *et al.* 2015). As a result, it is non-trivial to solve the denoising problem in practice.

2) If $\mathbf{H}$ is a square matrix representing blurring operations, e.g. motion blur and out-of-focus blur, we are faced with the *image deblurring* problem. The blur kernel associated with $\mathbf{H}$ is called the point spread function (PSF) (Biemond *et al.* 1990; Xu *et al.* 2013). Image deblurring with a known PSF (or known $\mathbf{H}$) is called *non-blind* deconvolution/deblurring; otherwise, when the PSF is unknown, we need to estimate it as well, in order to reconstruct $\mathbf{x}$. In this case, the problem is also called *blind* deconvolution in literature (Levin *et al.* 2009).

3) If some regions on the image $\mathbf{x}$ are missing, e.g. due to transmission, we need to perform *image inpainting* to get the original $\mathbf{x}$ back (Bertalmio *et al.* 2000). In this case, $\mathbf{H}$ is a square matrix zeroing out the pixels on $\mathbf{x}$. Note that by adding impulse noise to an image, the information of the affected pixels is completely "erased", which is equivalent to missing pixels. Hence, given that the location of the affected pixels is known, the removal of impulse noise now shares the same goal as image inpainting – both aim to fill out the missing pixels, although image inpainting typically means recovering larger missing regions.

4) When $\mathbf{H}$ corresponds to a down-sampling operator, e.g. bilinear and bicubic downsampling, we apply image *super-resolution* algorithms to the low-resolution

1 In some settings, e.g. super-resolution or deblurring with multiple images (Glasner *et al.* 2009; Cai *et al.* 2009), one may have several observations of $\mathbf{x}$.

image $\mathbf{y}$, in order to recover the high-resolution image $\mathbf{x}$ (Yang *et al.* 2010; Freeman *et al.* 2002). In this case, $m < n$ and $r = \sqrt{m/n}$ is the downsampling ratio.

Figure 6.1. *Several representative image restoration problems. The bottom-middle image is the original image. For a color version of this figure, see www.iste.co.uk/cheung/graph.zip*

In practice, different types of corruptions may appear in the same image simultaneously. For instance, the night view we captured with a hand-held camera is not only noisy, but is also blurry (Jia 2007). Consequently, image restoration in real-life often involves resolving several tasks jointly. Moreover, practical image restoration is not limited to denoising, deblurring, inpainting and super-resolution; other popular problems include haze removal (Fattal 2008), image colorization (Levin *et al.* 2004), image matting (Levin *et al.* 2007), flash/non-flash image pair enhancement (Petschnigg *et al.* 2004), and so on. Interestingly, under the plug-and-play framework (Romano *et al.* 2017; Venkatakrishnan *et al.* 2013), most restoration problems can be converted to the basic problem-removal of AWGN. As a result, one may leverage the superior performance of modern image denoisers to accomplish other restoration tasks.

6.1.2. *Restoration with signal priors*

Image restoration (equation [6.2]) is typically ill-posed, unless the noise $\mathbf{n}$ is zero and the matrix $\mathbf{H}$ is well conditioned, which rarely happens in practice. In other words, it is insufficient to use the information contained in $\mathbf{y}$ alone to reconstruct $\mathbf{x}$. Therefore, it is necessary to introduce extra prior knowledge describing the characteristics of the original image $\mathbf{x}$, so as to facilitate the recovery. Such prior knowledge is "encoded" in the form of *signal priors*. For example, if we know an original image contains mainly low-frequency contents, then the corresponding signal prior may state that $\|\mathbf{F}\mathbf{x}\|_2^2$ should be small for some high-frequency filter $\mathbf{F}$. In general, the formulation to recover $\mathbf{x}$ with a signal prior is given by the following optimization problem:

$$\mathbf{x}^\star = \arg\min_{\mathbf{x}} \operatorname{dist}(\mathbf{H}\mathbf{x}, \mathbf{y}) + \lambda \cdot \operatorname{prior}(\mathbf{x}), \qquad [6.2]$$

where prior$(\cdot)$ is the chosen signal prior, which serves as a *regularizer* to turn the restoration problem into a well-posed one. Popular priors in literature include total variation priors (Rudin *et al.* 1992), sparsity priors (Elad and Aharon 2006), autoregressive priors (Zhang and Wu 2008), low rank priors (Dong *et al.* 2014), and so on. In this chapter, we focus on *graph smoothness priors*, especially the simple-yet-effective *graph Laplacian regularizer* (Shuman *et al.* 2013; Cheung *et al.* 2018). In equation [6.2], the first term – $\operatorname{dist}(\mathbf{H}\mathbf{x}, \mathbf{y})$ – is some distance measure enforcing the proximity between $\mathbf{H}\mathbf{x}$ and $\mathbf{y}$, which is called the *data term* or *fidelity term*. As we will see shortly, different noise distributions of $\mathbf{n}$ induce different distance measure $\operatorname{dist}(\cdot,\cdot)$, e.g. AWGN gives rise to (squared) ℓ_2 norm–$\|\mathbf{H}\mathbf{x} - \mathbf{y}\|_2^2$. In equation [6.2], the regularization parameter $\lambda \in \mathbb{R}$ is a tunable parameter weighting the importance between the data term and the prior term.

The formulation equation [6.2] can be understood as seeking for the *maximum a posteriori* (MAP) estimation of $\mathbf{x}$ (Bishop 2006). With Bayes' theorem, the *posterior* distribution of $\mathbf{x}$, $P(\mathbf{x}|\mathbf{y})$, is given by

$$P(\mathbf{x}|\mathbf{y}) = \frac{P(\mathbf{y}|\mathbf{x}) \cdot P(\mathbf{x})}{P(\mathbf{y})}, \qquad [6.3]$$

where $P(\mathbf{x})$ is the *prior* distribution of $\mathbf{x}$, and $P(\mathbf{y}|\mathbf{x})$ is the *likelihood*, while the distribution of $\mathbf{y}$, $P(\mathbf{y})$, is called the *evidence*. The MAP estimation seeks for an optimal $\mathbf{x}$, denoted as $\mathbf{x}^\star$, that maximizes the posterior $P(\mathbf{x}|\mathbf{y})$, i.e.

$$\begin{aligned}\mathbf{x}^\star &= \arg\max_{\mathbf{x}} \frac{P(\mathbf{y}|\mathbf{x}) \cdot P(\mathbf{x})}{P(\mathbf{y})} \\ &= \arg\max_{\mathbf{x}} \log\left(P(\mathbf{y}|\mathbf{x})\right) + \log\left(P(\mathbf{x})\right) - \log\left(P(\mathbf{y})\right) \qquad \text{(take logarithm)} \\ &= \arg\min_{\mathbf{x}} -\log\left(P(\mathbf{y}|\mathbf{x})\right) - \log\left(P(\mathbf{x})\right) + \log\left(P(\mathbf{y})\right) \qquad \text{(take negation)}\end{aligned}$$

Since the last term $P(\mathbf{y})$ is irrelevant to $\mathbf{x}$, it is removed from the objective. Moreover, with our degradation model equation [6.1], $P(\mathbf{y}|\mathbf{x})$ is essentially the same as $P(\mathbf{y}|\mathbf{H}\mathbf{x})$. As a result, the MAP problem becomes

$$\mathbf{x}^\star = \arg\min_{\mathbf{x}} -\log\left(P(\mathbf{y}|\mathbf{H}\mathbf{x})\right) - \log\left(P(\mathbf{x})\right). \qquad [6.4]$$

By comparing the objectives of equations [6.2] and [6.4], solving problem equation [6.2] can be viewed as solving the MAP problem equation [6.4], since:

1) choosing a signal prior for $\mathbf{x}$ means "choosing" a prior distribution for $\mathbf{x}$ implicitly, where $P(\mathbf{x}) \propto \exp(-\text{prior}(\mathbf{x}))$;

2) the term $-\log(P(\mathbf{y}|\mathbf{Hx}))$ measures the "distance" of $\mathbf{y}$ and $\mathbf{Hx}$ by taking the noise model of $\mathbf{n}$ into account. In fact, when $\mathbf{n}$ is AWGN (its entries follow a i.i.d. Gaussian distribution), $-\log(P(\mathbf{y}|\mathbf{Hx})) \propto \|\mathbf{y} - \mathbf{Hx}\|_2^2$. That is why, in the literature on AWGN removal, squared ℓ_2-norm is usually employed as the data term.

Let us turn to an intuitive example. Consider formulating an optimization problem for AWGN removal using the graph Laplacian regularizer (see section I.5). Note that in this simple case, the $\mathbf{H}$ matrix is an identity, so $m = n$. Suppose we are equipped with some neighborhood graph $\mathcal{G}$ with n vertices, where each vertex represents a pixel on the image $\mathbf{x}$, while the graph weights connecting the pixels model the *similarity/affinity* among them. Then, based on the above discussions, our optimization problem becomes:

$$\mathbf{x}^\star = \arg\min_{\mathbf{x}} \|\mathbf{x} - \mathbf{y}\|_2^2 + \lambda \cdot \mathbf{x}^\top \mathbf{L}\mathbf{x}, \qquad [6.5]$$

where $\mathbf{L} \in \mathbb{R}^{n \times n}$ is the *graph Laplacian matrix* associated with the graph $\mathcal{G}$. In this case, equation [6.5] states that the graph Laplacian regularizer – $\mathbf{x}^\top\mathbf{L}\mathbf{x}$ – induced by $\mathcal{G}$ should be small for good candidates of $\mathbf{x}$. In other words, the proper $\mathbf{x}$'s should be smooth with respect to the chosen graph $\mathcal{G}$.

6.1.3. *Restoration via filtering*

Solving optimization problems can be formidable and sometimes even impractical. An alternative strategy to estimate the original $\mathbf{x}$ is to perform *image filtering* (Milanfar 2013), either in the *spatial domain* or in some *transform domain*. In fact, restoration in a transform domain is sometimes categorized as a different methodology. However, while spatial filtering can be loosely regarded as the approximation of filtering in a transform domain (Gonzales and Woods 2018; Pang *et al.* 2015), we still view transform-based methods as a type of filtering for simplicity. This chapter places more emphasis on approaches based on formulating optimization problems with signal priors. However, for completeness, we briefly discuss filter-based approaches for image restoration as follows. Interested readers are referred to section 1.3 for more discussions on filtering.

When filtering is performed in the spatial domain, each pixel on the corrupted image will be processed independently: its intensity will be updated by considering, not only its original intensity, but also its local/non-local neighborhood. Particularly, each pixel will be replaced by the weighted average of itself and other pixels on the corrupted image. This process can be modeled by a simple matrix multiplication:

$$\widehat{\mathbf{x}} = \mathbf{\Phi}\mathbf{y}, \qquad [6.6]$$

where $\mathbf{\Phi} \in \mathbb{R}^{m \times n}$ is the weight matrix representing the filter. Its (i, j)th entry, denoted as ϕ_{ij}, indicates how much the jth pixel on the corrupted image $\mathbf{y}$, is to be contributed to the ith pixel on the estimate $\widehat{\mathbf{x}}$. In some approaches, even information from other images (with similar characteristics) will be exploited for joint filtering (Kopf *et al.* 2007; He *et al.* 2013).

A simple approach of image filtering is Gaussian filtering (Wells 1986), which is a typical linear shift-invariant (LSI) filter. In this case, the filtering falls back to simple 2D convolution with a Gaussian kernel. A more advanced and representative approach is the bilateral filter (Tomasi and Manduchi 1998), which estimates the values of a pixel by taking not only the intensities of its neighboring pixels, but also its spatial distances to different neighbors into account. Its filter weight ϕ_{ij} is given by

$$\phi_{ij} = \exp\left(\frac{\|\mathbf{l}_i - \mathbf{l}_j\|_2^2}{\epsilon_l^2}\right) \exp\left(\frac{\|y_i - y_j\|_2^2}{\epsilon_i^2}\right), \qquad [6.7]$$

where $\mathbf{l}_i \in \mathbb{R}^2$ is the location of pixel i on the 2D image, y_i is the intensity of pixel i on the corrupted image $\mathbf{y}$, and ϵ_l and ϵ_i are two tunable parameters.

To go one step further, Buades *et al.* has proposed–perhaps one of the most important assumptions in image processing – the *non-local self-similarity* assumption (Buades *et al.* 2005b). It assumes that pixel patches with similar intensities appear on a natural image non-locally. With this assumption, Buades *et al.* introduced the non-local means (NLM) algorithm for denoising, where

$$\phi_{ij} = \exp\left(\frac{\|\text{patch}_{\mathbf{y}}(i) - \text{patch}_{\mathbf{y}}(j)\|_2^2}{\epsilon_i^2}\right), \qquad [6.8]$$

where $\text{patch}_{\mathbf{y}}(i)$ denotes a (vectorized) pixel patch on $\mathbf{y}$ centered at pixel i. One can see that equations [6.7] and [6.8] are closely related. Compared to the bilateral filter, NLM denoising has removed the term concerning spatial distance, and the filtering weight is determined solely by the similarity between pixel patches. It is worth noting that the principle of non-local self-similarity has lead to great success in research on image restoration (Dabov *et al.* 2007; Buades *et al.* 2005a). In fact, its rationale – similar patches recur across the whole image – are widely used to construct novel signal priors to formulate the problem of equation [6.2]. It is also imperative in applying GSP to image restoration, as will be seen later. Provided we have some proper graph $\mathcal{G}$ and its associated adjacency matrix $\mathbf{A}$, then a popular graph filter $\mathbf{K}$ to recover $\mathbf{x}$ is

$$\mathbf{K} = k_0\mathbf{A}^0 + k_1\mathbf{A} + \cdots + k_L\mathbf{A}^L, \qquad [6.9]$$

where the k_i's are constant coefficients. Equation [6.9] is called an L-th order polynomial in the matrix $\mathbf{A}$ (Cheung *et al.* 2018), then $\mathbf{x}$ is estimated by computing $\mathbf{Ky}$. Note that instead of the adjacency matrix $\mathbf{A}$, other suitable graph shift operators can also be used for filtering in equation [6.9] (e.g. Gavili and Zhang 2017).

When filtering in a transform domain, we assume the ground-truth $\mathbf{x}$ is either *sparse* or *smooth* in that domain (Dabov *et al.* 2007). When sparseness is assumed, we looked for an estimate $\widehat{\mathbf{x}}$, which on one hand is close to our observation $\mathbf{y}$, but on the other hand is sparse in the transform domain, i.e. only very few frequency components are non-zero. One may also assume that the signal is smooth in that transform domain, i.e. the spectral of the ground-truth $\mathbf{x}$ contains mainly *low-frequency* components. In this regard, one may suppress the magnitude of the high-frequency parts of $\mathbf{y}$ in order to get an estimate $\widehat{\mathbf{x}}$. While the transform being chosen is essential for the success of this type of method, the fixed transforms, such as Fourier transforms and discrete cosine transforms (DCT), often fail to capture the intrinsic image characteristics. Differently, the *graph Fourier transform* (GFT) adapts to the image content and provides a more tailored transformation to analyze images. Hence, provided that we have an appropriate graph $\mathcal{G}$ that describes the pixel similarities of the original image $\mathbf{x}$ well, it is undoubtedly beneficial to compute its corresponding GFT for subsequent filtering and restoration in the graph frequency domain.

Note that the filtering approaches presented here, and optimization with signal priors (equation [6.2]) have a profound relationship with each other. In fact, they both stem from the variational approaches based on diffusion with partial differential equations (PDEs) (Weickert 1998; Perona *et al.* 1994). While performing one step of image filtering can be regarded as discretizing a PDE in time, the optimization problems with signal priors can be viewed as the outcome of applying the *Euler–Lagrange* equation to a PDE. Since a deeper discussion of this aspect is out of the scope of this chapter, we refer interested readers to Sapiro (2006).

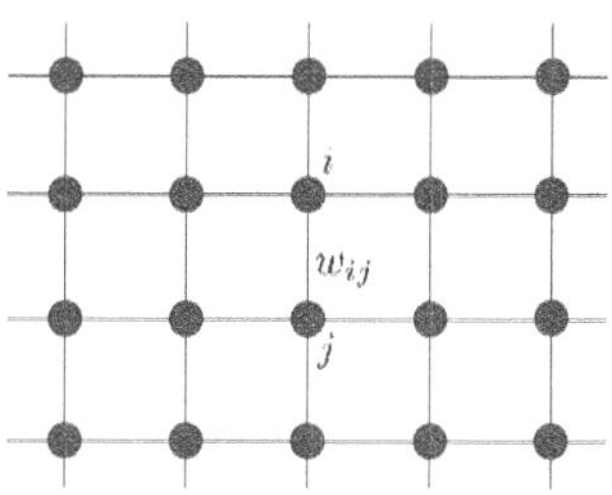

Figure 6.2. *A simple grid graph $\mathcal{G}$. This figure has only showed a part of the graph. For a color version of this figure, see www.iste.co.uk/cheung/graph.zip*

6.1.4. *GSP for image restoration*

From the above discussions, a central question to ask is: *how to construct an appropriate graph $\mathcal{G}$ given we only have the corrupted image* $\mathbf{y}$? Undoubtedly, building good graphs is vital to our recovery. A popular choice is to build a 4-connected grid graph, as shown in Figure 6.2, where each pixel i is connected to its four neighboring pixels, and the graph weight connecting pixel i and pixel j is given by the following Gaussian kernel:

$$w_{ij} = \exp\left(-\frac{\mathrm{dist}(i,j)^2}{2\epsilon^2}\right), \qquad [6.10]$$

where $\mathrm{dist}(i,j)$ is some distance metric between pixel i and pixel j, and ϵ is a constant (Shuman *et al.* 2013). A simple example of $\mathrm{dist}(i,j)$ is to just let it be $\|y_i - y_j\|$. Besides, one can see the filter weights in equations [6.7] and [6.8] are also examples of the generally defined equation [6.10].

According to the underlying modeling about the images, we categorize the works on image restoration with GSP into two types:

1) *Discrete-domain methods*: This type of method views each observation (e.g. an image patch) as a discrete signal in a discrete domain. To recover the ground truth, this type of work employs tools, such as graph spectral theory (Chung and Graham 1997) and Markov random fields (MRF) (Li 2009), where graph weights are viewed as correlations and a graph Laplacian matrix is viewed as a precision matrix. These "discrete-domain" tools inherently view the graph vertices as discrete objects.

2) *Continuous-domain methods*: This type of method views each observation as a discrete sample of an inherently continuous signal in a continuous domain. As a result, graphs are viewed as discrete counterparts of Riemannian manifolds (Osher *et al.* 2017; Pang and Cheung 2017; Elmoataz *et al.* 2008). This type of work has close connections to variational approaches (Sapiro 2006), and involves notions in the continuous-domain, e.g. Laplace-Beltrami operator, anisotropic diffusion, and so on.

We will present the discrete-domain methods and the continuous-domain methods in sections 6.2 and 6.3, respectively. Note that not all graph signals can be interpreted in the continuous domains, e.g. the voting pattern on a social network is inherently discrete. However, image signals can indeed be viewed as being captured continuously and then sampled at pixel grid locations; this enables a continuous interpretation of graph signals as discrete samples of continuous signals.

Both the discrete-domain and the continuous-domain methods are *model-based*, where a presumed model is applied for graph construction and image restoration. In contrast, *learning-based* approaches are data driven, i.e. the restoration is governed by the training data. By virtue of the developments of *deep learning* (LeCun *et al.*

2015; Kipf and Welling 2017), the capability of image processing has been escalated to an unprecedented level. Meanwhile, quite a few novel image restoration methods combining deep neural networks (DNN) and graph spectral signal processing have been proposed, leading to impressive restoration performance (Zeng *et al.* 2019; Su *et al.* 2020). These novel methods based on DNN will also be discussed in section 6.4.

6.2. Discrete-domain methods

A large body of research on image restoration with GSP can be classified as discrete-domain methods. We will first illustrate the discrete modeling of graph-based methods, and then discuss several representative approaches, e.g. non-local graph-based transforms (NLGBT) (Hu *et al.* 2013), doubly stochastic graph Laplacians (Kheradmand and Milanfar 2014), reweighted graph total variations (RGTV) (Bai *et al.* 2019) and left eigenvectors of the random walk graph Laplacian (LERaG) (Liu *et al.* 2017). We will also briefly talk about graph-based image filtering at the end.

6.2.1. *Non-local graph-based transform for depth image denoising*

As a concrete example, this section will discuss a simple yet representative method for depth image denoising, NLGBT (Hu *et al.* 2013).

6.2.1.1. *Non-local graph-based transform*

As introduced in section 6.1.3, non-local based methods are widely adopted in image denoising, among which the most representative works include non-local means (Buades *et al.* 2005a) and BM3D (Dabov *et al.* 2007). The basic assumption is that similar patches recur throughout an image, thus one patch can be restored by jointly averaging with non-local patches with high similarity.

The assumption also holds true for depth images where similar edge structures appear throughout an image. Moreover, the piece-wise smooth (PWS) property of depth image perfectly matches the GFT (referred to as graph-based transform in Hu *et al.* (2013)) representation, which preserves PWS signals well. In light of this, the authors jointly explore the local piece-wise smoothness and the non-local self-similarity of depth images with the proposed NLGBT.

To utilize non-local similarity, similar patches are clustered into one group, from which an average patch is computed, representing the average statistics of the patches in this group. In the average patch, a graph is constructed by connecting each pixel to its four neighbors, as shown in Figure 6.2. The graph weight is computed using equation [6.10], where $\mathrm{dist}(i, j)$ is set to be the intensity difference between pixel i and j, i.e. $|y_i - y_j|$. Then the GFT transform $\mathbf{U}$ is given by the eigenvectors of the combinatorial graph Laplacian matrix.

The averaging process enables the graph to represent non-local statistics, and also performs as a low-pass pre-filtering for the image patches, which contributes to the robustness of the transform matrix **U** to noise.

To preserve local piece-wise smoothness in the recovered image, these similar patches are transformed to the GFT domain, and denoising is performed by enforcing group sparsity in the GFT domain. Given the noisy patches $\mathbf{y}_i \in \mathbb{R}^n$, where i is the patch index, the formulation becomes:

$$\min_{\mathbf{U},\boldsymbol{\alpha}_i} \Sigma_{i=1}^{N} \|\mathbf{y}_i - \mathbf{U}\boldsymbol{\alpha}_i\|_2 + \lambda \Sigma_{i=1}^{N} \|\boldsymbol{\alpha}_i\|_0 \qquad [6.11]$$

where $\boldsymbol{\alpha}_i$ is the GFT coefficient for $\mathbf{y}_i$ and λ is the weighting parameter. The sparsity of $\boldsymbol{\alpha}_i$ is minimized as the smoothness prior to remove noise, as noise often appears as high-frequency components in the GFT domain.

In this way, NLGBT is able to utilize non-local similar geometry by using the same transform, **U**, for all patches and seeking sparse representation for the group, i.e. *group sparsity*. Furthermore, since $\boldsymbol{\alpha}_i$ is allowed to be different for each patch, NLGBT also preserves the textures of individual patches. In addition, the transform **U**, used in NLGBT, can be efficiently learned from the average patch, which simplifies the dictionary learning process.

6.2.1.2. *Algorithm implementation and performance demonstration*

The algorithm based on NLGBT is implemented as follows. First, similar patches are clustered into groups similar to BM3D. For each group, an average patch is computed and used for graph construction and GFT transform computation. Next, given the GFT, similar patches are jointly denoised by minimizing the l_0 norm of the GFT coefficients, as in equation [6.11], by hard-thresholding the coefficients similar to Donoho and Johnstone (1994). The optimization is solved iteratively, where graph edge weights and the corresponding GFT transform **U** are updated after each iteration. The optimization implementation is summarized in Algorithm 6.1.

The experiment results validate that NLGBT outperforms non-local-based schemes, such as non-local means (NLM) and BM3D. The example in Figure 6.3 shows that the result of NLGBT exhibits clean sharp edges and a smooth surface, while the one produced by BM3D is blurred along the edges to some extent, and the one produced by NLM still looks noisy.

6.2.2. *Doubly stochastic graph Laplacian*

NLGBT assumes the sparsity of GFT coefficients to preserve the PWS property of depth image, while variants of graph-based smoothness have been designed for different image restoration applications. In this section, we will introduce a smoothness prior proposed in Kheradmand and Milanfar (2014), based on a doubly stochastic graph Laplacian for non-blind image deblurring.

Algorithm 6.1. Depth image denoising with non-local graph-based transform

Require: Noisy depth image $\mathbf{y}$ and iteration number K

Initialize $\mathbf{y}^{(1)} = \mathbf{y}$
for $k = 1{:}K$ **do**
  Patch clustering
  GFT transform $\mathbf{U}$ computation
  Perform denoising by solving equation [6.11] via hard-thresholding coefficients
  Update $\mathbf{y}^{(k)}$ via inverse GFT
end for

Ensure: Denoised depth image

(a) NLM (b) BM3D (c) NLGBT

Figure 6.3. *Denoising results of different approaches, with depth map Teddy corrupted by AWGN (σ=10). © 2013 IEEE. Reprinted, with permission, from Hu* et al. *(2013)*

There are mainly three different definitions of the graph Laplacian commonly used in graph signal processing, i.e. combinatorial Laplacian, normalized Laplacian, and random walk Laplacian. However, these three Laplacian matrices are not ideal to be used in graph Laplacian regularization (GLR), due to the lack of desirable properties summarized in Table 6.1. The symmetry of the graph Laplacian is required so that the optimization can be solved efficiently with fast methods, such as conjugate gradient (CG) (Shewchuk *et al.* 1994). This is because, for CG to operate efficiently, the coefficient matrix is required to be sparse, symmetric and positive definite. The DC eigenvector of the graph Laplacian, i.e. the eigenvector for eigenvalue 0 being constant, is preferable because GLR preserves a constant signal, which is very common in natural images (Kheradmand and Milanfar 2014). The stochastic property of the adjacency matrix $\mathbf{W}$, i.e. $\mathbf{W1} = \mathbf{1}$ for row-stochasticity and $\mathbf{1W} = \mathbf{1}$ for column-stochasticity, keeps the mean value of the signal unchanged and leads to improved performance (Milanfar 2013).

Graph Laplacian	Combinatorial	Normalized	Random Walk	Doubly Stochastic
Math expression	$\mathbf{D}-\mathbf{W}$	$\mathbf{I}-\mathbf{D}^{-1/2}\mathbf{W}\mathbf{D}^{-1/2}$	$\mathbf{I}-\mathbf{D}^{-1}\mathbf{W}$	$\mathbf{I}-\mathbf{C}^{-1/2}\mathbf{W}\mathbf{C}^{-1/2}$
Symmetry	True	True	False	True
DC eigenvector	True	False	True	True
Stochastic property	False	False	Row-stochastic	Doubly stochastic

Table 6.1. *Property comparison of different graph Laplacians*

As seen in Table 6.1, none of the three Laplacian matrices fulfill all of the desirable properties. This motivated Kheradmand and Milanfar (2014) to propose a new doubly stochastic graph Laplacian with all three desirable properties. We will discuss how the doubly stochastic Laplacian is designed and applied to non-blind image deblurring.

6.2.2.1. *Doubly stochastic Laplacian and spectral interpretation*

In Kheradmand and Milanfar (2014), the adjacency matrix $\mathbf{W}$ is given by using equation [6.10] for edge weight computation, with $\text{dist}(i,j)$ set to be the pixel intensity difference. Then, the Sinkhorn matrix balancing procedure (Sinkhorn and Knopp 1967) is applied to $\mathbf{W}$, such that the resulting matrix $\mathbf{K} = \mathbf{C}^{-1/2}\mathbf{W}\mathbf{C}^{-1/2}$ is symmetric and row- and column-stochastic. The corresponding Laplacian $\mathbf{I}-\mathbf{K}$ is referred to as the *doubly stochastic graph Laplacian*, which is a symmetric and positive semi-definite (PSD) matrix, with constant eigenvector $\frac{1}{\sqrt{N}}\mathbf{1}$ corresponding to the zero eigenvalue. Being symmetric and PSD, the linear system of equations associated with $\mathbf{I}-\mathbf{K}$ can be solved with fast optimization methods like CG (Shewchuk *et al.* 1994). With a constant DC component, the graph Laplacian prior based on $\mathbf{I}-\mathbf{K}$ preserves a constant signal.

Given the blurry and noisy input $\mathbf{y}$ and the blur matrix $\mathbf{H}$ corresponding to the blur kernel, the deblurring result $\mathbf{x}$ is obtained by optimizing for

$$\min_{\mathbf{x}} \quad (\mathbf{y}-\mathbf{H}\mathbf{x})^{\top}\{\mathbf{I}+\beta(\mathbf{I}-\mathbf{K})\}(\mathbf{y}-\mathbf{H}\mathbf{x}) + \eta\mathbf{x}^{\top}(\mathbf{I}-\mathbf{K})\mathbf{x} \qquad [6.12]$$

where $\beta \geq -1$ and $\eta > 0$ are the parameters tuned based on the amount of blur and noise. The first term is a coupling of data fidelity and prior, and seeks to find a $\mathbf{x}$ that its blurred and filtered version is close to the filtered version of $\mathbf{y}$, where $\mathbf{I}+\beta(\mathbf{I}-\mathbf{K})$ implements the filtering. The frequency selectivity of the filter is determined by the parameter β. Meanwhile, the second term is to preserve the salient structure in $\mathbf{x}$ via the graph Laplacian prior based on $\mathbf{I}-\mathbf{K}$.

To analyze the spectral interpretation of the optimization, the objective function in equation [6.12] can be rewritten as:

$$E(\mathbf{x}) = \|\{\mathbf{I} + \beta(\mathbf{I} - \mathbf{K})\}^{1/2}(\mathbf{y} - \mathbf{H}\mathbf{x})\|^2 + \eta\|(\mathbf{I} - \mathbf{K})^{1/2}\mathbf{x}\|^2. \qquad [6.13]$$

Note that $\mathbf{I} + \beta(\mathbf{I} - \mathbf{K}) = \mathbf{V}\mathbf{\Lambda}\mathbf{V}^\top$ is symmetric and PSD, so $\{\mathbf{I} + \beta(\mathbf{I} - \mathbf{K})\}^{1/2} = \mathbf{V}\mathbf{\Lambda}^{1/2}\mathbf{V}^\top$ can be considered as a filter with similar filtering behavior to $\mathbf{I} + \beta(\mathbf{I} - \mathbf{K})$. Since $\mathbf{I} - \mathbf{K}$ is a high-pass filter, with $\beta > 0$, $\mathbf{I} + \beta(\mathbf{I} - \mathbf{K})$ behaves like a sharpening filter on the residuals $(\mathbf{y} - \mathbf{H}\mathbf{x})$, and so, is $\{\mathbf{I} + \beta(\mathbf{I} - \mathbf{K})\}^{1/2}$. In Shan *et al.* (2008), this is shown to be beneficial in modeling the underlying phenomenon for deblurring problems, by involving different derivatives of the residual in the data fidelity. Meanwhile, the second term $(\mathbf{I} - \mathbf{K})^{1/2}$ acts like a signal-adaptive high-pass filter. Hence, it penalizes high frequency to avoid artifacts due to noise amplification and ringing artifacts, while maintaining the fine details.

6.2.2.2. *Algorithm implementation and performance comparisons*

The solution to equation [6.12] is obtained by solving the following symmetric positive definite system of liner equations:

$$(\mathbf{H}^\top\{\mathbf{I} + \beta(\mathbf{I} - \mathbf{K})\}\mathbf{H} + \eta(\mathbf{I} - \mathbf{K}))\mathbf{x} = \mathbf{H}^\top\{\mathbf{I} + \beta(\mathbf{I} - \mathbf{K})\}\mathbf{y}. \qquad [6.14]$$

CG (Shewchuk *et al.* 1994) is used to solve the linear system. $\beta \geq -1$ is set to keep $\mathbf{I} + \beta(\mathbf{I} - \mathbf{K})$ PSD. The estimation for $\mathbf{x}$ is gradually improved through iterations, and the graph weight $\mathbf{K}$ is updated based on the last estimation of $\mathbf{x}$ in each iteration.

In Kheradmand and Milanfar (2014), a comparison with the non-local means regularization deconvolution algorithm in Zhang *et al.* (2010) is demonstrated in the example of the blurred noisy Cameraman image, with 9×9 box average blur kernel and additive white Gaussian noise $\sigma = 1$. The output of the doubly stochastic Laplacian produces less ringing artifacts and outperforms Zhang *et al.* (2010) by 0.59 dB in PSNR. Moreover, for nonlinear motion blur, Zhang *et al.* (2010) almost fail, while the doubly stochastic Laplacian shows superior performance in this case, outperforming Zhang *et al.* (2010) by 11.12 dB in PSNR. Additionally, Kheradmand and Milanfar (2014) perform a comparison between the combinatorial Laplacian and the doubly stochastic Laplacian, where the latter provides a much sharper result, validating the effectiveness of the proposed Laplacian.

6.2.3. *Reweighted graph total variation prior*

Kheradmand and Milanfar (2014) assume that the blur kernel is known. However, in most cases, the blur kernel is unknown, which is the case Bai *et al.* (2019) deal with, i.e. blind deblurring. Apart from deconvolving the blurry image to recover the latent image, as done in Kheradmand and Milanfar (2014), Bai *et al.* (2019) also need to estimate the blur kernel, making the problem highly ill-posed.

To solve kernel estimation, the authors propose to use a skeleton image as a PWS proxy for kernel estimation, and use the RGTV prior for the skeleton image reconstruction. The RGTV prior is demonstrated to promote sharpness better than the traditionally used GTV prior (Elmoataz *et al.* 2008), hence it is advantageous for the skeleton image recovery, which exhibits very sharp edges. With the estimated blur kernel from the skeleton image, the latent sharp image is recovered from the blurry input. Next, we will introduce how RGTV is designed and achieves edge preservation in the skeleton images.

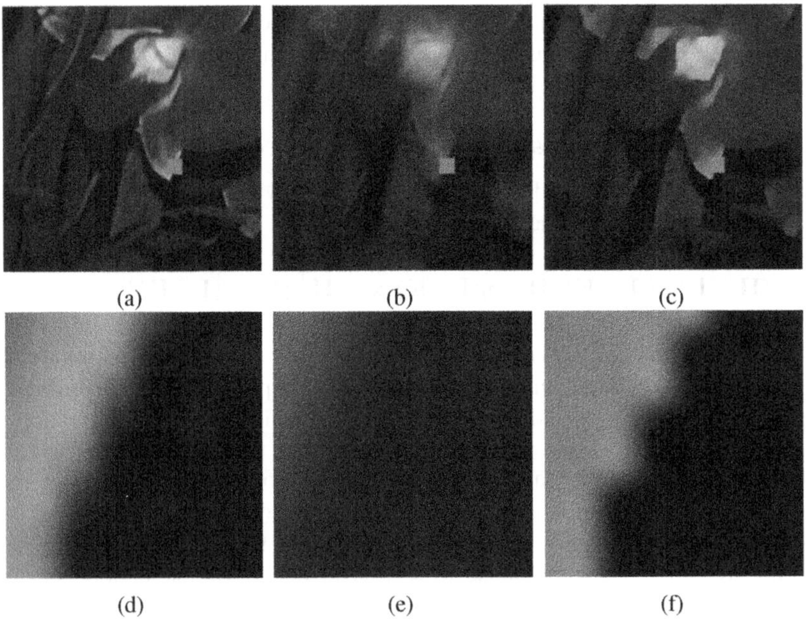

Figure 6.4. *Illustrations of different kinds of images. (a) A true natural image, (b) a blurry image, (c) a skeleton image, and (d), (e) and (f) are patches in the green squares of (a), (b) and (c), respectively. © 2018 IEEE. Reprinted, with permission, from Bai* et al. *(2019). For a color version of this figure, see www.iste.co.uk/cheung/graph.zip*

6.2.3.1. *Kernel estimation from skeleton image*

The skeleton image–a PWS version of the target image–is used as a proxy for the kernel estimation. The skeleton image smooths out the details and only retains the strong edge in the original image, as shown in Figure 6.4. It is demonstrated in Bai *et al.* (2019) that the skeleton image is as valuable as the natural image for kernel estimation. Since the skeleton image is much easier to estimate than the original natural image, given the blurring input, Bai *et al.* (2019) chose to use a skeleton image for kernel estimation.

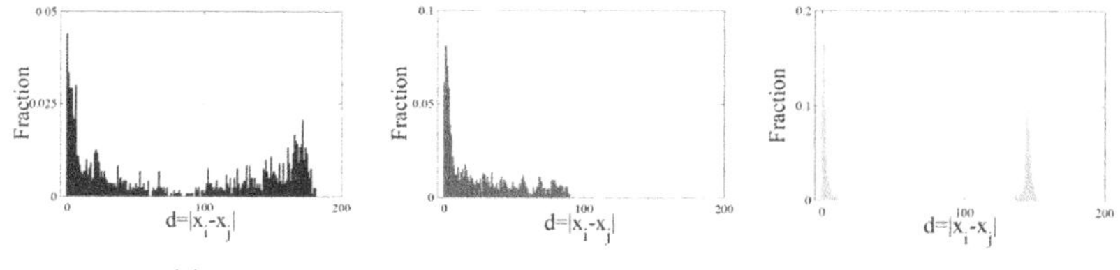

Figure 6.5. *Edge weight distribution around image edges. (a) A true natural patch, (b) a blurry patch, and (c) a skeleton patch. © 2018 IEEE. Reprinted, with permission, from Bai* et al. *(2019). For a color version of this figure, see www.iste.co.uk/cheung/graph.zip*

To distinguish the blurry image and the PWS image like the clean and skeleton image, Figure 6.5 demonstrates the edge weight distributions of the patches in Figures 6.4(d) – (f). A fully connected graph is constructed for each patch. x-axis is the inter-pixel difference $d = |x_i - x_j|$ for edge weight $w_{i,j}$ computation using equation [6.10], and $w_{i,j}$ is a monotonically decreasing function of d. The y-axis shows the fractions of weights, given d.

As shown in Figure 6.5, the edge weights of clean and skeleton patches are either very large or very small, following a *bi-modal distribution*, while the blurry patch does not. The authors take advantage of this statistic difference and design the RGTV prior to promote a bi-modal distribution of edge weights in the skeleton image, given the blurry input.

6.2.3.2. *Reweighted graph total variation and spectral analysis*

The RGTV prior is proposed to promote bi-modal edge weight distribution in image patches, which is defined as

$$\|x\|_{RGTV} = \Sigma_{i=1}^{N}\Sigma_{j=1}^{N} w_{i,j}(x_i, x_j)|x_j - x_i|, \qquad [6.15]$$

where $w_{i,j}(x_i, x_j)$ is computed using equation [6.10] with $d = |x_i - x_j|$, and is thus a function of the signal $\mathbf{x}$.

By reweighting the conventional graph total variation (GTV) (Elmoataz *et al.* 2008), the RGTV prior will not simply push d to 0 and smooth out the image like a fixed weight GTV does. Instead, minimizing RGTV will reduce d only when $d < \sigma/\sqrt{2}$, otherwise it will amplify d if $d > \sigma/\sqrt{2}$. In this way, the RGTV regularizer promotes the desirable bi-modal edge weight distribution of sharp images. In addition, by comparing with l_2 the GLR regularization $\mathbf{x}^\top \mathbf{L}\mathbf{x}$ in equation [6.5], when d approaches 0, the derivative of RGTV is 1 while GLR is 0, which means RGTV approaches the minimum faster than GLR.

In the form of l_1-norm, it is not obvious how GTV/RGTV preserves certain low frequencies like the GLR does. Since the Laplacian operator $\mathbf{L}$ can be deduced from the derivative of GLR $\mathbf{x}^\top \mathbf{L}\mathbf{x}$, the authors take the sub-derivative of GTV to deduce its corresponding Laplacian operator, so that the graph spectrum can be defined based on the new Laplacian. The new graph weight function is derived as

$$\gamma_{i,j} = \frac{w_{i,j}}{\max\{|x_i - x_j|, \epsilon\}}, \qquad [6.16]$$

where ϵ is a small constant (set to 0.01) to guarantee numerical stability around 0, and the resulting adjacency matrix is denoted as $\mathbf{\Gamma}$. The Laplacian operator $\mathbf{L}_\mathbf{\Gamma}$ is given as $\mathbf{L}_\mathbf{\Gamma} = \text{diag}(\mathbf{\Gamma}\mathbf{1}) - \mathbf{\Gamma}$, referred to as l_1-Laplacian.

The eigendecomposition of $\mathbf{L}_\mathbf{\Gamma}$ as a real symmetric PSD matrix on a one-dimensional signal example, experimentally demonstrates that GTV preserves the PWS property better than GLR. First, the second eigenvector (lowest AC frequency component) of GTV is more robust than the GLR, when noise and/or blur are presented in the original signal. Second, the relative eigenvalues λ_k/λ_2 of GTV are much larger than that of GLR when $k > 2$, which indicates that GTV has a larger penalization on high graph frequency components than GLR. RGTV inherits the desirable properties of GTV mentioned above, and can thus be interpreted as a low-pass graph filter that promotes edge-sharpness better than GLR.

6.2.3.3. *Algorithm implementation*

The blind image deblurring, given input $\mathbf{y}$, is solved via minimizing:

$$(\hat{\mathbf{x}}, \hat{\mathbf{k}}) = \arg\min_{\mathbf{x},\mathbf{k}} \frac{1}{2}||\mathbf{x} \otimes \mathbf{k} - \mathbf{y}||_2^2 + \beta||\mathbf{x}||_{RGTV} + \mu||\mathbf{k}||_2^2, \qquad [6.17]$$

where $\mathbf{k}$ is the blur kernel of size $h \times h$, $\otimes$ is the convolution operator, and β and μ are the parameters for the RGTV prior and the regularization for $\mathbf{k}$. The optimization in equation [6.17] is non-convex and non-differentiable. To solve equation [6.17], $\mathbf{x}$ and $\mathbf{k}$ are estimated iteratively.

Given $\hat{\mathbf{k}}$, $\mathbf{x}$ is solved with

$$\hat{\mathbf{x}} = \arg\min_{\mathbf{x}} \frac{1}{2}||\mathbf{x} \otimes \mathbf{k} - \mathbf{y}||_2^2 + \beta||\mathbf{x}||_{RGTV}. \qquad [6.18]$$

Since the RGTV prior is a function of $\mathbf{x}$, equation [6.18] is solved with an alternative scheme. First, the four-neighbor graph shown in Figure 6.2 is built with weight computed using equation [6.10]. Then, using the new graph weight function in equation [6.16], the RGTV prior is turned into the form of $\mathbf{x}^\top \mathbf{L}_\Gamma \mathbf{x}$. Therefore, $\mathbf{x}$ is optimized by solving a system of linear equations:

$$(\hat{\mathbf{H}}^\top \hat{\mathbf{H}} + 2\beta \mathbf{L}_\Gamma)\hat{\mathbf{x}} = \hat{\mathbf{H}}^\top \mathbf{y}, \qquad [6.19]$$

where $\hat{\mathbf{H}}$ is the matrix representation of convolving with $\hat{\mathbf{k}}$. Equation [6.19] can be solved via the CG method (Shewchuk *et al.* 1994) since the left-hand side matrix is sparse, symmetric and positive definite. Then, $\mathbf{L}_\Gamma$ is updated with $\mathbf{L}_\Gamma(\hat{\mathbf{x}})$ to optimize $\mathbf{x}$ in the next iteration. Note that $\mathbf{k}$ is initialized with delta function or the result from coarser scale, as in Fergus *et al.* (2006).

Given $\hat{\mathbf{x}}$, $\mathbf{k}$ is optimized in the gradient domain to avoid artifacts and is solved by

$$\hat{\mathbf{k}} = \arg\min_{\mathbf{k}} \frac{1}{2} \| \bigtriangledown \mathbf{x} \otimes \mathbf{k} - \bigtriangledown \mathbf{y} \|_2^2 + \mu \|\mathbf{k}\|_2^2, \qquad [6.20]$$

where $\bigtriangledown$ is the gradient operator. Equation [6.20] is a quadratic convex problem and also has a closed-form solution. Note that $\hat{\mathbf{x}}$ is the skeleton image, and to restore the natural sharp image, given $\hat{\mathbf{k}}$, the blurry image $\mathbf{y}$ is deblurred with recent non-blind image deblurring algorithms, as in Pan *et al.* (2016).

To sum up, RGTV is superior to GTV by reweighting the edge weight with the graph signal, which is advantageous to edge-sharpness preservation in the skeleton image. Further analysis of the spectral properties of GTV/RGTV shows that the prior is robust to noise/blur and promotes strong PWS filtering and sharpness. Moreover, thanks to the spectral interpretation of RGTV, an efficient algorithm is developed to solve for the skeleton image and blur kernel in the non-convex non-differentiable optimization problem.

6.2.3.4. *Performance demonstration*

RGTV is compared to state-of-the-art methods including Sun *et al.* (2013), Michaeli and Irani (2014), Lai *et al.* (2015) and Pan *et al.* (2016). The resulting comparison on a sample image from the database in Sun *et al.* (2013), is shown in Figure 6.6, where the image is blurred with the kernel, as shown in Figure 6.6(a), with added 1% white Gaussian noise. RGTV achieves more robust results compared to the competing algorithms, preserving the sharp edges with less ringing artifacts.

6.2.4. *Left eigenvectors of random walk graph Laplacian*

Apart from image denoising and deblurring, the graph-based regularization is also adopted in the soft decoding of JPEG[2] compressed images in Liu *et al.* (2017) to promote the PWS property in the reconstructed images. Liu *et al.* (2017) propose a new graph-signal smoothness prior based on the left eigenvectors of the random walk graph Laplacian matrix (LERaG), which is combined with the Laplacian prior for DCT coefficients and the sparsity prior for constructing a JPEG soft decoding algorithm.

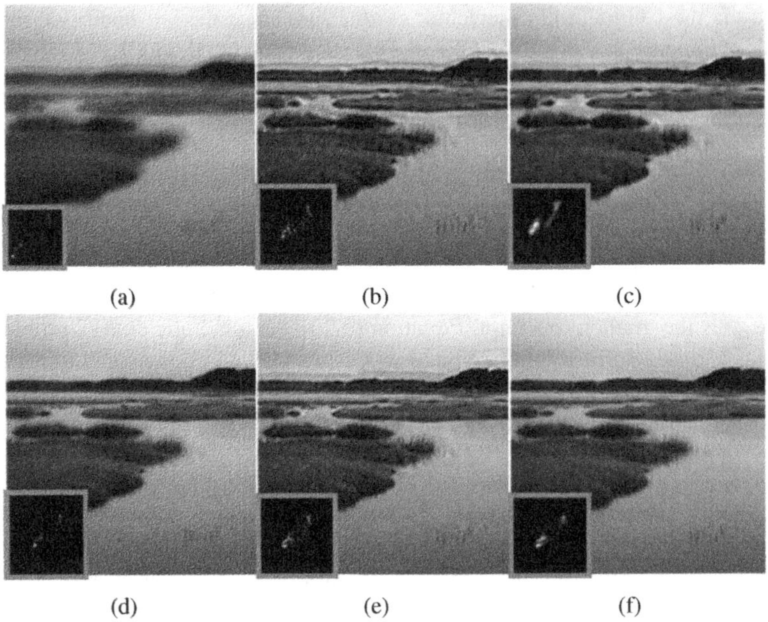

Figure 6.6. *Deblurring results comparison. (a) Blurry image. (b) Sun* et al. *(2013). (c) Michaeli and Irani (2014). (d) Lai* et al. *(2015). (e) Pan* et al. *(2016). (f) RGTV. The blur kernel is shown at the lower left corner. © 2018 IEEE. Reprinted, with permission, from Bai* et al. *(2019). For a color version of this figure, see www.iste.co.uk/cheung/graph.zip*

6.2.4.1. *JPEG soft decoding*

The problem setup of JPEG soft decoding is discussed first. In the JPEG encoding stage, for each non-overlapping 8×8 block $\mathbf{y} \in \mathbb{R}^{64}$ in an image, the image values are transformed via DCT into coefficients $\mathbf{Y} = \mathbf{T}\mathbf{y}$, where $\mathbf{T}$ is the transform matrix. $\mathbf{Y}$ is then lossily quantized using quantization parameter (QP) Q_i:

2 https://jpeg.org/

$$q_i = \text{round}(Y_i/Q_i), \qquad [6.21]$$

where q_i is the quantization bin (q-bin) index (q-index).

At the decoder, to recover Y_i from the q-index q_i, soft decoding is aimed at choosing the optimal value subject to the bin constraint:

$$q_i Q_i \leq Y_i \leq (q_i + 1)Q_i. \qquad [6.22]$$

The recovered Y is then transformed back to the pixel domain via inverse-DCT. Obviously, the dequantization is ill-posed and requires a prior.

6.2.4.2. *LERaG prior*

In Liu *et al.* (2017), three priors are combined for JPEG soft decoding: the Laplacian distribution prior of DCT coefficients, the sparse representation of image patches in dictionary domain and a new graph-signal smoothness prior – LERaG prior.

Given the bin constraint in equation [6.22] for individual DCT coefficients, DCT coefficients are assumed to follow *Laplacian distribution*, as used in Lam and Goodman (2000):

$$P_L(Y_i) = \frac{\mu_i}{2} \exp(-\mu_i |Y_i|), \qquad [6.23]$$

where μ_i is a parameter that increases with the frequency. The minimum mean square error (MMSE) solution using the Laplacian prior leads to a closed-form solution:

$$Y_i^* = \frac{(q_i Q_i + \mu_i) e^{\{\frac{-q_i Q_i}{\mu_i}\}} - ((q_i + 1)Q_i + \mu_i) e^{\{\frac{-(q_i+1)Q_i}{\mu_i}\}}}{e^{\{\frac{-q_i Q_i}{\mu_i}\}} - e^{\{\frac{-(q_i+1)Q_i}{\mu_i}\}}}. \qquad [6.24]$$

The optimal DCT coefficients are then transformed into image values $\mathbf{y}^*$ via inverse DCT.

Since the solution is optimized for each block separately, it does not handle block artifacts at block boundaries. Therefore, the *sparsity prior* is employed to optimize, at a larger patch level, $\mathbf{x} \in \mathbb{R}^n$, which encloses the smaller block $\mathbf{y} \in \mathbb{R}^{64}$, where $n > 64$, as illustrated in Figure 6.7(a). Mathematically, $\mathbf{y} = \mathbf{M}\mathbf{x}$, where $\mathbf{M} \in \{0,1\}^{64 \times n}$ extracts pixels in $\mathbf{x}$ corresponding to $\mathbf{y}$. Assuming a sparsity model, x is projected onto the overcomplete dictionary $\mathbf{\Phi} \in \mathbb{R}^{n \times M}$, where $M \gg n$, given as

$$\mathbf{x} = \mathbf{\Phi}\boldsymbol{\alpha} + \xi, \qquad [6.25]$$

where $\boldsymbol{\alpha} \in \mathbb{R}^M$ are the coefficients corresponding to the atoms of $\boldsymbol{\Phi}$, and $\xi \in \mathbb{R}^n$ is a small error term. Dictionary $\boldsymbol{\Phi}$ is learned offline with a large set of training patches using K-SVD (Aharon *et al.* 2006). Given $\mathbf{x}$, the optimal $\boldsymbol{\alpha}$ is given by minimizing the l_0-norm of $\boldsymbol{\alpha}$:

$$\boldsymbol{\alpha}^* = \arg\min_{\boldsymbol{\alpha}} \|\mathbf{x} - \boldsymbol{\Phi\alpha}\|_2^2 + \lambda\|\boldsymbol{\alpha}\|_0, \qquad [6.26]$$

where λ is the weighting factor to trade-off the data fidelity term and the sparsity prior. The optimization in equation [6.26] is solved via the orthogonal matching pursuit (OMP) (Aharon *et al.* 2006).

To reduce the complexity of OMP, the dictionary size M is required to be small, resulting in a poor recovery of high DCT frequencies. The problem is that a small dictionary size undermines the performance of sparsity-based soft decoding, as seen in Figure 6.7(b).

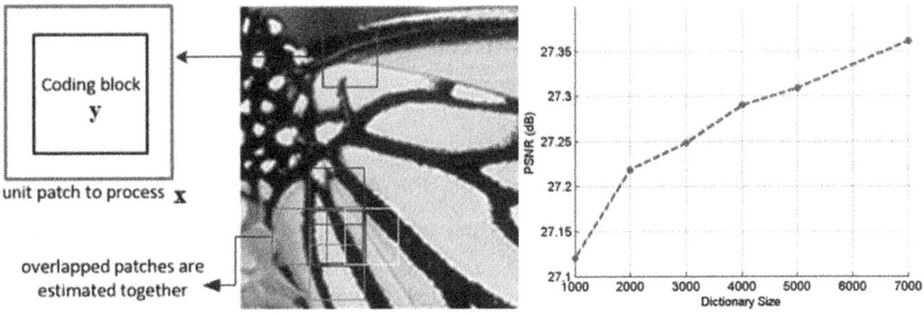

(a) (b)

Figure 6.7. *(a) A patch being optimized encloses a smaller code block. Boundary discontinuity is removed by averaging overlapping patches. (b) The relationship between the dictionary size and the restoration performance. © 2016 IEEE. Reprinted, with permission, from Liu* et al. *(2017). For a color version of this figure, see www.iste.co.uk/cheung/graph.zip*

To recover high DCT frequencies when the dictionary is small, the *LERaG* prior is proposed to promote the PWS property in the reconstructed image. For each block $\mathbf{x}$, a fully connected graph is constructed with a graph weight similar to the bilateral filter weight defined in equation [6.7]. LERaG is proposed as $\mathbf{x}^\top \mathcal{L}_r^\top \mathcal{L}_r \mathbf{x}$, where $\mathcal{L}_r = \mathbf{D}^{-1}\mathbf{L}$ is the random walk Laplacian and $\mathbf{L}$ is the combinatorial graph Laplacian.

6.2.4.3. *Property of LERaG*

As discussed in section 6.2.2, the doubly stochastic Laplacian is designed to have matrix symmetry for fast implementation, DC eigenvector for constant signal preservation, and stochasticity to keep the mean value of the signal unchanged. LERaG is also designed to have the three properties and is more computationally efficient than Kheradmand and Milanfar (2014), which requires the performance of matrix balancing, taking non-trivial computational cost.

First, using $\mathcal{L}_r^\top \mathcal{L}_r$ instead of $\mathcal{L}_r$ keeps the matrix symmetric. Second, $\mathcal{L}_r$ can be obtained from the normalized Laplacian $\mathcal{L}_n$ via similarity-transform $\mathcal{L}_r = \mathbf{D}^{-1/2}\mathcal{L}_n\mathbf{D}^{1/2} = \mathbf{D}^{-1}\mathbf{L}$. Therefore, $\mathcal{L}_r$ has the same eigenvalues as $\mathcal{L}_n$, and has the left eigenvectors:

$$\begin{aligned} &\mathcal{L}_n = \mathbf{V}\boldsymbol{\Lambda}\mathbf{V}^\top, \\ &\mathbf{V}^\top\mathbf{D}^{1/2}\mathcal{L}_r = \mathbf{V}^\top\mathbf{D}^{1/2}\mathbf{D}^{-1/2}\mathbf{V}\boldsymbol{\Lambda}\mathbf{V}^\top\mathbf{D}^{1/2} = \boldsymbol{\Lambda}\mathbf{V}^\top\mathbf{D}^{1/2}, \end{aligned} \qquad [6.27]$$

where $\mathbf{V}^\top\mathbf{D}^{1/2}$ are the left eigenvectors of $\mathcal{L}_r$. Since $\mathbf{D}^{1/2}\mathbf{1}$ is the first eigenvector corresponding to eigenvalue 0 of $\mathcal{L}_n$, when projecting $\mathbf{1}$ to the $\mathbf{V}^\top\mathbf{D}^{1/2}$, the left eigenbasis coefficients of $\mathcal{L}_r$, i.e. $\beta = \mathbf{V}^\top\mathbf{D}^{1/2}\mathbf{1}$, is only non-zero in the first element. Therefore, unlike $\mathbf{x}^\top\mathcal{L}_n\mathbf{x}$, LERaG is based on the left eigenvectors of $\mathcal{L}_r$ and preserves constant signal.

Lastly, unlike $\mathbf{x}^\top\mathbf{L}\mathbf{x}$, LERaG is based on the normalized Laplacian $\mathcal{L}_n$, thus the filtering is insensitive to the vertex degrees of particular graphs, resolving the need to tune weighting parameters for each patch.

6.2.4.4. *Algorithm implementation*

The LERaG prior is computed as $(d_{min}^{-1})\mathbf{x}^\top\mathbf{L}\mathbf{D}^{-1}\mathbf{L}\mathbf{x}$ in Liu *et al.* (2017), and the formulation is given as

$$\begin{aligned} &\min_{\mathbf{x},\boldsymbol{\alpha}} \|\mathbf{x} - \boldsymbol{\Phi}\boldsymbol{\alpha}\|_2^2 + \lambda_1\|\boldsymbol{\alpha}\|_0 + \lambda_2(d_{min}^{-1})\mathbf{x}^\top\mathbf{L}\mathbf{D}^{-1}\mathbf{L}\mathbf{x}, \\ &\text{s.t.} \quad \mathbf{q}\mathbf{Q} \preceq \mathbf{T}\mathbf{M}\mathbf{x} \preceq (\mathbf{q}+1)\mathbf{Q}, \end{aligned} \qquad [6.28]$$

where $\mathbf{q}$ and $\mathbf{Q}$ are the quantization bin indices and QP's. Optimization in equation [6.28] is not convex, so an alternating optimization scheme is adopted to solve for $\mathbf{x}$ and $\boldsymbol{\alpha}$ iteratively. First, the initialization of $\mathbf{x}$ is given by the Laplacian prior based MMSE solution in equation [6.24]. Then in the tth iteration, $\mathbf{x}$ is fixed to solve for $\boldsymbol{\alpha}$, as in equation [6.26] via OMP. Then with fixed $\boldsymbol{\alpha}$, $\mathbf{x}$ is solved using quadratic programming (QP), and the output solution $\mathbf{x}^t$ serves as the initial for the $(t+1)$th iteration, and the graph is also updated based on $\mathbf{x}^t$. The iterative algorithm terminates when both $\mathbf{x}$ and $\boldsymbol{\alpha}$ converge.

6.2.4.5. *Performance demonstration*

LERaG is compared with competing schemes including BM3D (Dabov *et al.* 2007), KSVD (Aharon *et al.* 2006), ANCE (Zhang *et al.* 2013), DicTV (Zhang *et al.*

2013) and SSRQC (Zhao *et al.* 2016). In Figure 6.8, a sample image Butterfly is compressed with QF = 5 and decoded with different methods. LERaG outperforms competing schemes in terms of blocking artifact removal and edge sharpness preservation, and achieves the highest PSNR and SSIM.

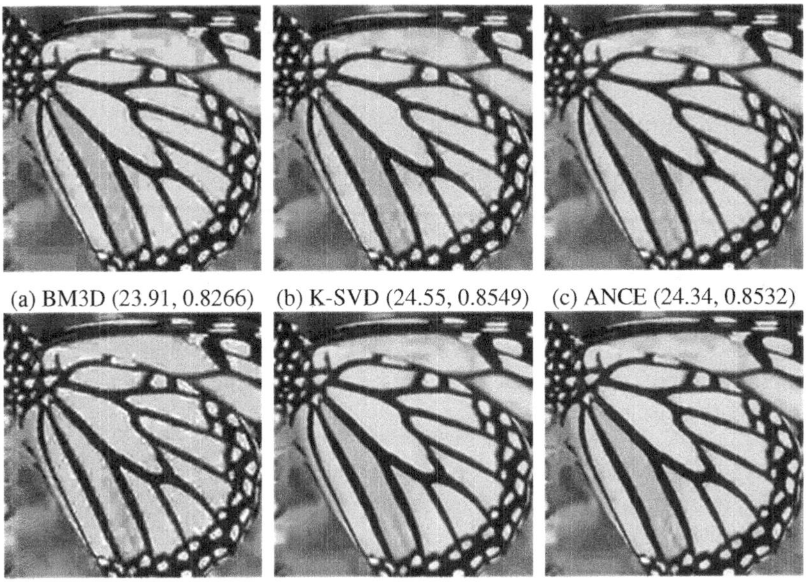

(a) BM3D (23.91, 0.8266) (b) K-SVD (24.55, 0.8549) (c) ANCE (24.34, 0.8532)

(d) DicTV (23.42, 0.8176) (e) SSRQC (25.31, 0.8764) (f) LERaG (25.82, 0.8861)

Figure 6.8. *Comparison with competing schemes on Butterfly at QF = 5. The corresponding PSNR and SSIM values are shown © 2016 IEEE. Reprinted, with permission, from Liu* et al. *(2017)*

Additionally, LERaG is compared with other graph-based priors, including the combinatorial graph Laplacian regularizer, normalized graph Laplacian regularizer and the doubly stochastic Laplacian. The comparison result in PSNR is shown in Table 6.2, with eight widely used testing images of size 256 × 256. It can be seen that LERaG outperforms the other three in the application of JPEG soft decoding.

Priors	Combinatorial	Normalized	Doubly stochastic	LERaG
Average	20.36	27.43	24.36	**30.10**

Table 6.2. *Comparison with different graph variants in PSNR (dB) at QF = 5*

6.2.5. *Graph-based image filtering*

Graph-based image filtering can be applied for image denoising without solving optimization problems (section 6.1.3). One can build edge-aware filters by smartly constructing the graphs (Milanfar 2013).

For example, in Gadde *et al.* (2013), with the bilateral filter as a simple building block, an efficient method is provided to implement graph spectral filtering with the desirable spectral property. By analyzing the bilateral filter as a graph spectral filter, a family of more general bilateral-like filters is designed with desired spectral responses for different applications. These spectral filters are implemented as iterative bilateral filtering operations without the need for expensive Laplacian diagonalization.

In Talebi and Milanfar (2016), the computational cost of applying graph-based filtering is achieved by simplifying the graph weight computation of Laplacian operators. The filtering procedure is as follows. The input is first decomposed into multi-scale detail layers by filtering with different Laplacian operators. Each layer is then mapped by a nonlinear function to boost or suppress the associated detail, and finally, these manipulated layers are blended into the output to achieve a range of filtering applications from detail enhancement to denoising. The computational cost is reduced by approximating the graph weights of the Laplacian operators. With precomputed graph filter weights using the NLM and bilateral kernels, different versions of the filter are computed by the simple direct product of the weights. The implementation is further sped up via the approximation of the Laplacian normalization to provide a computationally simplified un-normalized filtering paradigm. In this way, the implementation is fast and more suitable to be applied to mobile devices.

6.3. Continuous-domain methods

We go on with our journey of image restoration via GSP, with a focus on the *continuous-domain* methods. Different from the previous section, we now view neighborhood graphs as discrete approximations of *Riemannian manifolds* (Singer 2006; Ting *et al.* 2010). In this scenario, each vertex on a graph represents a sample point on its associated manifold, and a graph signal is a collection of samples from a continuous function on the Riemannian manifold. Intuitively, as the number of vertices on a graph increases, the graph gets closer and closer to a Riemannian manifold. The natural relationship between neighborhood graphs and the Riemannian manifold enables us to process and understand images with tools from differential geometry and variational methods (Sochen *et al.* 1998), using spectral graph theory as a bridge (Shuman *et al.* 2013).

Typical methods in this category include Zhang and Hancock (2008), Elmoataz *et al.* (2008), Pang and Cheung (2017), Osher *et al.* (2017), and so on. It is also combined with the recent blossom of deep learning, leading to hybrid approaches such as Pang *et al.* (2018); Zeng *et al.* (2019) and Zhu *et al.* (2018a). In this section, we look into two threads of representative works (Pang and Cheung 2017; Pang *et al.* 2014) (Osher *et al.* 2017; Zhu *et al.* 2018b), where the first thread still recovers an image on a patch-by-patch basis (similar to Dabov *et al.* (2007); Hu *et al.* (2013), *etc.*) and the second thread recovers an image as a whole in one optimization problem.

6.3.1. *Continuous-domain analysis of graph Laplacian regularization*

Despite the success of graph Laplacian regularization in image restoration, e.g. the discrete-domain approaches discussed in the previous section, there is still a lack of fundamental understanding of how it works well. In particular:

1) How does the graph Laplacian regularizer promote a correct solution to restore a corrupted image effectively?

2) What is the optimal graph, and hence the optimal graph Laplacian regularizer, for inverse imaging?

The series of works (Pang *et al.* 2014, 2015; Pang and Cheung 2017) view neighborhood graphs of pixel patches as discrete counterparts of Riemannian manifolds (Hein 2006; Ting *et al.* 2010) and analyze them in the continuous domain, which shed light on the above open questions. From the analysis in the continuous domain, an iterative algorithm, *optimal graph Laplacian regularization* (OGLR) for denoising, is proposed to verify the developed understanding.

6.3.1.1. *Interpreting the graph Laplacian regularizer*

We first construct an underlying graph $\mathcal{G}$ that supports a graph-signal $\mathbf{u}$ on top, where $\mathbf{u}$ is also a pixel patch in an image. We then demonstrate the convergence of the graph Laplacian regularizer $S_{\mathcal{G}}(\mathbf{u}) = \mathbf{u}^{\top}\mathbf{L}\,\mathbf{u}$ to the anisotropic Dirichlet energy S_{Ω} (Flucher 1999) – a quadratic functional in the continuous domain. We then analyze, in detail, the functional S_{Ω} to understand its discrete counterpart $S_{\mathcal{G}}$.

To facilitate understanding, we describe the construction of our discrete graph $\mathcal{G}$ and define corresponding continuous quantities in parallel. We first define Ω, a bounded region in $\mathbb{R}^2$, as the *domain* on which a continuous image (or image patch) lives. In practice, Ω takes a rectangular shape; see Figure 6.9 for an illustration. Denote by $\Gamma = \{\mathbf{s}_i = (x_i, y_i) \mid \mathbf{s}_i \in \Omega, 1 \leq i \leq M\}$ a set of M random coordinates uniformly distributed on Ω (e.g. orange crosses in Figure 6.9). Since pixel coordinates are uniformly distributed on an image, we interpret the collection of pixel coordinates as one possible set Γ.

For any location $\mathbf{s} = (x, y) \in \Omega$, we denote $f_n(\mathbf{s}) : \Omega \mapsto \mathbb{R}$, $1 \leq n \leq N$ as a set of N continuous functions defined on Ω; we call them *exemplar functions*.

These functions, which can be freely chosen by the users, are critical in determining graph connections and edge weights, as will be seen shortly. One possible choice for the exemplar functions is the estimates/observations of the desired ground-truth signal. For example, in image denoising where a noisy patch is given, the f_n's can be the noisy patch itself and another $K-1$ non-local similar patch due to non-local self-similarity of natural images (Dabov *et al.* 2007). Hence, in this case, there are $N = K$ exemplar functions. However, this selection turns out to be sub-optimal. We will develop a methodology to choose f_n's optimally in the coming section.

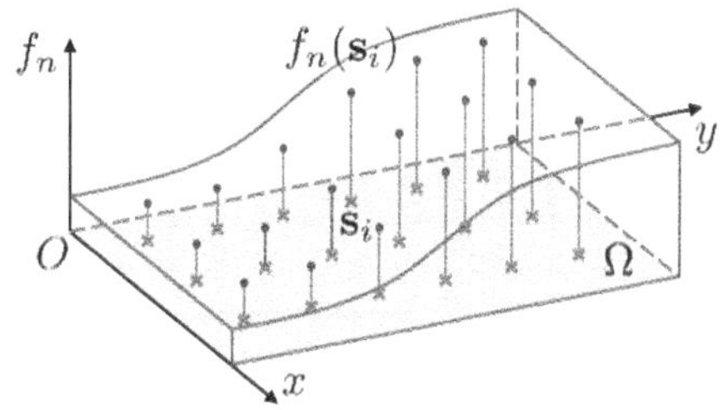

Figure 6.9. *Sampling the exemplar function f_n at pixel locations in domain Ω. © 2017 IEEE. Reprinted, with permission, from Pang and Cheung (2017). For a color version of this figure, see www.iste.co.uk/cheung/graph.zip*

By sampling the exemplar functions $\{f_n\}_{n=1}^{N}$ at coordinates Γ, N *discrete* exemplar functions of length M are obtained:

$$\mathbf{f}_n = [f_n(\mathbf{s}_1)\; f_n(\mathbf{s}_2)\; \ldots\; f_n(\mathbf{s}_M)]^{\mathrm{T}}, \qquad [6.29]$$

where $1 \le n \le N$. Figure 6.9 illustrates the sampling process of an exemplar function f_n – a simple ramp in Ω. The blue dots are samples of f_n and collectively form the vector $\mathbf{f}_n$. For each pixel location $\mathbf{s}_i \in \Gamma$, we construct a length N vector $\mathbf{v}_i$ $(1 \le i \le M)$ using the previously defined $\mathbf{f}_n$

$$\mathbf{v}_i = [\mathbf{f}_1(i)\; \mathbf{f}_2(i)\; \ldots\; \mathbf{f}_N(i)]^{\mathrm{T}}, \qquad [6.30]$$

where we write the ith entry of $\mathbf{f}_n$ as $\mathbf{f}_n(i)$, so $\mathbf{f}_n(i) = f_n(\mathbf{s}_i)$. With vectors $\{\mathbf{v}_i\}_{i=1}^{M}$, we build a weighted neighborhood graph $\mathcal{G}$ with M vertices, where each pixel location $\mathbf{s}_i \in \Gamma$ is represented by a vertex $V_i \in \mathcal{V}$. The weight w_{ij} between two different vertices V_i and V_j is computed as a thresholded Gaussian function:

$$w_{ij} = \begin{cases} \exp\left(-\dfrac{\|\mathbf{v}_i - \mathbf{v}_j\|_2^2}{2\epsilon^2}\right) & \text{if } \|\mathbf{v}_i - \mathbf{v}_j\|_2 \le r, \\ 0 & \text{otherwise,} \end{cases} \qquad [6.31]$$

where constant ϵ controls the sensitivity of the graph weights. Under these settings, $\mathcal{G}$ is an r-*neighborhood graph*, i.e. there is no edge connecting two vertices with a distance greater than r. With the constructed graph, we can now compute its combinatorial graph Laplacian matrix $\mathbf{L}$ (section I.3). We note that graphs employed in many recent works (e.g. Kheradmand and Milanfar (2014); Hu *et al.* (2014); Liu *et al.* (2014); Wan *et al.* (2014)) are special cases of our more generally defined graph $\mathcal{G}$.

Now let us formally introduce the counterpart of the graph Laplacian regularizer in the continuous domain. Denote by $u(x, y) : \Omega \mapsto \mathbb{R}$ a smooth[3] *candidate function* defined in domain Ω. Sampling u at positions of Γ leads to its discretized version, $\mathbf{u} = [u(\mathbf{s}_1)\ u(\mathbf{s}_2)\ \ldots\ u(\mathbf{s}_M)]^{\mathrm{T}}$. Using $\mathbf{L}$, the graph Laplacian regularizer for $\mathbf{u}$ can now be written as $S_{\mathcal{G}}(\mathbf{u}) = \mathbf{u}^{\top}\mathbf{L}\,\mathbf{u}$. The continuous counterpart of regularizer $S_{\mathcal{G}}(\mathbf{u})$ is given by a functional $S_{\Omega}(u)$ for function u, defined in domain Ω

$$S_{\Omega}(u) = \int_{\Omega} \nabla u^{\top} \mathbf{G}^{-1} \nabla u \left(\sqrt{\det \mathbf{G}}\right)^{-1} ds, \qquad [6.32]$$

where $\nabla u = [\partial_x u\ \partial_y u]^{\top}$ is the gradient of continuous function u, and $\mathbf{s} = (x, y)$ is a location in Ω. $S_{\Omega}(u)$ is also called the *anisotropic Dirichlet energy* in literature (Alliez *et al.* 2007; Flucher 1999), and $\mathbf{G}$ is a $2{\times}2$ matrix:

$$\mathbf{G} = \sum_{n=1}^{N} \begin{bmatrix} (\partial_x f_n)^2 & \partial_x f_n \cdot \partial_y f_n \\ \partial_x f_n \cdot \partial_y f & (\partial_y f_n)^2 \end{bmatrix} = \sum\nolimits_{n=1}^{N} \nabla f_n \cdot \nabla f_n^{\top}. \qquad [6.33]$$

$\mathbf{G} : \Omega \mapsto \mathbb{R}^{2\times 2}$ is a matrix-valued function of location $\mathbf{s} \in \Omega$. It can be viewed as the *structure tensor* (Knutsson *et al.* 2011) of the gradients $\{\nabla f_n\}_{n=1}^{N}$. Note that the exemplar functions $\{f_n\}_{n=1}^{N}$ exactly determine the functional S_{Ω} and the graph Laplacian regularizer $S_{\mathcal{G}}$. Loosely speaking, as the number of samples $M \to \infty$ and the neighborhood size $r \to 0$, $S_{\mathcal{G}}(\mathbf{u})$ approaches $S_{\Omega}(u)$ uniformly. Interested readers are referred to Pang and Cheung (2017) for a rigorous elaboration.

The convergence of the graph Laplacian regularizer $S_{\mathcal{G}}$ to the anisotropic Dirichlet energy S_{Ω} allows us to understand the mechanisms of $S_{\mathcal{G}}$ by analyzing S_{Ω}. From equation [6.32], the quadratic term $\nabla u^{\top}\mathbf{G}^{-1}\nabla u$ measures the length of gradient ∇u in a *metric space* determined by matrix $\mathbf{G}$; it is also the *Mahalanobis distance* between the point ∇u and the distribution of the points $\{\nabla f_n\}_{n=1}^{N}$ (Mahalanobis 1936). In the

3 Here "smooth" means a function with derivatives of all orders.

following, we slightly abuse the notation and call $\mathbf{G}$ the *metric space*, and perform eigendecomposition to $\mathbf{G}$ to analyze S_Ω (Takeda *et al.* 2007):

$$\mathbf{G} = \alpha \mathbf{U} \boldsymbol{\Lambda} \mathbf{U}^\top, \quad \mathbf{U} = \begin{bmatrix} \cos\theta & -\sin\theta \\ \sin\theta & \cos\theta \end{bmatrix}, \quad \boldsymbol{\Lambda} = \begin{bmatrix} \mu & 0 \\ 0 & \mu^{-1} \end{bmatrix}, \qquad [6.34]$$

where $\mu \geq 1$, $\theta \in [0, \pi)$ and $\alpha > 0$. One can verify that the *unit-distance ellipse* – the set of points having distance 1 from the origin of metric space $\mathbf{G}$ is an ellipse with semi-major axis $\sqrt{\alpha\mu}$ and semi-minor axis $\sqrt{\alpha/\mu}$. For the same length $|\nabla u|$, $\nabla u^\top \mathbf{G}^{-1} \nabla u$ computes to different values for ∇u with different directions. The Euclidean space is a special case of $\mathbf{G}$ by letting $\alpha = \mu = 1$, whose unit-distance ellipse is a unit circle.

From the convergence result, using the graph Laplacian regularizer in the discrete domain corresponds to using the functional S_Ω as a regularizer in the continuous domain. From the expression of S_Ω (equation [6.32]), this boils down to using the metric norm $\nabla u^\top \mathbf{G}^{-1} \nabla u$ as a regularizer on a point-by-point basis, throughout the image domain Ω. Figure 6.10 shows different scenarios of applying the metric norm $\nabla u^\top \mathbf{G}^{-1} \nabla u$ as a "point-wise" regularizer. Denote by $\mathbf{g}$ the point-wise ground-truth gradient of the original image, which is marked with a red dot in each plot. We also draw the contour lines of the metric spaces, where the most inner (bold) ones are the unit-distance ellipses. We see that both metric spaces in Figure 6.10(a) and Figure 6.10(b) have major directions aligned with $\mathbf{g}$, Figure 6.10(b) is more skewed, and hence more *discriminant*, i.e. a small Euclidean distance away from $\mathbf{g}$ along the minor direction of $\mathbf{G}$ results in a large metric distance. It is *desirable* for a regularizer to distinguish between good image patch candidates (close to ground-truth) and bad candidates (far from ground-truth). However, if the metric space is skewed, but its major direction does not align with $\mathbf{g}$ (Figure 6.10(c)), it is *undesirable* because bad image patch candidates will have a smaller cost than good candidates.

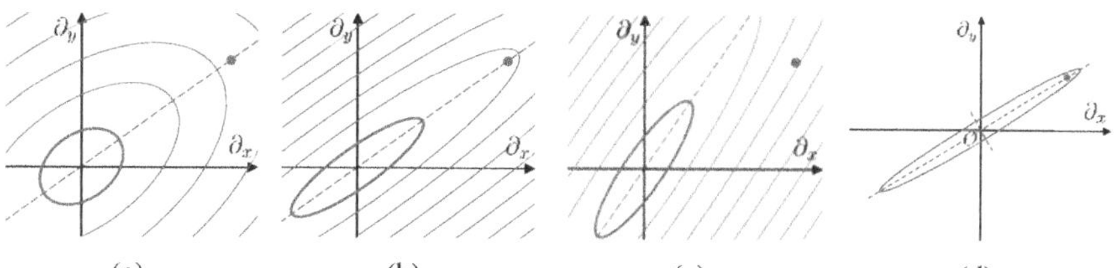

Figure 6.10. *(a) – (c) Different scenarios of using the metric norm as a "pointwise" regularizer. (d) The ideal metric space. The red dots mark the ground-truth gradient. © 2017 IEEE. Reprinted, with permission, from Pang and Cheung (2017). For a color version of this figure, see www.iste.co.uk/cheung/graph.zip*

As a result, for inverse imaging problems where g is unknown, one should design a *robust* metric space $\mathbf{G}$ based on an initial estimate of $\mathbf{g}$, such that:

1) $\mathbf{G}$ has a major direction aligned with the estimate, i.e. it is discriminant with respect to the estimate;

2) the metric space $\mathbf{G}$ is discriminant only to the extent that the estimate is reliable.

6.3.1.2. *Optimal graph for image denoising*

Equipped with the analysis of S_Ω, let us now derive the *optimal* graph Laplacian regularizer for image denoising via a non-local patch-based approach (Hu *et al.* 2013; Buades *et al.* 2005b). We consider denoising an input image on a patch-by-patch basis, hence, the domain Ω is a square region accommodating continuous image patches in this approach.

If the ground-truth gradient $\mathbf{g}(\mathbf{s})$ at location $\mathbf{s}$, $\mathbf{s} \in \Omega$ is known, we can build the following ideal metric space $\mathbf{G}_I$, for regularization:

$$\mathbf{G}_I(\mathbf{g}) = \mathbf{g}\mathbf{g}^\top + \beta \mathbf{I}, \qquad [6.35]$$

where β is a small positive constant. The quantity $\beta\mathbf{I}$ is included in equation [6.35] to ensure that the metric space $\mathbf{G}_I$ is well defined – i.e. $\mathbf{G}_I$ is invertible, so $\nabla u^\top \mathbf{G}_I^{-1} \nabla u$ is computable. In fact, when $\mathbf{g}(\mathbf{s}) = \mathbf{0}$, e.g. in flat regions, $\mathbf{G}_I = \beta\mathbf{I}$, corresponding to a scaled Euclidean space. By performing eigendecomposition, as done in the previous section, we can see that $\mathbf{G}_I$ has a major direction aligned with $\mathbf{g}$. Since the ground truth $\mathbf{g}$ is known, it is desirable to have a very discriminative metric space, implying that β should be very small. For illustration, Figure 6.10(d) shows an ideal metric space with an elongated unit-distance ellipse.

As mentioned in section 6.1, self-similarity plays a key role in modern image restoration (Buades *et al.* 2005b; Dabov *et al.* 2007). In this proposal, similar pixel patches recurring throughout an image is also assumed. Specifically, given a $\sqrt{M} \times \sqrt{M}$ noisy target patch $\mathbf{z}_0 \in \mathbb{R}^M$, we assume that there exists a set of $K - 1$ non-local patches in the noisy image that are similar to $\mathbf{z}_0$ in terms of *gradients*. Together with $\mathbf{z}_0$, the K patches $\{\mathbf{z}_k\}_{k=0}^{K-1}$ are collectively called a *cluster* in the sequel. We denote the continuous counterpart of patch $\mathbf{z}_k$ as $z_k(\mathbf{s}) : \Omega \mapsto \mathbb{R}$, $0 \le k \le K-1$, and represent $\nabla z_k(\mathbf{s})$ – the gradient of z_k at location $\mathbf{s} \in \Omega$—as $\mathbf{g}_k(\mathbf{s})$. The variable $\mathbf{s}$ is omitted hereafter for simplicity.

For simplicity, herein we introduce AWGN in the gradient domain, as done in Karaçali and Snyder (2004). With the cluster $\{\mathbf{z}_k\}_{k=0}^{K-1}$, we model the noisy gradients $\{\mathbf{g}_k\}_{k=0}^{K-1}$ at a location $\mathbf{s} \in \Omega$ as

$$\mathbf{g}_k = \mathbf{g} + \mathbf{e}_k, \; 0 \le k \le K-1, \qquad [6.36]$$

where $\mathbf{g}$ is the ground-truth (noiseless) gradient at $\mathbf{s}$, to be recovered. $\{\mathbf{e}_k\}_{k=0}^{K-1}$ are i.i.d. noise terms in the gradient domain, which follow a 2D Gaussian distribution with zero-mean and covariance matrix $\sigma_g^2\mathbf{I}$ ($\mathbf{I}$ is the 2×2 identity matrix). So the probability density function (PDF) of $\mathbf{g}_k$, given $\mathbf{g}$, is

$$Pr(\mathbf{g}_k|\mathbf{g}) = \frac{1}{2\pi\sigma_g^2} \exp\left(-\frac{1}{2\sigma_g^2} \|\mathbf{g} - \mathbf{g}_k\|_2^2\right). \quad [6.37]$$

We assume that σ_g^2 is constant over Ω. Given a patch $\mathbf{z}_0$, one may search for similar patches and empirically estimate the value of σ_g^2 (Pang and Cheung 2017).

Given the noisy gradients $\{\mathbf{g}_k\}_{k=0}^{K-1}$, we can now seek the optimal metric space $\mathbf{G}^\star$ in the MMSE sense. We consider the following minimization problem:

$$\mathbf{G}^\star = \arg\min_{\mathbf{G}} \int_{\mathbb{R}^2} \|\mathbf{G} - \mathbf{G}_I(\mathbf{g})\|_F^2 \cdot Pr\left(\mathbf{g}\middle|\{\mathbf{g}_k\}_{k=0}^{K-1}\right) d\mathbf{g}, \quad [6.38]$$

where the differences between metric spaces are measured by the *Frobenius* norm. By assuming the prior $Pr(\mathbf{g})$ follows a 2D zero-mean Gaussian with constant covariance $\sigma_p^2\mathbf{I}$, the optimal metric space equation [6.38] can be derived in closed form:

$$\mathbf{G}^\star = \widetilde{\mathbf{g}}\widetilde{\mathbf{g}}^\top + \beta_{\mathcal{G}}\mathbf{I}, \quad [6.39]$$

where we denote the constant $\beta_{\mathcal{G}} = \widetilde{\sigma}^2 + \beta$, and

$$\widetilde{\mathbf{g}} = \frac{1}{K + \sigma_g^2/\sigma_p^2} \sum_{k=0}^{K-1} \mathbf{g}_k, \quad \widetilde{\sigma}^2 = \frac{\sigma_g^2}{K + \sigma_g^2/\sigma_p^2}. \quad [6.40]$$

Here, $\widetilde{\mathbf{g}}$ averages the noisy gradients, and it can be viewed as an *estimate* of the ground truth $\mathbf{g}$. $\widetilde{\sigma}^2$ is a constant in domain Ω, and it decreases as the number of observations K increases. From equation [6.39], $\mathbf{G}^\star$ has a major direction aligned with the estimate $\widetilde{\mathbf{g}}$. It has an intuitive interpretation: when the noise variance $\widetilde{\sigma}^2$ is small, the first term dominates and the metric space is *skewed and discriminant*; when $\widetilde{\sigma}^2$ is large, i.e. the estimated gradient $\widetilde{\mathbf{g}}$ is unreliable, the second term dominates and the metric space is not skewed and is close to a *non-discriminant Euclidean space*. Such properties of the optimal metric space $\mathbf{G}^\star$ are consistent with the previous analysis of designing robust metric spaces (Figure 6.10).

Given equation [6.33], which relates exemplar functions to a metric space, there exists a natural assignment of $N = 3$ discrete exemplar functions leading to the optimal metric space equation [6.39]. Let

$$\mathbf{f}_1^\star(i) = \sqrt{\beta_{\mathcal{G}}} \cdot x_i, \quad \mathbf{f}_2^\star(i) = \sqrt{\beta_{\mathcal{G}}} \cdot y_i, \quad \mathbf{f}_3^\star = \frac{1}{K + \sigma_g^2/\sigma_p^2} \sum_{k=0}^{K-1} \mathbf{z}_k. \quad [6.41]$$

Recall that (x_i, y_i) are the coordinates of pixel i. According to equation [6.33], $\mathbf{f}_1^\star(i)$ and $\mathbf{f}_2^\star(i)$ correspond to the term $\beta_\mathcal{G}\mathbf{I}$ in equation [6.39], and $\mathbf{f}_3^\star$ averages the whole cluster $\{\mathbf{z}_k\}_{k=0}^{K-1}$. From the expression of $\widetilde{\mathbf{g}}$ (equation [6.40]), $\mathbf{f}_3^\star$ corresponds to the term $\widetilde{\mathbf{g}}\widetilde{\mathbf{g}}^\top$ in equation [6.39]. With the defined $\mathbf{f}_1^\star$, $\mathbf{f}_2^\star$ and $\mathbf{f}_3^\star$, we can obtain the neighborhood graph $\mathcal{G}^\star$, and hence its graph Laplacian $\mathbf{L}^\star$ and graph Laplacian regularizer $S_{\mathcal{G}^\star}$.

6.3.1.3. *Experimentation*

With the derived optimal graph, an iterative patch-based image denoising algorithm is developed. Given a noisy patch $\mathbf{z}_0$, we first search for its similar patches via block-matching and Euclidean distance as the metric. We then compute the optimal exemplar functions, leading to the optimal graph Laplacian $\mathbf{L}^\star$. The denoised patch can then be obtained by solving equation [6.5]. Similar to Dabov *et al.* (2007), the denoised patches are aggregated to form the updated denoised image at the end of each iteration. We also denoise the given image iteratively to gradually enhance the image quality.

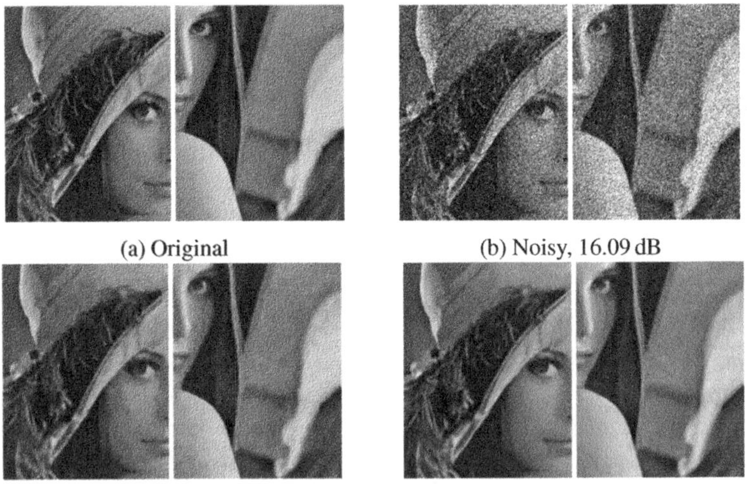

(a) Original (b) Noisy, 16.09 dB

(c) BM3D, 29.86 dB (d) OGLR, 30.04 dB

Figure 6.11. *Denoising of the image Lena, where the original image is corrupted by AWGN with $\sigma_\mathcal{I} = 40$. Two cropped fragments of each image are presented for comparison. © 2017 IEEE. Reprinted, with permission, from Pang and Cheung (2017)*

For illustration, we present a simple denoising experiment. We add i.i.d. AWGN (with standard deviation $\sigma_\mathcal{I} = 40$) to the image *Lena*, and then apply the OGLR algorithm to it. We also employ the BM3D algorithm (Dabov *et al.* 2007) for comparison. Cropped fragments from the *Lena* image are shown in Figure 6.11, along with the corresponding PSNR values. As can be seen, OGLR effectively denoised the image, leading to pleasing visual quality. The OGLR algorithm also effectively performs on piece-wise smooth images (e.g. depth images); interested readers may refer to Pang and Cheung (2017) for more details.

6.3.2. *Low-dimensional manifold model for image restoration*

In a series of works (Shi *et al.* 2016; Osher *et al.* 2017; Shi *et al.* 2018), the authors proposed a novel continuous-domain model called the low-dimensional manifold model (LDMM). The LDMM model is very effective for image restoration and its interesting formulation also has a deep connection to the graph Laplacian regularization. We provide the intuition and the core methodology of LDMM, without diving into its rigorous derivations with differential geometry.

6.3.2.1. *The low-dimensional manifold model*

The LDMM prior assumes all pixel patches extracted from a clean image lie on a smooth low-dimensional manifold. This property is studied in Peyré (2008) and Peyré (2009), while several other works (Shi *et al.* 2016; Osher *et al.* 2017; Shi *et al.* 2018) explicitly exploit this property to perform image restoration. In the following, we first formulate a general image restoration problem with LDMM as the signal prior, then take a graph-based treatment to understand the problem. We assume the same corruption model as illustrated in equation [6.1], but adopt a similar set of notations as in section 6.3.1.2. Particularly, the corruption model is given by $\mathbf{z} = \mathbf{Hu}+\mathbf{e}$, where $\mathbf{u}$ is the original image, $\mathbf{z}$ is the corrupted image, $\mathbf{e}$ is the additive noise and $\mathbf{H}$ is the corruption matrix. We assume $\mathbf{u}, \mathbf{z}, \mathbf{e} \in \mathbb{R}^{HW}$, they are vectorized from images of size H-by-W, while $\mathbf{H} \in \mathbb{R}^{HW \times HW}$.

To begin with, we define a set of coordinates regularly located on the image $\mathbf{u}$,

$$\Gamma = \Big\{\mathbf{s}_i \in \mathbb{Z}_+^2 \Big| \mathbf{s}_i \in \{1+t, 1+2t, \cdots, H-\sqrt{N}+1\} \times \\ \{1+t, 1+2t, \cdots, W-\sqrt{N}+1\}\Big\}, \tag{6.42}$$

where $\sqrt{N}$ is an integer, and $t \in \mathbb{Z}_+$ is the step size. We also denote the cardinality of Γ, i.e. $|\Gamma|$, as M. An image patch $\mathbf{p}(\mathbf{s}_i) \in \mathbb{R}^N$ is the (vectorized) $\sqrt{N}$-by-$\sqrt{N}$ pixel patch extracted from image $\mathbf{u}$, such that the coordinate of the top-left corner of $\mathbf{p}(\mathbf{s}_i)$ is $\mathbf{s}_i \in \Gamma$. Let us define the set of pixel patches as

$$\mathcal{P} = \{\mathbf{p}(\mathbf{s}_i) | \mathbf{s}_i \in \Gamma, 1 \leq i \leq M\}, \tag{6.43}$$

which contains M pixel patches extracted from image $\mathbf{u}$. For simplicity, we denote $\mathbf{p}(\mathbf{s}_i)$ by $\mathbf{p}_i$ in the sequel. The LDMM assumes that the patches in $\mathcal{P}$ are point samples from a smooth manifold $\mathcal{M}(\mathbf{u})$, where $\mathcal{M}(\mathbf{u})$ is a *low-dimensional manifold* embedded in high-dimensional space $\mathbb{R}^N$, i.e. $\dim(\mathcal{M}(\mathbf{u}))$ is small. With a certain way to sample the pixel patches, e.g. one patch is sampled every t pixels horizontally and vertically (equation [6.42]), the patches in $\mathcal{P}$ are determined by image $\mathbf{u}$, while the patches further define the underlying manifold $\mathcal{M}$. Therefore, we view the manifold $\mathcal{M}$ as a function of the image $\mathbf{u}$. Hence, $\mathcal{M}(\mathbf{u})$ is also termed as a *patch manifold*. By assuming $\mathcal{M}$ is low dimensional, ideally, we aim at minimizing the following optimization problem:

$$\mathbf{u}^\star = \arg\min_{\mathbf{u}} \ \dim(\mathcal{M}(\mathbf{u})) + \lambda \left\|\mathbf{z} - \mathbf{H}\mathbf{u}\right\|_2^2, \qquad [6.44]$$

where the dimension $\dim(\mathcal{M}(\mathbf{u}))$ directly serves as the regularization term, while $\|\mathbf{z} - \mathbf{H}\mathbf{u}\|_2^2$ is the fidelity term. Despite the attractive concept of minimizing the dimensionality, the problem equation [6.44] is highly nonlinear and non-convex, which is difficult to solve.

To make the problem of equation [6.44] tractable, the work of Osher *et al.* (2017) introduces a set of N continuous functions, which are called *coordinate functions*, $f_i(\mathbf{x}) : \mathcal{M} \subset \mathbb{R}^N \mapsto \mathbb{R}$, $1 \leq i \leq N$, which map a point $\mathbf{x} = [x_1 \ x_2 \cdots x_N]^\top \in \mathcal{M}$ on the manifold $\mathcal{M}$ to a real number:

$$f_n(\mathbf{x}) = x_n. \qquad [6.45]$$

In other words, f_i simply maps a sample on $\mathcal{M}$ to its ith coordinate. Hence,

$$f_n(\mathbf{p}_i) = \mathbf{p}_i(n), \qquad [6.46]$$

where $\mathbf{p}_i(n)$ is the nth pixel of patch $\mathbf{p}_i$. With tools in differential geometry, Osher *et al.* (2017) prove that, $\forall \mathbf{x} \in \mathcal{M} \subset \mathbb{R}^N$,

$$\dim(\mathcal{M}) = \sum_{i=1}^{N} \|\nabla_{\mathcal{M}} f_i(\mathbf{x})\|_2^2, \qquad [6.47]$$

where $\nabla_{\mathcal{M}}$ denotes the gradient operator on manifold $\mathcal{M}$. With this property, equation [6.44] can be reformulated as

$$\mathbf{u}^\star = \arg\min_{\mathbf{u},\mathcal{M}} \sum_{i=1} \int_{\mathcal{M}} \|\nabla_{\mathcal{M}} f_i(\mathbf{x})\|_2^2 \, d\mathbf{x} + \lambda \left\|\mathbf{z} - \mathbf{H}\mathbf{u}\right\|_2^2. \qquad [6.48]$$

Osher *et al.* (2017) introduce an alternating direction iteration to solve equation [6.48]. First, given the manifold $\mathcal{M}$ (represented by the patch samples $\mathcal{P}$), we update the image $\mathbf{u}$. Then the image $\mathbf{u}$ is fixed and we update the manifold $\mathcal{M}$ by sampling pixel patches $\mathcal{P}$ from $\mathbf{u}$. While the second step is straightforward, the first one is non-trivial. As pointed out in Osher *et al.* (2017), it is common to solve this type of problem by linking it with graph Laplacian. Osher *et al.* also employ a solution based on graph Laplacian, which is called the *point integral method* (PIM) (Li *et al.* 2017). Compared to directly approximating the Laplace–Beltrami operator with graph Laplacian, PIM essentially incorporates another boundary correction term for more accurate numerical results. Instead of presenting the PIM in Osher *et al.* (2017), we analyze the problem of equation [6.48] in a way similar to section 6.3.1.1, to better understand the link between LDMM and graph-based methods.

6.3.3. *LDMM as graph Laplacian regularization*

We begin by building a graph $\mathcal{G} = (V, E)$, where V and E denote the set of vertices and edges on $\mathcal{G}$, respectively. On the graph $\mathcal{G}$, each patch in $\mathcal{P}$ (or each coordinate in Γ) will be represented by a vertex, so $|V| = M$. The weight w_{ij} connecting vertices (or patches) V_i and V_j is defined in the same way as presented in equation [6.31], where the squared distance is now computed by the patch distance $\|\mathbf{p}_i - \mathbf{p}_j\|_2^2$. We denote the graph Laplacian matrix of the constructed graph $\mathcal{G}$ by $\mathbf{L} \in \mathbb{R}^{M\times M}$. Define $h(\cdot)$ as some continuous function on the manifold $\mathcal{M}$, i.e. $h(\cdot) : \mathcal{M} \subset \mathbb{R}^N \mapsto \mathbb{R}$; and its discretized version $\mathbf{h}$ sampled at locations of Γ, i.e. $\mathbf{h} = [h(\mathbf{p}_1)\ h(\mathbf{p}_2)\ \cdots\ h(\mathbf{p}_M)]^\top$. From the convergence results in Hein (2006), we have

$$\lim_{\substack{M\to\infty \\ r\to 0}} \mathbf{h}^\top \mathbf{L}\mathbf{h} \sim \int_{\mathcal{M}} \|\nabla_{\mathcal{M}} h\|_2^2\, q\, dV, \tag{6.49}$$

where dV is the natural volume element of manifold M. q is the underlying PDF on $\mathcal{M}$ describing the distribution of the patches of $\mathcal{P}$ on $\mathcal{M}$, $\int_{\mathcal{M}} q\, dV = 1$. We now apply equation [6.49] to the coordinate functions $\{f_n\}_{=1}^N$:

$$\begin{aligned}\lim_{\substack{M\to\infty \\ r\to 0}} \sum\nolimits_{i=1}^{N} \mathbf{f}_i^\top \mathbf{L}\mathbf{f}_i \sim \sum\nolimits_{i=1}^{N} \int_{\mathcal{M}} \|\nabla_{\mathcal{M}} f_i\|_2^2\, q\, dV = \\ \int_{\mathcal{M}} \dim(\mathcal{M}) q\, dV = \dim(\mathcal{M}),\end{aligned} \tag{6.50}$$

where the relationship of equation [6.47] is applied, and $\mathbf{f}_i$ is the discretization of f_i, given by

$$\mathbf{f}_i = [f_i(\mathbf{p}_1)\ f_i(\mathbf{p}_2) \cdots f_i(\mathbf{p}_N)]^\top = [\mathbf{p}_1(i)\ \mathbf{p}_2(i) \cdots \mathbf{p}_N(i)]^\top. \tag{6.51}$$

Consequently, one can approximate the dimension of $\mathcal{M}(\mathbf{u})$ with the $\mathbf{f}_i$'s and the constructed graph Laplacian $\mathbf{L}$. Moreover, from equation [6.51], $\mathbf{f}_i$ is essentially the *subsampled* version of image $\mathbf{u}$, which is formed by the ith pixel of all patches in $\mathcal{P}$.

Let us construct the matrices $\mathbf{S}_i \in \mathbb{R}^{M \times HW}$, $1 \leq i \leq N$, such that $\mathbf{f}_i = \mathbf{S}_i\mathbf{u}$. In other words, $\mathbf{S}_i$ is the matrix extracting $\mathbf{f}_i$ from image $\mathbf{u}$. Then

$$\sum_{i=1}^{N} \mathbf{f}_i^\top \mathbf{L}\mathbf{f}_i = \sum_{i=1}^{N} (\mathbf{S}_i\mathbf{u})^\top \mathbf{L}\,(\mathbf{S}_i\mathbf{u}) = \mathbf{u}^\top \left(\sum_{i=1}^{N} \mathbf{S}_i^\top \mathbf{L}\mathbf{S}_i\right) \mathbf{u}. \qquad [6.52]$$

By denoting $\mathbf{L}' = \sum_{i=1}^{N} \mathbf{S}_i^\top \mathbf{L}\mathbf{S}_i$ and applying equation [6.50], we finally have

$$\lim_{\substack{M\to\infty \\ r\to 0}} \mathbf{u}^\top \mathbf{L}'(\mathbf{u})\mathbf{u} \sim \dim(\mathcal{M}). \qquad [6.53]$$

Note that $\mathbf{L}' \in \mathbb{R}^{HW \times HW}$ is essentially a matrix-valued function of image $\mathbf{u}$, i.e. $\mathbf{L}' : \mathbb{R}^{HW} \mapsto \mathbb{R}^{HW \times HW}$. It can be viewed as an overall graph Laplacian matrix about the whole image $\mathbf{u}$. Using equation [6.53], the optimization problem of equation [6.44] can be rewritten as

$$\mathbf{u}^\star = \arg\min_{\mathbf{u}} \mathbf{u}^\top \mathbf{L}'(\mathbf{u})\mathbf{u} + \lambda \left\|\mathbf{z} - \mathbf{H}\mathbf{u}\right\|_2^2. \qquad [6.54]$$

Though equation [6.54] is still non-convex, its solution can be approximated iteratively, similar to the aforementioned alternating direction iteration to solve equation [6.48]. First, given an initial $\mathbf{u}$, we can compute the corresponding matrix $\mathbf{L}'$. Then equipped with $\mathbf{L}'$, we solve the QP problem

$$\mathbf{u}^\star = \arg\min_{\mathbf{u}} \mathbf{u}^\top \mathbf{L}'\mathbf{u} + \lambda \left\|\mathbf{z} - \mathbf{H}\mathbf{u}\right\|_2^2, \qquad [6.55]$$

to update $\mathbf{u}$. We iteratively construct the matrix $\mathbf{L}'$ and update $\mathbf{u}$ using equation [6.55] until $\mathbf{u}$ converges. Consequently, by interpreting graphs as discrete counterparts of Riemannian manifolds, we resolve the difficult problem of equation [6.44].

We now look at a simple experiment of applying LDMM for image inpainting. In this experiment, 90% of pixels are randomly removed from an original image, leading to its corrupted version. We aim to recover the original image from the corrupted observation. For the LDMM model, we let the patch size be 10×10, i.e. $N = 100$, and $\mathbf{u}$ is initialized by random numbers. We showcase the restoration results for the image *Peppers* in Figure 6.12, where LDMM gradually improves the results, leading to effective restoration. The methodology of LDMM is also extended for point cloud denoising in Zeng *et al.* (2020) (Chapter 7).

Compared to the discrete-domain approaches introduced in section 6.2, the continuous-domain methods, OGLR (section 6.3.1) and LDMM (section 6.3.2), view discrete graphs as proxies of Riemannian manifolds. Therefore mathematical tools from functional analysis (Kreyszig 1978) and differential geometry (Helgason 1979) can be applied, leading to novel insights and promising restoration results.

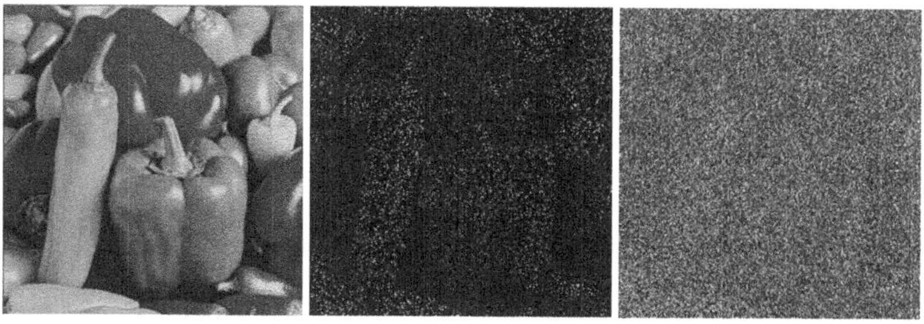

(a) Original (b) Corrupted (c) Initialization

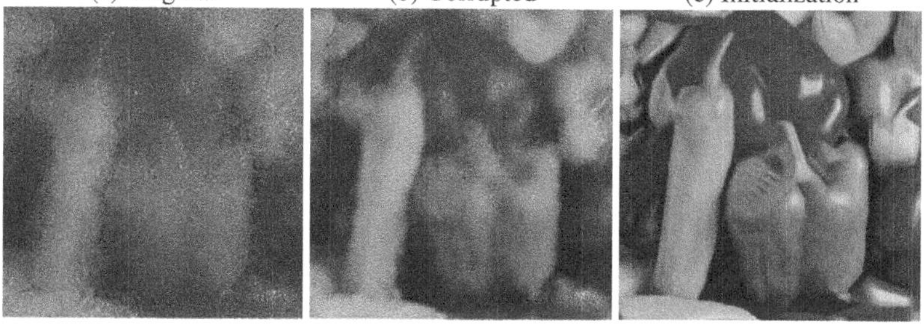

(d) LDMM, iteration 1, 16.51 dB (e) LDMM, iteration 2, 19.70 dB (f) LDMM, iteration 15, 25.10 dB

Figure 6.12. *Inpainting of the image* Peppers *with the low-dimensional manifold model (LDMM), where the corrupted image only keeps 10% of the pixels from the original image*

6.4. Learning-based methods

Recent developments in deep learning have revolutionized the paradigm of image restoration, leading to state-of-the-art performance in various tasks, including image denoising (Zhang *et al.* 2017), image deconvolution (Xu *et al.* 2014), image super-resolution (Dong *et al.* 2015), image inpainting (Iizuka *et al.* 2017), and so on. Endowed with deep learning, a camera is able to provide pleasing images under challenging capture conditions, e.g. extreme low light (Liba *et al.* 2019) and large zoom-in factor (Wronski *et al.* 2019). Recent trials even propose to replace the whole image signal processing (ISP) pipeline with an end-to-end deep neural network, encompassing a sequence of low-level vision tasks including demosaicing, denoising, color correction and so on (Schwartz *et al.* 2018; Chen *et al.* 2018).

Deep learning differs from the model-based techniques discussed above, in that they are typically data driven and automatically learn representations from images without involving hand-coded priors. However, when testing images exhibit a statistical mismatch from the training images, data-driven approaches may fail to generalize to testing images (McCann *et al.* 2017). For example, in real image denoising, it is non-trivial to obtain ground-truth images, leading to a limited amount of training data (Chen *et al.* 2018). In this case, a pure data-driven approach is prone to overfit to the particular characteristics of the training data and fail on test images, e.g. Figure 6.13(b) showcases the result of a pure data-driven approach trained for a different domain.

Model-based approaches, as illustrated above, rely on the priors about the original images. Without the notion of training, the performance of model-based approaches is generally more robust than data-driven approaches when facing the heterogeneity of natural images, e.g. the result in Figure 6.13(a) provides a relatively smoother result than Figure 6.13(b).

In this section, we focus on the recent trials of incorporating graph signal processing tools into convolutional neural networks (CNN). These methods combine the robustness merit of model-based approaches and the learning power of data-driven approaches (Zeng *et al.* 2019; Su *et al.* 2020), and are applied in image denoising to achieve higher robustness when training and testing data exhibit statistical mismatch; e.g. Figure 6.13(c) illustrates the effectiveness of incorporating GSP into CNN.

Apart from the drawback of generalization ability, deep learning is also known to be difficult to apply to data with irregular structure and utilize non-local correlations among pixels. The two drawbacks are overcome by the GCNN approaches. We will not discuss the GCNN works in this chapter; for details please refer to Chapter 10.

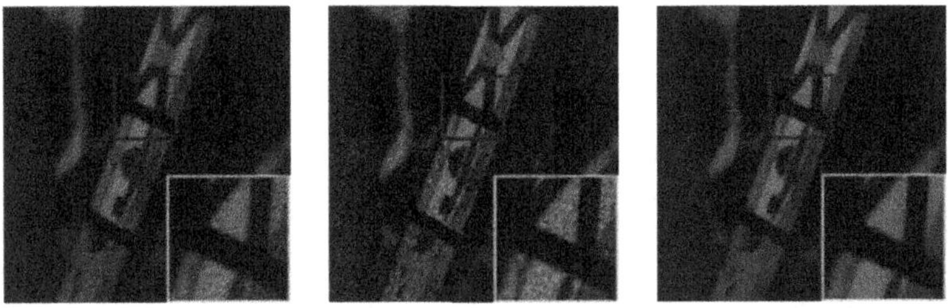

(a) Noise clinic (b) CDnCNN (c) DeepGLR

Figure 6.13. *Results of real image denoising. (a) Noise clinic (model-based) (Lebrun* et al. *2015); (b) CDnCNN (data-driven) (Zhang* et al. *2017); (c) DeepGLR. CDnCNN and DeepGLR are trained for Gaussian denoising. ©2019 IEEE. Reprinted, with permission, from Zeng* et al. *(2019). For a color version of this figure, see www.iste.co.uk/cheung/graph.zip*

6.4.1. *CNN with GLR*

In Zeng *et al.* (2019), a recent work called *DeepGLR* is the first in literature to incorporate the GLR into deep neural networks as a fully differentiable layer, extracting underlying features of the input noisy images and boosting the performance of the subsequent restoration.

6.4.1.1. *Combining advantages of GLR and CNN*

Section 6.3.1 discussed the theoretical reasoning for GLR and the approach to compute edge weights that leads to optimal GLR for denoising. Nevertheless, the analysis is based on the assumption that the graph is a discrete approximation of the Riemannian manifold, which may not hold for real images. Therefore, DeepGLR utilizes the powerful representation ability of CNN to uncover the correlation between pixels, for a more sophisticated graph construction that better fits the underlying structure of the input image.

On the other hand, the GLR is integrated as a fully differentiable layer into CNN to restrict its search space, achieving higher robustness to overfitting with strong cross-domain generalization ability. This is achieved by coupling the graph Laplacian regularization layer – an adaptive low-pass linear filter regardless of the training data – with a light-weight CNN for pre-filtering. Moreover, by constraining the regularization weight to prevent a steep local minimum, the pipeline is provably numerical stable.

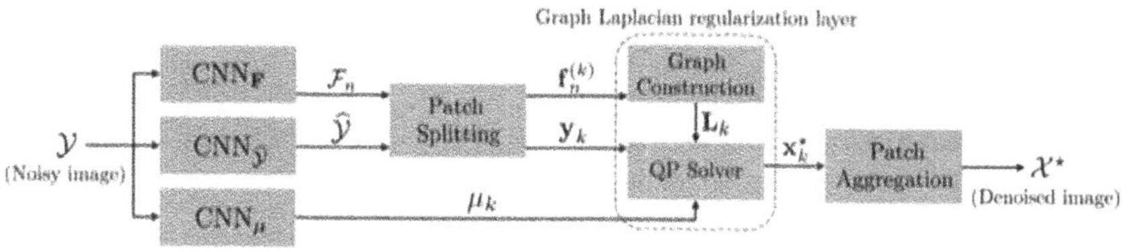

Figure 6.14. *Block diagram of the proposed GLRNet that employs a graph Laplacian regularization layer for image denoising. © 2019 IEEE. Reprinted, with permission, from Zeng* et al. *(2019)*

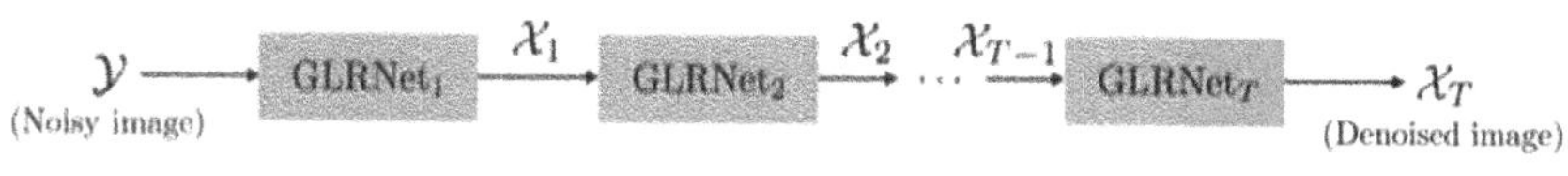

Figure 6.15. *Block diagram of the overall DeepGLR framework. © 2019 IEEE. Reprinted, with permission, from Zeng* et al. *(2019)*

6.4.1.2. *DeepGLR Framework*

Image denoising is achieved through GLRNet, which is the key component of the DeepGLR framework, with its structure shown in Figure 6.14. GLRNet takes a real noisy image $\mathcal{Y}$ as input and feeds into $\text{CNN}_{\mathbf{F}}$, and outputs N feature maps $\{\mathbf{f}_n\}_{n=1}^{N}$, which function as the exemplars, in section 6.3.1, for edge weight computation. Meanwhile, the noisy image goes through a lightweight CNN ($\text{CNN}_{\widehat{\mathcal{Y}}}$) to give the pre-filtered $\widehat{\mathcal{Y}}$.

$\{\mathbf{f}_n\}_{n=1}^{N}$ and $\widehat{\mathcal{Y}}$ are then divided into K overlapping patches, giving $\mathbf{f}_n^{(k)}$ and $\{\mathbf{y}_k\}_{k=1}^{K}$, $1 \leq k \leq K$ so that the denoising is performed patch by patch, in order to reduce computational complexity. That is, an eight-connected graph with Laplacian matrix $\mathbf{L}_k$ is built using the features for each patch, then the corresponding patch is denoised by solving an unconstrained QP problem, following the same optimization formulation in equation [6.5]. The denoised patches $\mathbf{x}_k^{\star}$ are aggregated to output the final denoising result $\mathcal{X}^{\star}$.

Besides the denoising modules, a light-weight CNN (CNN_{μ}) takes the noisy image as the input and estimates the weighting parameters $\{\mu_k\}_{k=1}^{K}$ to balance the fidelity term and the GLR prior. Similar to previous approaches, DeepGLR iteratively performs denoising on the input to gradually enhance the output image quality, and the overall framework is composed of T blocks of GLRNet, as shown in Figure 6.15. Each GLRNet is supposed to remove noise components following a Gaussian distribution so that the whole framework removes a mixture of Gaussian that can approximate any distribution arbitrarily well (Reynolds 2009). The network parameters are shared among all the GLRNets.

During training, a loss penalizing differences between the recovered image $\mathcal{X}_T$ and the ground-truth $\mathcal{X}^{(\text{gt})}$ is defined as the mean-square-error (MSE) between $\mathcal{X}^{(\text{gt})}$ and $\mathcal{X}_T$, i.e.

$$L_{\text{res}}\left(\mathcal{X}^{(\text{gt})}, \mathcal{X}_T\right) = \frac{1}{HW}\sum_{i=1}^{H}\sum_{j=1}^{W}\left(\mathcal{X}^{(\text{gt})}(i,j) - \mathcal{X}_T(i,j)\right)^2, \tag{6.56}$$

where H and W are the height and width of the images, respectively. $\mathcal{X}^{(\text{gt})}(i,j)$ is the (i,j)th pixel of $\mathcal{X}^{(\text{gt})}$, the same for $\mathcal{X}_T(i,j)$. Details of the network structures can be found in Zeng *et al.* (2019) and are not discussed here.

6.4.1.3. *Result demonstration*

By using CNN for graph learning, DeepGLR can extract features that better fit the input noisy image than model-based OGLR. For example, when tested for Gaussian noise removal on the BSDS500 dataset, DeepGLR outperforms OGLR by 0.13 dB at noise level $\sigma = 15$, 0.24 dB at noise level $\sigma = 25$ and 0.02 dB at noise level $\sigma = 50$. Moreover, when trained on a small dataset, where five images from RENOIR (Anaya and Barbu 2018) are adopted for training, DeepGLR provides a more stable performance than pure data-driven CDnCNN (Zhang *et al.* 2017). For example, even if CDnCNN outperforms DeepGLR by 0.54 dB during training, DeepGLR exceeds CDnCNN by 0.17 dB during testing, which indicates CDnCNN overfits to the training images, while DeepGLR is more robust to a small training set because of the restrictions from GLR. More importantly, to investigate the cross-domain generalization ability, DeepGLR and CDnCNN are trained on BSDS500 for blind Gaussian noise removal, with a noise level range $\sigma \in [0, 55]$, and are then tested on the RENOIR dataset for real image removal. By comparing the results in Figures 6.13(b) and (c), we can see DeepGLR successively generalizes to the real noise removal while CDnCNN fails to do so.

In summary, by incorporating GLR into a deep learning framework, DeepGLR not only provides a sophisticated graph construction, but also demonstrates high immunity to overfitting when trained with a small dataset and strong cross-domain generalization ability.

6.4.2. *CNN with graph wavelet filter*

In DeepGLR, the denoising is performed via the QP solver optimizing for GLR. This requires solving a system of linear equations, which is computational complex, even if a fast method like CG (Shewchuk *et al.* 1994) is used. Therefore, Su *et al.* propose a Deep Analytical Graph Filter (DeepAGF) (Su *et al.* 2020) that implements denoising through analytical graph wavelet filters, which is fast in execution.

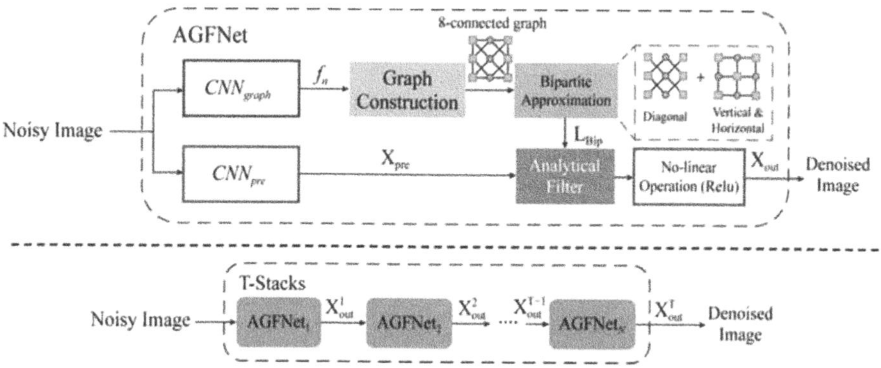

Figure 6.16. *DeepAGF framework. Top: Block diagram of the AGFNet, using analytical graph filter for image denoising. Bottom: Block diagram of the N stacks DeepAGF framework. © 2020 IEEE. Reprinted, with permission, from Su* et al. *(2020). For a color version of this figure, see www.iste.co.uk/cheung/graph.zip*

6.4.2.1. *DeepAGF framework*

DeepAGF is inspired by Hadji and Wildes (2017), who used *fixed* Gaussian filters combined with point-wise nonlinearity and pooling operators for each convolutional layer, so that the filter in the network is "explainable" and nonetheless achieves state-of-the-art performance in image texture recognition.

In DeepAGF, the bi-orthogonal graph wavelet called GraphBio (Narang and Ortega 2013) (discussed in Chapter 1) is adopted as the analytical graph filter to build the graph convolutional layer. GraphBio designs low-pass/high-pass filters to decompose the input graph signal into low-/high-frequency components and is implemented efficiently. The low-pass filter is used in the graph convolutional layer to remove noise so that the network is explainable and of a lower complexity than DeepGLR. Meanwhile, similar to DeepGLR, the graph is learned through CNN so that the underlying graph is optimized in a data-driven manner. In DeepAGF, image denoising is implemented by AGFNet, which is the building block shown in Figure 6.16. The difference from GLRNet is that, instead of using a QP solver, AGFNet employs GraphBio as the analytical filter, followed by a nonlinear operation for denoising. Additionally, to apply GraphBio on the eight-connected graph, the graph needs to be decomposed into bipartite subgraphs. This is implemented by separating the edges into two bipartite graphs: one with vertical and horizontal edges, and the other with diagonal edges.

For effective denoising, the input image is also iteratively denoised to gradually enhance the output image quality. Therefore, the overall framework of DeepAGF is composed of T blocks of AGFNet, as depicted in Figure 6.16, where the blocks share the same parameters. MSE Loss is applied at the final output for training.

6.4.2.2. *Result demonstration*

DeepAGF provides similar performance as DeepGLR, with strong robustness to training/testing data mismatch. For example, in Figure 6.17, the pure data-driven DnCNN (Zhang *et al.* 2017) and DeepAGF are both trained on noise level $\sigma = 50$ with the dataset (Zhang *et al.* 2017), but are tested on $\sigma = 70$. For the Starfish image, DeepAGF outperforms DnCNN by 1.48 dB in PSNR and better preserves the edges in the areas highlighted by red boxes, demonstrating that DeepAGF is more robust to statistical mismatch.

6.5. Concluding remarks

Image restoration aims at recovering a clean image from the corrupted one, which is critical for both aesthetic purposes and algorithmic processing tasks. Although images are essentially entities defined on regular grids, by modeling them as irregular graph signals on the underlying graphs, it is possible to reveal the intrinsic correlation within images, therefore achieving effective reconstructions (section 6.1).

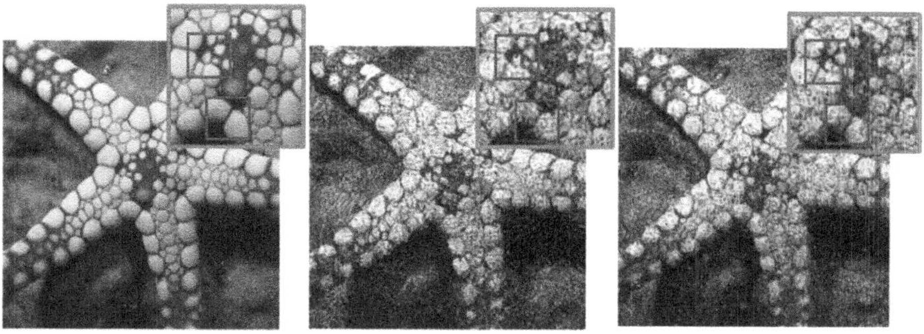

(a) Original (b) PSNR = 18.24 dB (c) PSNR = 19.72 dB

Figure 6.17. *Denoising result comparison on the Starfish image with input noisy level $\sigma = 70$ (a) Original, (b) DnCNN (Zhang* et al. *2017) and (c) DeepAGF. DnCNN and DeepAGF are trained with noise level $\sigma = 50$. © 2020 IEEE. Reprinted, with permission, from Su* et al. *(2020). For a color version of this figure, see www.iste.co.uk/cheung/graph.zip*

To model images as graph signals, one may view the graph signal itself as inherently discrete, which typically regards a graph as a statistical object describing the correlations among image pixels. This thread of works usually adopt tools from graph spectral theory (Chung and Graham 1997), Markov random field (Li 2009), etc. and are also related to probabilistic graphical models (Koller and Friedman 2009) (section 6.2). Differently, a graph can also be viewed as a proxy of an underlying Riemannian manifold. In this way, we establish its connection to differential geometry, and therefore, a different realm of tools can be applied for analysis (Hein *et al.* 2007) (section 6.3). The aforementioned continuous-domain and discrete-domain methods are both model-based, where prior knowledge about the ground-truth image is employed for formulation. The recent advance of data-driven approaches, especially those based on deep learning (LeCun *et al.* 2015), has further pushed the state of the art and has achieved synergies with model-based methods (section 6.4).

6.6. References

Aharon, M., Elad, M., Bruckstein, A. (2006). K-SVD: An algorithm for designing overcomplete dictionaries for sparse representation. *IEEE Transactions on Signal Processing*, 54(11), 4311–4322.

Alliez, P., Cohen-Steiner, D., Tong, Y., Desbrun, M. (2007). Voronoi-based variational reconstruction of unoriented point sets. *Symposium on Geometry Processing*, 7, 39–48.

Anaya, J. and Barbu, A. (2018). RENOIR – A dataset for real low-light noise image reduction. *Journal of Visual Communication and Image Representation*, 51, 144–154.

Aubert, G. and Vese, L. (1997). A variational method in image recovery. *SIAM Journal on Numerical Analysis*, 34(5), 1948–1979.

Bai, Y., Cheung, G., Liu, X., Gao, W. (2019). Graph-based blind image deblurring from a single photograph. *IEEE Transactions on Image Processing*, 28(3), 1404–1418.

Bertalmio, M., Sapiro, G., Caselles, V., Ballester, C. (2000). Image inpainting. *ACM Annual Conference on Computer Graphics and Interactive Techniques*, 417–424.

Biemond, J., Lagendijk, R.L., Mersereau, R.M. (1990). Iterative methods for image deblurring. *Proceedings of the IEEE*, 78(5), 856–883.

Bishop, C.M. (2006). *Pattern Recognition and Machine Learning*, Springer-Verlag, New York.

Buades, A., Coll, B., Morel, J.-M. (2005a). A non-local algorithm for image denoising. *IEEE Conference on Computer Vision and Pattern Recognition*, 2, 60–65.

Buades, A., Coll, B., Morel, J.-M. (2005b). A review of image denoising algorithms, with a new one. *Multiscale Modeling and Simulation*, 4(2), 490–530.

Cai, J.-F., Ji, H., Liu, C., Shen, Z. (2009). Blind motion deblurring using multiple images. *Journal of Computational Physics*, 228(14), 5057–5071.

Chen, C., Chen, Q., Xu, J., Koltun, V. (2018). Learning to see in the dark. *IEEE Conference on Computer Vision and Pattern Recognition*, 3291–3300.

Cheung, G., Magli, E., Tanaka, Y., Ng, M.K. (2018). Graph spectral image processing. *Proceedings of the IEEE*, 106(5), 907–930.

Chung, F.R. and Graham, F.C. (1997). Spectral graph theory. *American Mathematical Soc.*, 92.

Dabov, K., Foi, A., Katkovnik, V., Egiazarian, K. (2007). Image denoising by sparse 3-D transform-domain collaborative filtering. *IEEE Transactions on Image Processing*, 16(8), 2080–2095.

Dong, W., Zhang, L., Shi, G., Li, X. (2012). Nonlocally centralized sparse representation for image restoration. *IEEE Transactions on Image Processing*, 22(4), 1620–1630.

Dong, W., Shi, G., Li, X., Ma, Y., Huang, F. (2014). Compressive sensing via nonlocal low-rank regularization. *IEEE Transactions on Image Processing*, 23(8), 3618–3632.

Dong, C., Loy, C.C., He, K., Tang, X. (2015). Image super-resolution using deep convolutional networks. *IEEE Transactions on Pattern Analysis and Machine Intelligence*, 38(2), 295–307.

Donoho, D.L. and Johnstone, J.M. (1994). Ideal spatial adaptation by wavelet shrinkage. *Biometrika*, 81(3), 425–455.

Elad, M. and Aharon, M. (2006). Image denoising via learned dictionaries and sparse representation. *IEEE Conference on Computer Vision and Pattern Recognition*, 1, 895–900.

Elmoataz, A., Lezoray, O., Bougleux, S., (2008). Nonlocal discrete regularization on weighted graphs: A framework for image and manifold processing. *IEEE Transactions on Image Processing*, 17(7), 1047–1060.

Egilmez, H.E., Chao, Y.H., Ortega, A., Lee, B., Yea, S. (2016). GBST: Separable transforms based on line graphs for predictive video coding. *IEEE International Conference on Image Processing (ICIP)*, 2375–2379.

Fattal, R. (2008). Single image dehazing. *ACM Transactions on Graphics*, 27(3), 72.

Fergus, R., Singh, B., Hertzmann, A., Roweis, S.T., Freeman, W.T. (2006). Removing camera shake from a single photograph. *ACM SIGGRAPH*, 787–794.

Flucher, M. (1999). Compactness criteria. In *Variational Problems With Concentration*, Flucher, M. Springer, Birkhauser, Basel .

Freeman, W.T., Jones, T.R., Pasztor, E.C. (2002). Example-based super-resolution. *IEEE Computer Graphics and Applications*, 22(2), 56–65.

Gadde, A., Narang, S.K., Ortega, A. (2013). Bilateral filter: Graph spectral interpretation and extensions. *IEEE International Conference on Image Processing*, 1222–1226.

Gavili, A. and Zhang, X.-P. (2017). On the shift operator, graph frequency, and optimal filtering in graph signal processing. *IEEE Transactions on Signal Processing*, 65(23), 6303–6318.

Gonzales, R. and Woods, R. (2018). *Digital Image Processing*. Pearson, New York.

Glasner, D., Bagon, S., Irani, M. (2009). Super-resolution from a single image. *IEEE International Conference on Computer Vision*, 349–356.

Gunturk, B.K. and Li, X. (2012). *Image Restoration: Fundamentals and Advances*, CRC Press, Boca Raton, Florida.

Hadji, I. and Wildes, R.P. (2017). A spatiotemporal oriented energy network for dynamic texture recognition. *IEEE International Conference on Computer Vision*, 3066–3074.

He, K., Sun, J., Tang, X. (2013). Guided image filtering. *IEEE Transactions on Pattern Analysis and Machine Intelligence*, 35(6), 1397–1409.

Hein, M. (2006). Uniform convergence of adaptive graph-based regularization. In *International Conference on Computational Learning Theory*, Lugosi, G. and Simon, H.U. (eds). Springer-Verlag, Berlin, Heidelberg.

Hein, M., Audibert, J.-Y., Luxburg, U.V. (2007). Graph Laplacians and their convergence on random neighborhood graphs. *Journal of Machine Learning Research*, 8(June), 1325–1368.

Helgason, S. (1979). *Differential Geometry, Lie Groups, and Symmetric Spaces*. Academic Press, Massachusetts.

Hu, W., Li, X., Cheung, G., Au, O. (2013). Depth map denoising using graph-based transform and group sparsity. *IEEE International Workshop on Multimedia Signal Processing*, 001–006.

Hu, W., Cheung, G., Li, X., Au, O.C. (2014). Graph-based joint denoising and super-resolution of generalized piecewise smooth images. *IEEE International Conference on Image Processing (ICIP)*, 2056–2060.

Iizuka, S., Simo-Serra, E., Ishikawa, H. (2017). Globally and locally consistent image completion. *ACM Transactions on Graphics*, 36(4), 107.

Jia, J. (2007). Single image motion deblurring using transparency. *IEEE Conference on Computer Vision and Pattern Recognition*, 1–8.

Karaçali, B. and Snyder, W. (2004). Noise reduction in surface reconstruction from a given gradient field. *International Journal of Computer Vision*, 60(1), 25–44.

Katsaggelos, A.K. (2012). *Digital Image Restoration*. Springer-Verlag, Berlin, Heidelberg.

Kheradmand, A. and Milanfar, P. (2014). A general framework for regularized, similarity-based image restoration. *IEEE Transactions on Image Processing*, 23(12), 5136–5151.

Kipf, T.N. and Welling, M. (2017). Semi-supervised classification with graph convolutional networks. In *International Conference on Learning Representations*. arXiv:1609.02907.

Knutsson, H., Westin, C.-F., Andersson, M. (2011). Representing local structure using tensors II. In *Scandinavian Conference on Image Analysis*, Heyden, A. and Kahl, F. (eds). Springer, Berlin, Heidelberg.

Koller, D. and Friedman, N. (2009). *Probabilistic Graphical Models: Principles and Techniques*. MIT Press, Massachusetts.

Kopf, J., Cohen, M.F., Lischinski, D., Uyttendaele, M. (2007). Joint bilateral upsampling. *ACM Transactions on Graphics*, 26(3), 96–es.

Kreyszig, E. (1978). *Introductory Functional Analysis with Applications*. Wiley, New York.

Kwon, Y., Kim, K.I., Tompkin, J., Kim, J.H., Theobalt, C. (2015). Efficient learning of image super-resolution and compression artifact removal with semi-local Gaussian processes. *IEEE Transactions on Pattern Analysis and Machine Intelligence*, 37(9), 1792–1805.

Lai, W.-S., Ding, J.-J., Lin, Y.-Y., Chuang, Y.-Y. (2015). Blur kernel estimation using normalized color-line prior. *IEEE Conference on Computer Vision and Pattern Recognition*, 64–72.

Lam, E.Y. and Goodman, J.W. (2000). A mathematical analysis of the DCT coefficient distributions for images. *IEEE Transactions on Image Processing*, 9(10), 1661–1666.

Lebrun, M., Colom, M., Morel, J.-M. (2015). The noise clinic: A blind image denoising algorithm. *Image Processing On Line*, 5, 1–54.

LeCun, Y., Bengio, Y., Hinton, G. (2015). Deep learning. *Nature*, 521(7553), 436.

Levin, A., Lischinski, D., Weiss, Y. (2004). Colorization using optimization. *ACM Transactions on Graphics*, 23(3), 689–694.

Levin, A., Lischinski, D., Weiss, Y. (2007). A closed-form solution to natural image matting. *IEEE Transactions on Pattern Analysis and Machine Intelligence*, 30(2), 228–242.

Levin, A., Weiss, Y., Durand, F., Freeman, W.T. (2009). Understanding and evaluating blind deconvolution algorithms. *IEEE Conference on Computer Vision and Pattern Recognition*, 1964–1971.

Li, S.Z. (2009). *Markov Random Field Modeling in Image Analysis*. Springer-Verlag, London.

Li, Z., Shi, Z., Sun, J. (2017). Point integral method for solving Poisson-type equations on manifolds from point clouds with convergence guarantees. *Communications in Computational Physics*, 22(1), 228–258.

Liba, O., Murthy, K., Tsai, Y.-T., Brooks, T., Xue, T., Karnad, N., He, Q., Barron, J.T., Sharlet, D., Geiss, R., Hasinoff, S.W., Pritch, Y., Levoy, M. (2019). Handheld mobile photography in very low light. *ACM Transactions on Graphics*, 38(6), 1–16.

Liu, X., Zhai, D., Zhao, D., Zhai, G., Gao, W. (2014). Progressive image denoising through hybrid graph Laplacian regularization: A unified framework. *IEEE Transactions on Image Processing*, 23(4), 1491–1503.

Liu, X., Cheung, G., Wu, X., Zhao, D. (2017). Random walk graph Laplacian-based smoothness prior for soft decoding of JPEG images. *IEEE Transactions on Image Processing*, 26(2), 509–524.

Mahalanobis, P.C. (1936). On the generalized distance in statistics. *Proceedings of National Institute of Sciences*, India, 2(1), 49–55.

McCann, M.T., Jin, K.H., Unser, M. (2017). Convolutional neural networks for inverse problems in imaging: A review. *IEEE Signal Processing Magazine*, 34(6), 85–95.

Michaeli, T. and Irani, M. (2014). Blind deblurring using internal patch recurrence. In *Computer Vision – ECCV 2014: 13th European Conference on Computer Vision*, Fleet, D., Pajdla, T., Schiele, B., Tuytelaars, T. (eds). Springer, Cham.

Milanfar, P. (2013). A tour of modern image filtering. *IEEE Signal Processing Magazine*, 30(1), 106–128.

Milanfar, P. (2013). Symmetrizing smoothing filters. *SIAM Journal on Imaging Sciences*, 6(1), 263–284.

Narang, S.K. and Ortega, A. (2013). Compact support biorthogonal wavelet filterbanks for arbitrary undirected graphs. *IEEE Transactions on Signal Processing*, 61(19), 4673–4685.

Osher, S., Shi, Z., Zhu, W. (2017). Low dimensional manifold model for image processing. *SIAM Journal on Imaging Sciences*, 10(4), 1669–1690.

Pan, J., Hu, Z., Su, Z., Yang, M.-H. (2016). *l*0-regularized intensity and gradient prior for deblurring text images and beyond. *IEEE Transactions on Pattern Analysis and Machine Intelligence*, 39(2), 342–355.

Pan, J., Sun, D., Pfister, H., Yang, M.-H. (2016). Blind image deblurring using dark channel prior. *IEEE Conference on Computer Vision and Pattern Recognition*, 1628–1636.

Pang, J. and Cheung, G. (2017). Graph Laplacian regularization for image denoising: Analysis in the continuous domain. *IEEE Transactions on Image Processing*, 26(4), 1770–1785.

Pang, J., Cheung, G., Hu, W., Au, O.C. (2014). Redefining self-similarity in natural images for denoising using graph signal gradient. *Asia-Pacific Signal and Information Processing Association Annual Summit and Conference*, 1–8.

Pang, J., Cheung, G., Ortega, A., Au, O.C. (2015). Optimal graph Laplacian regularization for natural image denoising. *IEEE International Conference on Acoustics, Speech and Signal Processing*, 2294–2298.

Pang, J., Fang, L., Zeng, J., Guo, Y., Tang, K. (2015). Subpixel-based image scaling for grid-like subpixel arrangements: A generalized continuous-domain analysis model. *IEEE Transactions on Image Processing*, 25(3), 1017–1032.

Pang, J., Sun, W., Yang, C., Ren, J., Xiao, R., Zeng, J., Lin, L. (2018). Zoom and learn: Generalizing deep stereo matching to novel domains. *IEEE Conference on Computer Vision and Pattern Recognition*, 2070–2079.

Perona, P., Shiota, T., Malik, J. (1994). Anisotropic diffusion. In *Geometry-driven Diffusion in Computer Vision*, Haar Romeny, B.M. (ed.). Springer, Netherlands.

Petschnigg, G., Szeliski, R., Agrawala, M., Cohen, M., Hoppe, H., Toyama, K. (2004). Digital photography with flash and no-flash image pairs. *ACM Transactions on Graphics*, 23(3), 664–672.

Peyré, G. (2008). Image processing with nonlocal spectral bases. *Multiscale Modeling & Simulation*, 7(2), 703–730.

Peyré, G. (2009). Manifold models for signals and images. *Computer Vision and Image Understanding*, 113(2), 249–260.

Reynolds, D. (2009). Gaussian mixture models. *Encyclopedia of Biometrics*, 741, 659–663.

Romano, Y., Elad, M., Milanfar, P. (2017). The little engine that could: Regularization by denoising (RED). *SIAM Journal on Imaging Sciences*, 10(4), 1804–1844.

Rudin, L.I., Osher, S., Fatemi, E. (1992). Nonlinear total variation based noise removal algorithms. *Physica D: Nonlinear Phenomena*, 60(1–4), 259–268.

Sapiro, G. (2006). *Geometric Partial Differential Equations and Image Analysis*. Cambridge University Press, Cambridge.

Schwartz, E., Giryes, R., Bronstein, A.M. (2018). DeepISP: Toward learning an end-to-end image processing pipeline. *IEEE Transactions on Image Processing*, 28(2), 912–923.

Shan, Q., Jia, J., Agarwala, A. (2008). High-quality motion deblurring from a single image. *ACM Transactions on Graphics*, 27(3), 1–10.

Shewchuk, J.R. (1994). An introduction to the conjugate gradient method without the agonizing pain. *CMU Technical Report*, 1–64.

Shi, Z., Osher, S., Zhu, W. (2016). Low dimensional manifold model with semi-local patches. UCLA CAM report, 16-63.

Shi, Z., Osher, S., Zhu, W. (2018). Generalization of the weighted nonlocal Laplacian in low dimensional manifold model. *Journal of Scientific Computing*, 75(2), 638–656.

Shuman, D., Narang, S.K., Frossard, P., Ortega, A., Vandergheynst, P. (2013). The emerging field of signal processing on graphs: Extending high-dimensional data analysis to networks and other irregular domains. *IEEE Signal Processing Magazine*, 30(3), 83–98.

Singer, A. (2006). From graph to manifold Laplacian: The convergence rate. *Applied and Computational Harmonic Analysis*, 21(1), 128–134.

Sinkhorn, R. and Knopp, P. (1967). Concerning nonnegative matrices and doubly stochastic matrices. *Pacific Journal of Mathematics*, 21(2), 343–348.

Sochen, N., Kimmel, R., Malladi, R. (1998). A general framework for low level vision. *IEEE Transactions on Image Processing*, 7(3), 310–318.

Su, W.-T., Cheung, G., Wildes, R., Lin, C.-W. (2020). Graph neural net using analytical graph filters and topology optimization for image denoising. *IEEE International Conference on Acoustics, Speech and Signal Processing*, 8464–8468.

Sun, L., Cho, S., Wang, J., Hays, J. (2013). Edge-based blur kernel estimation using patch priors. *IEEE International Conference on Computational Photography*, 1–8.

Takeda, H., Farsiu, S., Milanfar, P. (2007). Kernel regression for image processing and reconstruction. *IEEE Transactions on Image Processing*, 16(2), 349–366.

Talebi, H. and Milanfar, P. (2016). Fast multilayer Laplacian enhancement. *IEEE Transactions on Computational Imaging*, 2(4), 496–509.

Ting, D., Huang, L., Jordan, M. (2010). An analysis of the convergence of graph Laplacians. *International Conference on Machine Learning*, 1079–1086.

Tomasi, C. and Manduchi, R. (1998). Bilateral filtering for gray and color images. *IEEE International Conference on Computer Vision*, 839–846.

Venkatakrishnan, S.V., Bouman, C.A., Wohlberg, B. (2013). Plug-and-play priors for model based reconstruction. *IEEE Global Conference on Signal and Information Processing*, 945–948.

Wan, P., Cheung, G., Florencio, D., Zhang, C., Au, O.C. (2014). Image bit-depth enhancement via maximum-a-posteriori estimation of graph AC component. *IEEE International Conference on Image Processing*, 4052–4056.

Weickert, J. (1998). *Anisotropic Diffusion in Image Processing*. Teubner, Stuttgart.

Wells, W.M. (1986). Efficient synthesis of Gaussian filters by cascaded uniform filters. *IEEE Transactions on Pattern Analysis and Machine Intelligence*, 8(2), 234–239.

Wronski, B., Garcia-Dorado, I., Ernst, M., Kelly, D., Krainin, M., Liang, C.-K., Levoy, M., Milanfar, P. (2019). Handheld multi-frame super-resolution. *ACM Transactions on Graphics*, 38(4), 1–18.

Xiong, Z., Orchard, M.T., Zhang, Y.-Q. (1997). A deblocking algorithm for JPEG compressed images using overcomplete wavelet representations. *IEEE Transactions on Circuits and Systems for Video Technology*, 7(2), 433–437.

Xu, L., Zheng, S., Jia, J. (2013). Unnatural *l*0 sparse representation for natural image deblurring. *IEEE Conference on Computer Vision and Pattern Recognition*, 1107–1114.

Xu, L., Ren, J.S., Liu, C., Jia, J. (2014). Deep convolutional neural network for image deconvolution. *Advances in Neural Information Processing Systems*, 1790–1798.

Yang, J., Wright, J., Huang, T.S., Ma, Y. (2010). Image super-resolution via sparse representation. *IEEE Transactions on Image Processing*, 19(11), 2861–2873.

Yuan, L., Sun, J., Quan, L., Shum, H.-Y. (2007). Image deblurring with blurred/noisy image pairs. *ACM Transactions on Graphics*, 26(3), 1.

Zeng, J., Pang, J., Sun, W., Cheung, G. (2019). Deep graph Laplacian regularization for robust denoising of real images. *IEEE Conference on Computer Vision and Pattern Recognition Workshops*, 16–17 June, Long Beach, CA.

Zeng, J., Cheung, G., Ng, M., Pang, J., Yang, C. (2020). 3D point cloud denoising using graph Laplacian regularization of a low dimensional manifold model. *IEEE Transactions on Image Processing*, 29, 3474–3489.

Zhang, F. and Hancock, E.R. (2008). Graph spectral image smoothing using the heat kernel. *Pattern Recognition*, 41(11), 3328–3342.

Zhang, X. and Wu, X. (2008). Image interpolation by adaptive 2-D autoregressive modeling and soft-decision estimation. *IEEE Transactions on Image Processing*, 17(6), 887–896.

Zhang, X., Burger, M., Bresson, X., Osher, S. (2010). Bregmanized nonlocal regularization for deconvolution and sparse reconstruction. *SIAM Journal on Imaging Sciences*, 3(3), 253–276.

Zhang, X., Xiong, R., Fan, X., Ma, S., Gao, W. (2013). Compression artifact reduction by overlapped-block transform coefficient estimation with block similarity. *IEEE Transactions on Image Processing*, 22(12), 4613–4626.

Zhang, K., Zuo, W., Chen, Y., Meng, D., Zhang, L. (2017). Beyond a Gaussian denoiser: Residual learning of deep CNN for image denoising. *IEEE Transactions on Image Processing*, 26(7), 3142–3155.

Zhao, C., Zhang, J., Ma, S., Fan, X., Zhang, Y., Gao, W. (2016). Reducing image compression artifacts by structural sparse representation and quantization constraint prior. *IEEE Transactions on Circuits and Systems for Video Technology*, 27(10), 2057–2071.

Zhu, W., Qiu, Q., Huang, J., Calderbank, R., Sapiro, G., Daubechies, I. (2018a). LDMNet: Low dimensional manifold regularized neural networks. *IEEE Conference on Computer Vision and Pattern Recognition*, 2743–2751.

Zhu, W., Shi, Z., Osher, S. (2018b). Scalable low dimensional manifold model in the reconstruction of noisy and incomplete hyperspectral images. *IEEE Workshop on Hyperspectral Image and Signal Processing: Evolution in Remote Sensing (WHISPERS)*, 1–5.

7

Graph Spectral Point Cloud Processing

Wei HU[1], Siheng CHEN[2] and Dong TIAN[3]

[1] *Peking University, Beijing, China*

[2] *Shanghai Jiao Tong University, China*

[3] *InterDigital Inc., New York, USA*

7.1. Introduction

Images provide 2D visual projection of the 3D world and could have geometric information embedded implicitly. In contrast, 3D point clouds form omnidirectional representations of the explicit geometric structure and visual information of the 3D world. A point cloud consists of a set of 3D points that are *irregularly* sampled from the surface of objects or scenes, as demonstrated in Figure 7.1. Each point typically contains geometry information (i.e. its 3D coordinate) and may be additionally associated with other attribute information, such as RGB colors and/or reflection intensity, depending on the scanning device. The maturity of depth sensing and laser scanning techniques[1] makes the acquisition of 3D point clouds more affordable. A point cloud is gradually becoming a popular 3D representation format and has a critical role in a wide range of applications, including immersive tele-presence (Mekuria *et al.* 2016), autonomous driving (Chen *et al.* 2021) and digital preservation of cultural heritage (Gomes *et al.* 2014).

---

1 Commercial products include Microsoft Kinect (2010–2014), Intel RealSense (2015–), Velodyne LiDAR (2007–2020), the LiDAR scanner of the Apple iPad Pro (2020), and so on.

Graph Spectral Image Processing,
coordinated by Gene CHEUNG and Enrico MAGLI.

By their very nature, point clouds have a profound connection with graphs. A point cloud is typically a representation (signal) of some underlying surface or 2D manifold over a set of quantized (sampled) nodes, which reside on an *unstructured* grid. More specifically, we define every 3D point as a graph node, and capture the relationship between a pair of 3D points by a graph edge. For example, two 3D points can be connected when their Euclidean distance is smaller than a certain threshold. Finally, we treat the 3D coordinates (and additional attributes) of each point as graph signals. Hence, graphs serve as a natural representation of point clouds, and this makes it particularly suitable to apply graph signal processing (GSP) techniques to point cloud data.

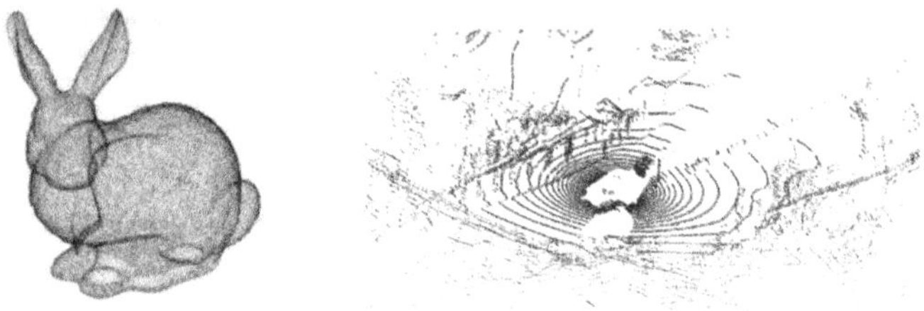

(a) The "Stanford Bunny" object (b) Street view from LiDAR

Figure 7.1. *Examples of 3D point clouds. The 3D point cloud in plot shows (a) a 3D* Bunny *model obtained by range scanners with some postprocessing and (b) one LiDAR sweep directly collected by Velodyne HDL-64E for autonomous driving*

In principle, there exists a link between point clouds and graphs through Riemannian manifolds in a continuous domain. On the one hand, point clouds are discrete samples of functions on Riemannian manifolds, representing the geometry of 3D objects. On the other hand, under certain conditions, graph operators converge to operators on Riemannian manifolds (Ting *et al.* 2010b; Hein *et al.* 2007), while graph-based regularizers converge to smoothness functionals on Riemannian manifolds (Pang and Cheung 2017; Hein 2006). Hence, as discrete counterparts of operators on Riemannian manifolds, GSP tools are naturally advantageous to point cloud processing by representing their underlying geometry on graphs.

In practice, graphs and GSP have already demonstrated great benefits for point cloud processing, from low-level processing, such as restoration, resampling and compression, etc., to high-level understanding, such as supervised or unsupervised feature learning for segmentation, classification and recognition.

The outline of this chapter is as follows. We start from the definition of graphs and graph signals in point cloud applications. Then we introduce two families of methodologies for point cloud processing, namely, traditional GSP methods and geometric deep learning methods. Both approach families benefit from the insights provided by graph spectral analysis. Depending on a specific algorithm design, the computation may be performed mainly in the spectral-domain or the nodal-domain. Finally, we discuss several typical applications and how they benefit from the GSP approaches.

7.2. Graph and graph-signals in point cloud processing

Point clouds describe the geometry of objects or scenes with a set of irregularly sampled 3D points. To denote a 3D point cloud, we consider an unordered set of points, which ignores any order of 3D points. Let $\mathbf{S} = \{(\mathbf{p}_i, \mathbf{a}_i)\}_{i=1}^N$ be a set of N 3D points, whose ith element $\mathbf{p}_i = [x_i, y_i, z_i] \in \mathbb{R}^3$ represents the 3D coordinate of the ith point and $\mathbf{a}_i$ represents other associated attributes, such as RGB colors, reflection intensity and surface normals. To ease mathematical computation, we could also represent a 3D point cloud as a matrix,

$$\mathbf{X} = \begin{bmatrix} \mathbf{x}_1^T \\ \mathbf{x}_2^T \\ \vdots \\ \mathbf{x}_N^T \end{bmatrix} \in \mathbb{R}^{N \times d}, \qquad [7.1]$$

where the ith row vector $\mathbf{x}_i^T = [\mathbf{p}_i^T \ \mathbf{a}_i^T] \in \mathbb{R}^{1 \times d}$ records the information from the ith point.

To capture the relationships among 3D points, one can introduce a graph, where each node is a 3D point and each edge reflects the pairwise relationship of 3D points. This graph can be viewed as a discrete approximation of the Riemannian manifold that describes the continuous surface of an object (or a scene). One mathematical representation of a graph with N nodes is via an adjacency matrix $\mathbf{W} \in \mathbb{R}^{N \times N}$, whose (i, j)th element indicates the pairwise relationship between the ith and the jth 3D points. As a 3D point, each node is associated with a 3D coordinate and, possibly, with more attributes. The collection of such associated variables forms a d-dimensional *graph signal* (see Figure 7.2).

There are two typical methods of constructing a graph:

1) model-based graph construction, which builds graphs with models from domain knowledge;

2) learning-based graph construction, which infers/learns the underlying graph from geometric data.

Model-based graph construction for point clouds often assumes edge weights are inversely proportional to the affinities in coordinates, such as a K-nearest-neighbor graph (K-NN graph) and an ϵ-neighborhood graph (ϵ-N graph). A K-NN graph is a directed graph in which two nodes are connected by an edge when their Euclidean distance is among the Kth smallest Euclidean distances from one 3D point to all the other 3D points. An ϵ-N graph is a graph in which two nodes are connected by an edge when their Euclidean distance is smaller than a given threshold ϵ. A K-NN graph intends to maintain a consistent node degree in the graph that may lead to a more stable algorithm implementation, while an ϵ-N graph intends to make the node degree reflect local point density, leading to more physical interpretation. Both K-NN graphs and ϵ-N graphs can be efficiently implemented by using efficient data structures, such as octrees (Peng and Kuo 2005; Hornung *et al.* 2013). Though these graphs exhibit manifold convergence properties (Ting *et al.* 2010b; Hein *et al.* 2007), it still remains challenging to find an efficient estimation of the sparsification parameters such as K and ϵ, given finite and non-uniformly sampled data.

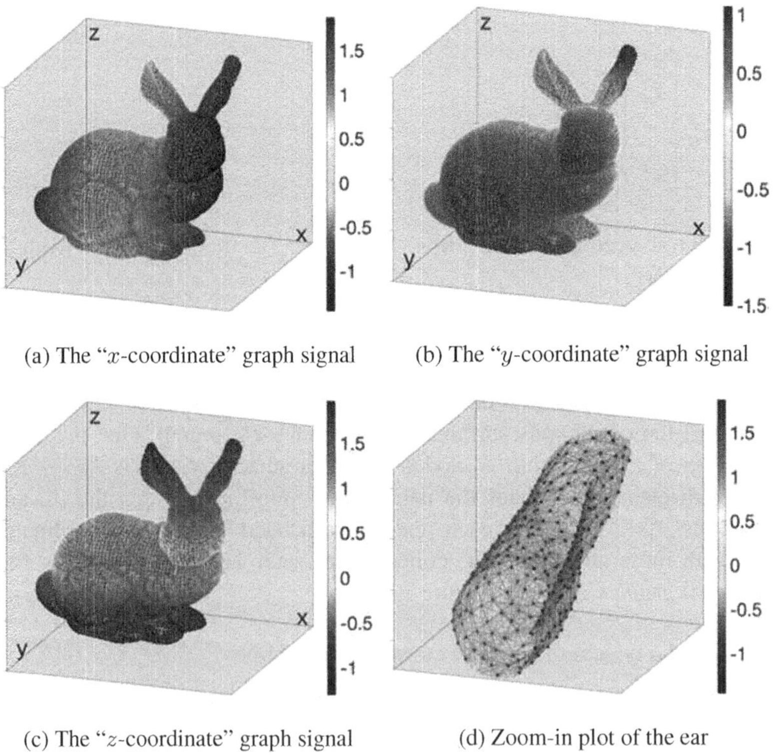

(a) The "x-coordinate" graph signal (b) The "y-coordinate" graph signal

(c) The "z-coordinate" graph signal (d) Zoom-in plot of the ear

Figure 7.2. *Graph and graph signals for 3D point clouds. A K-nearest-neighbor graph is constructed to capture the pairwise spatial relationships among 3D points. The values of graph signals are reflected via color. For a color version of this figure, see www.iste.co.uk/cheung/graph.zip*

In learning-based graph construction, the underlying graph is inferred or optimized from point clouds, in terms of a certain optimization criterion. For example, given a single or partial observation, Hu *et al.* (2020a) optimize a distance metric from relevant feature vectors on nodes by minimizing the graph Laplacian regularizer, leading to learned edge weights. Besides, edge weights could be trainable in an end-to-end learning architecture (Wang *et al.* 2019b). General graph learning methodologies can also be applied to the graph construction of point clouds. Please refer to Chapter 2 for a detailed discussion.

7.3. Graph spectral methodologies for point cloud processing

In this section, we elaborate on graph spectral methods for point cloud processing. Based on how we infer graph filter parameters, we classify graph spectral methods into two classes: *model-based graph spectral methods* and *learning-based graph spectral methods*. Model-based methods are built upon a deep understanding and assumptions of the characteristics of point clouds, such as *sparsity* models in the spectral domain or *smoothness* models in the nodal domain. The learning-based paradigm (i.e. geometric deep learning (Bronstein *et al.* 2017)) learns filter parameters end-to-end in an unsupervised or supervised fashion, such as the recently developed graph neural networks (GNNs) that extend classical neural networks to graph data via graph convolution in the spectral domain or nodal domain.

A detailed description of GSP has been introduced in Part 1. We revisit them here in the context of point clouds and emphasize how we deploy graph spectral methods to process point clouds.

7.3.1. *Spectral-domain graph filtering for point clouds*

As introduced in the Introduction and Chapter 1, graph spectral filtering is defined in the graph frequency domain, such as the graph Fourier transform (GFT) domain. Given a point cloud $\mathbf{X}$, suppose we have constructed a graph over $\mathbf{X}$, whose connectivity is encoded in a graph Laplacian $\mathbf{L}$ computed from $\mathbf{W}$. Let us denote $\mathbf{U}$ as the eigenvector matrix of $\mathbf{L}$ and $\{\lambda_0 = 0 \leq \lambda_1 \leq ... \leq \lambda_{N-1}\}$ as the set of eigenvectors of $\mathbf{L}$. Assuming the graph frequency response of the graph spectral filtering is $\hat{h}(\lambda_k)$ $(k = 0, \dots, N-1)$, then this filtering can be written as

$$\mathbf{Y} = \mathbf{U} \begin{bmatrix} \hat{h}(\lambda_0) & & \\ & \ddots & \\ & & \hat{h}(\lambda_{N-1}) \end{bmatrix} \mathbf{U}^\top \mathbf{X}. \qquad [7.2]$$

This graph spectral filtering essentially first converts the point cloud data into the GFT domain, performs filtering on each eigenvalue (i.e. the spectrum of the graph) and finally projects back to the spatial domain via the inverse GFT. The key is to specify N graph frequency responses in $\hat{h}(\lambda_k)$ tailored for each point cloud processing task. Next, we will discuss *low-pass* graph spectral filtering and *high-pass* graph spectral filtering separately.

As in image processing, we can use a low-pass graph filter to capture the rough shape of a point cloud and attenuate noise/outliers, under the assumption that signals on point clouds are smooth. In practice, a point cloud signal (e.g. coordinates, normals, RGB colors) is naturally smooth, with respect to the graph we construct from the signal. Thus, we can perform point cloud smoothing by a low-pass graph filter, where high-frequency components are likely to be generated by noise, as the energy of a smooth point cloud concentrates on low-frequency components. This essentially leads to a smoothed discrete representation of the underlying manifold.

One intuitive realization is an ideal low-pass graph filter, which completely eliminates all graph frequencies above a given bandwidth while keeping those below unchanged. The graph frequency response of an ideal low-pass graph filter with bandwidth b is

$$\hat{h}(\lambda_k) = \begin{cases} 1, & k \leq b, \\ 0, & k > b, \end{cases} \qquad [7.3]$$

which projects the input point cloud onto a bandlimited subspace by removing components corresponding to large eigenvalues (i.e. high-frequency components). The smoothed result provides a bandlimited approximation of the original point cloud. Figure 7.3 demonstrates an example of the bandlimited approximation of the 3D coordinates of a sampled point cloud of *Bunny* with 10, 100 and 1000 graph frequencies, respectively. We see that the first 10 low frequency components are able to represent the rough shape of *Bunny*, while the shape gets more accurate with more graph frequencies (see Rosman *et al.* (2013) and Chen *et al.* (2018) for more examples).

(a) Original (b) 10 frequencies (c) 100 frequencies (d) 1000 frequencies

Figure 7.3. *Low-pass approximation of the point cloud* Bunny*. Plot (a) is the original sampled point cloud with 1,797 points. Plots (b)–(d) show the low-pass approximations with 10, 100 and 1,000 graph frequencies*

Another simple choice is Haar-like low-pass graph filter, as discussed in Chen *et al.* (2018), where the graph frequency response is:

$$\hat{h}(\lambda_k) = 1 - \frac{\lambda_k}{\lambda_{max}}, \qquad [7.4]$$

where $\lambda_{max} = \lambda_{N-1}$ is the maximum eigenvalue for normalization. As $\lambda_{k-1} \leq \lambda_k$, we have $\hat{h}(\lambda_{k-1}) \geq \hat{h}(\lambda_k)$. As such, low-frequency components are preserved while high-frequency components are attenuated.

In contrary to the low-pass filtering, high-pass filtering eliminates low-frequency components, which enables the detection of large variations in a point cloud, such as geometric contours or texture variations. A simple design is a Haar-like high-pass graph filter with the following graph frequency response

$$\hat{h}(\lambda_k) = \frac{\lambda_k}{\lambda_{max}}. \qquad [7.5]$$

As $\lambda_{k-1} \leq \lambda_k$, we have $\hat{h}(\lambda_{k-1}) \leq \hat{h}(\lambda_k)$. This indicates that lower frequency responses are attenuated, whereas high-frequency responses are preserved. An application example is contour detection, as discussed in Chen *et al.* (2018).

In addition, we can design a desired spectral distribution and then use graph filter coefficients to fit this distribution. For example, an P-length graph filter is

$$\hat{h}(\Lambda) = \begin{bmatrix} \sum_{p=0}^{P-1} \hat{h}_p \lambda_1^p & & \\ & \ddots & \\ & & \sum_{p=0}^{P-1} \hat{h}_p \lambda_N^p \end{bmatrix},$$

where Λ is a diagonal matrix with diagonal entries as the eigenvalues of the graph Laplacian $\mathbf{L}$ and $\hat{h}_p$'s are the filter coefficients. If a desirable response of the ith graph frequency is c_i, we set

$$\hat{h}(\lambda_i) = \sum_{p=0}^{P-1} \hat{h}_p \lambda_i^p = c_i,$$

and solve a set of linear equations to obtain the graph filter coefficients $\hat{h}_p$. It is also possible to use the Chebyshev polynomial to construct graph filter coefficients (Hammond *et al.* 2011).

However, since spectral-domain graph filtering entails high computational complexity due to the full eigendecomposition of the graph Laplacian, this class of methods are either dedicated to *small-scale* point clouds or applied to point clouds in a divide-and-conquer manner. For instance, one may divide a point cloud into regular cubes, and perform graph spectral filtering for each cube separately. One may also deploy a fast algorithm of GFT implementation (e.g. fast GFT in Le Magoarou *et al.* (2017)) to accelerate the spectral filtering process.

7.3.2. Nodal-domain graph filtering for point clouds

This class of methods performs filtering on point clouds in the nodal domain, which is often computationally efficient and thus amenable to *large-scale* data. Intuitively, the functionality of nodal-domain graph filtering is to fuse the feature (attributes) at each 3D point with the feature from its neighboring 3D points in the graph vertex domain. Simple fusion might be implemented by a weighted linear combination.

A *graph shift operator* is an elementary but non-trivial graph filter. Often a graph Laplacian matrix $\mathbf{L}$ is used as a graph shift operator, although this is controversial when energy-preserving is a property of interest (Gavili and Zhang 2017).

Every linear, shift-invariant graph filter is a polynomial in the graph shift,

$$h(\mathbf{L}) = \sum_{p=0}^{P-1} h_p \mathbf{L}^p = h_0 \mathbf{I} + h_1 \mathbf{L} + \ldots + h_{P-1} \mathbf{L}^{P-1}, \qquad [7.6]$$

where h_p, $p = 0, 1, \ldots, P-1$ are filter coefficients and P is the graph filter length. A higher order corresponds to a larger support size on the graph vertex domain. The output of graph filtering is given by the matrix-vector product. For example, let $\mathbf{X} \in \mathbb{R}^{N \times 3}$ be the RGB values of a 3D point cloud, the filtering response $\mathbf{X}_{\text{out}} = h(\mathbf{L})\mathbf{X} \in \mathbb{R}^{N \times 3}$ is the weighted averaged RGB values. The output of the ith point, $\sum_{p=0}^{P-1} h_p \left(\mathbf{L}^p \mathbf{X}\right)_i$ is a weighted average of the attributes of points that are within P hops away from the ith point. The pth graph filter coefficient, h_p, quantifies the contribution from the pth-hop neighbors. We design the filter coefficients to change the weights in local averaging.

Similarly to spectral-domain graph filtering, we now consider two typical types of nodal-domain graph filters as follows:

– Low-pass graph filtering: similarly to classical signal processing, we can use a low-pass graph filter to smooth feature values over a 3D point cloud. Figure 7.4 shows the effect of low-pass graph filtering: plot (a) shows a graph signal activating the right ear of the *Bunny*, where the activated points are marked in red. Plot (b) shows that the boundary between activated points and inactivated points gets blurred after low-pass graph filtering. This indicates that the functionality of low-pass graph filtering is to reduce transient change and capture the main shape.

– High-pass graph filtering: in image processing, a high-pass filter is used to extract edges and contours; similarly, we can use a high-pass graph filter to extract contours in a point cloud. Figure 7.4(c) shows that the boundary between activated points and inactivated points gets highlighted after high-pass graph filtering. This indicates that the functionality of high-pass graph filtering is to detect transient changes of feature values over a 3D point cloud.

(a) Original graph signal (b) Low-pass graph filtering (c) High-pass graph filtering

Figure 7.4. *Graph filtering for 3D point clouds. Low-pass graph filtering smooths the sharp transition in a graph signal, while high-pass graph filtering highlights the sharp transition. For a color version of this figure, see www.iste.co.uk/cheung/graph.zip*

Graph filtering can be used in various processing tasks. For example, Chen *et al.* (2018) design a high-pass graph filter to extract contour-related features from a 3D point cloud with applications to 3D point cloud downsampling; Zeng *et al.* (2019) design a graph filter to solve the graph Laplacian-regularized optimization problem with applications for 3D point cloud denoising. Graph filtering also lays the mathematical foundation for graph convolutional neural networks (CNNs), where the key operation can be assumed to be a graph filter.

7.3.3. *Learning-based graph spectral methods for point clouds*

Another class of methods is based on neural networks, which infer filter parameters end-to-end. As introduced in Chapter 3, CNNs can only process data that is defined on regular grids, where a simple spatial relationship between data samples is defined. For example, for an image pixel, its top, bottom, left and right pixels are all uniquely defined. Unfortunately, point clouds reside on irregular domains, where the neighborhood of a point scatters in 3D space, and no such simple spatial relationship exists. In contrast, GSP provides efficient filtering and sampling of such data with elegant spectral interpretation, which is able to generalize the key operations (e.g. convolution and pooling) in a neural network for irregular point clouds. For example, graph convolution can be defined as graph filtering either in the spectral or the spatial domain, while pooling can be treated as the sampling operation of a graph signal. This leads to the recently developed graph convolutional neural networks (GCNNs) (see Bronstein *et al.* (2017) and references therein), which generalize CNNs to unstructured data. Please refer to Chapter 3 for a detailed description of GCNNs, in this chapter we focus on GCNN methods for point clouds.

While some previous works represent irregular point clouds with regular 3D voxel grids or collections of 2D images before feeding them to a neural network or impose certain symmetrizations in the net computation (e.g. PointNet; Qi *et al.* (2017b)), many recent efforts represent point clouds on graphs naturally and develop various GCNN methods.

Among many inspiring works, Wang *et al.* (2019b) introduced two useful techniques: the edge convolution operation and learnable graphs (see details in Chapter 3). Subsequent research has proposed novel graph neural networks for point cloud understanding, such as recognition, segmentation and classification. For example, Wang *et al.* (2018a) proposed local spectral graph convolution, where filter coefficients represented in the graph spectral domain are learnable, and Wang *et al.* (2019a) proposed graph attention convolution, which selects the most relevant points from all the neighbors by trainable attentional weights. It also uses graph coarsening to promote multiscale analysis on graphs. Te *et al.* (2018) propose a regularized graph convolutional neural network (RGCNN), which regularizes each layer by a graph-signal smoothness prior, with spectral smoothing functionality. Such regularization leads to robustness to both low density and noise in point clouds. As one of the most recent works in this area, Li *et al.* (2019a) construct the deepest yet graph convolution network (GCN) architecture, which has 56 layers. It transplants a series of techniques from CNNs, such as residual and dense connections, as well as dilated graph convolutions, to the graph domain. Some other works do not rely on graphs, but have based their computations on local neighbors defined by either K-NN graphs (Zhao *et al.* 2019; Liu *et al.* 2019b; Pan *et al.* 2019) or ϵ-graphs (Thomas *et al.* 2019).

7.4. Low-level point cloud processing

In this section, we discuss the applications of graph spectral methods in low-level point cloud processing, including restoration, resampling, compression and so on. Please refer to Chapter 5 for a detailed discussion on point cloud compression, in this chapter we focus on the restoration and resampling of point clouds. We also introduce popular benchmarks for point clouds and commonly used evaluation metrics.

Similar to image restoration introduced in Chapter 6, point cloud restoration is an inverse problem to reconstructing point clouds from degraded versions. As point clouds are discrete samples of functions on Riemannian manifolds, point cloud restoration essentially reconstructs the underlying manifold from a corrupted observation. Due to various causes, such as the inherent limitations of acquisition sensors and capturing viewpoints, point clouds often suffer from noise, holes, compression artifacts and so on. Consequently, many attempts have been made at point cloud restoration, with, as a representative problem, point cloud denoising. For those who are interested in point cloud inpainting, please see Hu *et al.* (2019) and references therein.

7.4.1. *Point cloud denoising*

Point clouds are often perturbed by noise, which comes from hardware, software or environmental causes. Hardware-wise, noise occurs due to the inherent limitations of the acquisition equipment. Software-wise, in the case of generating point clouds with existing algorithms, points may be located somewhere completely wrong due to imprecise triangulation (e.g. a false epipolar matching). Environmentally, outliers may appear due to the surrounding contamination, such as dust in the air. Noise corruption directly affects the subsequent applications of point clouds. We show a point cloud corrupted by Gaussian noise and its denoised result from one of the latest graph spectral denoising methods (Hu *et al.* 2020a) in Figure 7.5.

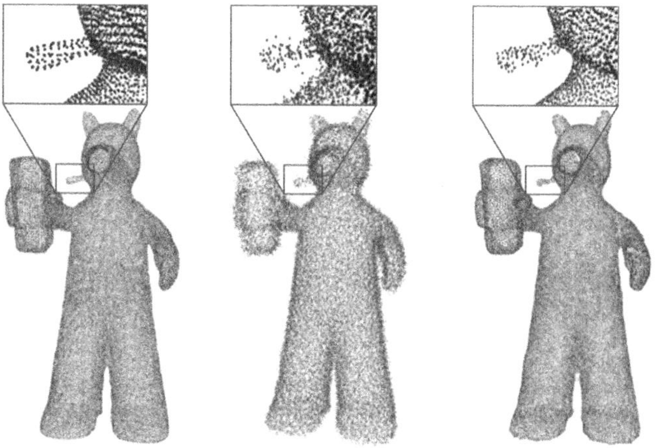

(a) The ground truth (b) The noisy point cloud (c) The denoised result

Figure 7.5. *A synthetic noisy point cloud with Gaussian noise* $\sigma = 0.04$ *for* Quasimoto *and one denoised result: (a) the ground truth; (b) the noisy point cloud; (c) the denoised result by Hu* et al. *(2020a)*

Assuming an additive noise model for a point cloud, namely

$$\mathbf{P} = \mathbf{Y} + \mathbf{E}, \qquad [7.7]$$

where $\mathbf{P} \in \mathbb{R}^{N\times 3}$ denotes the observed noise-corrupted 3D coordinates of the target point cloud with N points, $\mathbf{Y} \in \mathbb{R}^{N\times 3}$ is the ground truth 3D coordinates of the point cloud and $\mathbf{E} \in \mathbb{R}^{N\times 3}$ is an additive noise. The problem of point cloud denoising is then to reconstruct $\mathbf{Y}$ from $\mathbf{P}$.

Modeling of $\mathbf{E}$ depends on the actual point cloud acquisition mechanism. There exist a wide range of point cloud acquisition systems at different price points – from consumer-level depth sensors like Intel RealSense[1], costing 150 USD, to high-end outdoor scanners like Teledyne Optech[2], which cost up to 250,000 USD – and defining accurate noise models for all of them is difficult. One may select the most common Gaussian noise model, which has been shown to be reasonably accurate for popular depth cameras like Microsoft Kinect (Nguyen *et al.* 2012; Sun *et al.* 2008). Hence, $\mathbf{E}$ in equation [7.7] represents zero-mean *additive white Gaussian noise* (AWGN) with standard deviation σ, i.e.

$$\mathbf{E} \sim \mathcal{N}(\mathbf{0}, \sigma^2\mathbf{I}), \tag{7.8}$$

where $\mathbf{I}$ is an identity matrix.

Graph-based methods interpret a point cloud as a graph signal and perform denoising by designing spectral/spatial graph filters, as discussed in sections 7.3.1 and 7.3.2. For example, in accordance with the ideal low-pass spectral graph filtering discussed in section 7.3.1, hard thresholding of GFT coefficients was proposed in Rosman *et al.* (2013) to remove spectrum components with small energy, assuming they were introduced by noise. Due to the computation complexity of eigendecomposition in spectral filtering, many methods resort to spatial graph filters in the form of the following optimization problem:

$$\hat{\mathbf{Y}} = \underset{\mathbf{Y}}{\operatorname{argmin}} \|\mathbf{Y} - \mathbf{P}\|_2^2 + \gamma g(\mathbf{Y}). \tag{7.9}$$

Here, $\hat{\mathbf{Y}}$ is the denoised point cloud, $g(\cdot)$ is a graph-signal prior assumed for $\mathbf{Y}$ and γ is a weighting parameter to strike a balance between the data fidelity term and the graph-signal prior. If the prior $g(\cdot)$ is differentiable, equation [7.9] admits a closed-form solution that corresponds to a spatial graph filter.

Various graph-signal priors have been designed under different models. (Schoenenberger *et al.* 2015) assume smoothness in the gradient $\nabla_{\mathcal{G}}\mathbf{Y}$ of the point cloud $\mathbf{Y}$ on the graph $\mathcal{G}$, leading to a Tikhonov regularization $g(\mathbf{Y}) = \|\nabla_{\mathcal{G}}\mathbf{Y}\|_2^2$. Besides, assuming the manifold underlying the point cloud to be piecewise smooth instead of smooth, they propose to replace it by a total variation (TV) regularization $g(\mathbf{Y}) = \|\nabla_{\mathcal{G}}\mathbf{Y}\|_1$.

In Dinesh *et al.* (2018), a reweighted graph Laplacian regularizer for surface normals was designed, i.e. the graph Laplacian (or the edge weights) is a function of the graph signal:

$$g(\mathbf{Y}) = \sum_{(i,j)\in\mathcal{E}} w_{i,j}(\mathbf{n}_i, \mathbf{n}_j)\|\mathbf{n}_i - \mathbf{n}_j\|_2^2, \tag{7.10}$$

1 https://www.intelrealsense.com/.

2 https://www.teledyneoptech.com/en/products/static-3d-survey/.

where $\mathbf{n}_i$ and $\mathbf{n}_i$ are the normal vectors at point i and j, respectively. Here, the edge weight between nodes i and j is a function of the normals, i.e. $w_{i,j}(\mathbf{n}_i, \mathbf{n}_j)$. Moreover, they established a linear relationship between normals and 3D point coordinates via bipartite graph approximation for ease of optimization.

Zeng *et al.* (2019) assume a low-dimensional manifold model and approximates the manifold dimension computation defined in continuous domain with a patch-based graph Laplacian regularizer, i.e.

$$g(\mathbf{Y}) = \mathrm{tr}\left(\mathcal{P}_{\mathbf{Y}}^{\top}\mathbf{L}\mathcal{P}_{\mathbf{Y}}\right), \qquad [7.11]$$

where $\mathcal{P}_{\mathbf{Y}}$ denotes patches in $\mathbf{Y}$. They seek self-similar patches among $\mathcal{P}_{\mathbf{Y}}$ for simultaneous denoising.

Instead of constructing the underlying graph with pre-defined edge weights from hand-crafted parameters, (Hu *et al.* 2020b) proposed feature graph learning by minimizing graph Laplacian regularizer using the Mahalanobis distance metric matrix $\mathbf{M}$ as a variable, assuming feature vector per node is available. The graph Laplacian $\mathbf{L}$ in equation [7.11] then becomes a function of $\mathbf{M}$, i.e. $\mathbf{L}(\mathbf{M})$. A fast algorithm was presented and applied to point cloud denoising, where the graph for each set of self-similar patches is computed from 3D coordinates and surface normals as features.

Instead of directly smoothing the coordinates or normals of 3D points, Duan *et al.* (2018) estimated a local tangent plane at each 3D point based on a graph and then reconstructed each 3D point by weighted averaging of its projections on multiple tangent planes. Duan *et al.* (2019) further consider a deep-neural-network-based model for denoising 3D point clouds. Please refer to Hu *et al.* (2020b) for a thorough description of previous point cloud denoising approaches. Here, we will focus on the graph spectral methods to address the point cloud denoising problem.

7.4.2. *Point cloud resampling*

Resampling of 3D point clouds considers changing the point density of a 3D point cloud to benefit downstream tasks. Depending on the demands of a downstream task, resampling could either be downsampling, which reduces the point density, or upsampling, which increases the point density.

7.4.2.1. *Downsampling*

The goal of downsampling is to select a subset of 3D points in an original 3D point cloud while preserving representative information (see Figure 7.6). As a typical task of 3D point cloud processing, downsampling is potentially useful to data storage and visualization. For example, autonomous-driving systems usually create a large-scale point-cloud map to precisely represent a 3D environment; however, handling a large number of 3D points is challenging and expensive. Therefore, to concisely represent a 3D scene, one can select representative 3D points from a point-cloud map through downsampling, leading to faster and better map-creation and localization performances (Chen *et al.* 2018).

Figure 7.6. *Dual problems of 3D point cloud downsampling and upsampling. For a color version of this figure, see www.iste.co.uk/cheung/graph.zip*

The most straightforward downsampling is uniform sampling, where we select downsampled indices from a uniform distribution; however, uniform sampling does not consider any geometric property of a 3D point cloud. For example, a 3D point cloud could include a densely sampled region and a sparsely sampled region, and both are semantically important to reflect the 3D scene; however, when we sample uniformly, there might be only a few points from the sparsely sampled region.

To overcome this issue, a simple and popular technique is the farthest point sampling (Qi *et al.* 2017a). It randomly chooses the first 3D point and then iteratively chooses the next 3D point that has the largest distance from all the points in the downsampled set. It is nothing but the deterministic version of K-means++ (Arthur and Vassilvitskii 2007). Compared with uniform sampling, it has better coverage of the entire 3D point cloud, given the same number of samples; however, farthest point sampling is agnostic to a subsequent application, such as localization and recognition.

To better preserve the geometric properties of a 3D point cloud during downsampling, we can design a non-uniform distribution, which actively reflects the demanded properties in a subsequent task (Chen *et al.* 2018). Once the non-uniform distribution is fixed and given, we can efficiently draw samples as downsampled indices. The design mechanism is based on extracting geometric-sensitive features from a 3D point cloud. Depending on the application, those features can be edges, key points and flatness (Daniels *et al.* 2007; Weber *et al.* 2010). Here, we consider extracting features from a point cloud by using a high-pass graph filter, as discussed in section 7.3.2. In image processing, a high-pass filter is used to extract edges and contours. Intuitively, a graph captures the spatial relationship of 3D points and a high-pass graph filter reflects the discontinuity in the spatial domain, which are, essentially, 3D edges or contours.

We use the response of high-pass graph filtering to measure local variation on graphs to detect contour points; local variation at the ith point is given as

$$f_i(\mathbf{X}) = \|(h(\mathbf{L})\mathbf{X})_i\|_2^2, \qquad [7.12]$$

where $h(\mathbf{L})$ is a high-pass graph filter. The simplest choice is $h(\mathbf{L}) = \mathbf{L}$. The local variation $f(\mathbf{X}) \in \mathbb{R}^N$ quantifies the energy of the response after high-pass graph filtering; when the local variation at a point is high, its 3D coordinates cannot be approximated well from those of its neighboring points, and the point is thus likely to be a contour point. In the graph vertex domain, the response of the ith point is

$$(h(\mathbf{L})\mathbf{X})_i = (\mathbf{L}\mathbf{X})_i = \sum_{j \in \mathcal{N}_i} \mathbf{W}_{i,j} (\mathbf{x}_i - \mathbf{x}_j). \qquad [7.13]$$

We see that each element compares the difference between each point and its neighbors, reflecting how much information we can glean about a point from its neighbors.

Next we design the non-uniform sampling distribution based on that local variation. We resample a point cloud M times. At the jth step, we independently choose a point $\mathcal{M}_j = i$ with probability π_i. Let $\Psi \in \mathbb{R}^{M \times N}$ be the resampling operator and $\mathbf{S} \in \mathbb{R}^{N \times N}$ be a diagonal rescaling matrix with $\mathbf{S}_{i,i} = 1/(M\pi_i)$. We quantify the performance of a resampling operator through

$$D_{f(\mathbf{X})}(\Psi) = \left\|\mathbf{S}\Psi^T \Psi f(\mathbf{X}) - f(\mathbf{X})\right\|_F^2, \qquad [7.14]$$

where $\|\cdot\|_F$ is the Frobenius norm. $\Psi^T\Psi \in \mathbb{R}^{N \times N}$ is a zero-padding operator, a diagonal matrix with diagonal elements $(\Psi^T\Psi)_{i,i} > 0$ when the ith point is sampled and 0 otherwise, which ensures that the resampled and the original point clouds are of the same size. $\mathbf{S}$ is used to compensate for non-uniform weights during resampling. $\mathbf{S}\Psi^T$ is the naive interpolation operator that reconstructs the original feature $f(\mathbf{X})$ from its resampled version $\Psi f(\mathbf{X})$, and $\mathbf{S}\Psi^T\Psi f(\mathbf{X})$ represents the preserved features after resampling in a zero-padded form. We consider the interpolation operator $\mathbf{S}\Psi^T$ for two reasons. (1) The formulation is simple and leads to a closed-form solution. (2) To ensure the robustness of the resampling operator, we simply use zero-padding to be an interpolation operator. Note that practical interpolation operators may consider reconstructing the underlying surface, instead of original 3D points (which may not be exactly known).

We next take the expectation of equation [7.14] and obtain

$$\mathbb{E}_{\Psi\sim\pi} D_{f(\mathbf{X})}(\Psi) \ = \ \operatorname{Tr}\left(f(\mathbf{X})\mathbf{Q}f(\mathbf{X})^T\right), \tag{7.15}$$

where $\mathbf{Q} \in \mathbb{R}^{N\times N}$ is a diagonal matrix with the ith element $\mathbf{Q}_{i,i} = (1-\pi)/(M\pi)$; see detailed derivations in Chen *et al.* (2018). Finally, the optimal resampling distribution π^* is

$$\pi_i^* \ \propto \ \|f_i(\mathbf{X})\|_2 = \|(\mathbf{LX})_i\|_2^2, \tag{7.16}$$

which is proportional to the local variation. The points with higher local variations have a higher probability of being selected, while points with a smaller local variations have a smaller probability of being selected. The intuition is that the response after applying high-pass graph filtering reflects the information contained in each 3D point and determines the resampling probability for each. Figure 7.7 illustrates the local variation-based downsampling of 3D point clouds. The first row shows the original point clouds; the second row shows the resampled versions, which is 10 % of points in the original point cloud.

Downsampling of 3D point clouds can also be designed based on deep neural networks. For example, simplification network (S-NET) (Dovrat *et al.* 2018) is a deep neural network-based downsampling system. It takes a 3D point cloud and produces a downsampled 3D point cloud that is optimized for a subsequent task. The architecture is similar to latentGAN, used for 3D point cloud reconstruction and generation (Achlioptas *et al.* 2018). The difference is that S-NET does not reconstruct all the 3D points, but only reconstructs a fixed number of 3D points. The loss function includes a reconstruction loss, such as the Earth mover's distance and Chamfer distance, as will be discussed in section 7.4.3, and a task-specific loss, such as classification loss. Since the reconstructed 3D point cloud is not a subset of the original 3D point cloud any more, S-NET matches each reconstructed 3D point to its nearest neighbor in the original 3D point cloud; however, it is non-trivial to apply S-NET to train and operate on large-scale 3D point clouds, which makes it less practical in autonomous driving.

7.4.2.2. *Upsampling*

As an inverse procedure of downsampling, the goal of upsampling is to generate a dense (high-resolution) 3D point cloud from a sparse (low-resolution) 3D point cloud, to describe the underlying geometry of an object or a scene (see Figure 7.6). 3D point cloud upsampling is similar in nature to the super-resolution of 2D images.

Classical 3D point cloud upsampling algorithms are based on super-resolution algorithms for 2D images. For example, (Alexa *et al.* 2003) constructs surfaces with the moving least squares algorithm and generates new points at the vertices of the

Voronoi diagram to upsample a 3D point cloud; to avoid over-smoothing, Huang *et al.* (2013) apply an anisotropic locally optimal projection operator to preserve sharp edges by pushing 3D points away from the edges, and achieves the edge-aware 3D point cloud upsampling; Wu *et al.* (2015a) combine the smoothness of surfaces and the sharpness of edges through an extracted meso-skeleton. The meso-skeleton consists of a mixture of skeletal curves and sheets to parameterize the underlying surfaces. It then generates new 3D points by jointly optimizing both the surface and 3D points residing on the meso-skeleton; however, these classical upsampling algorithms usually depend heavily on local geometry priors, such as the normal vectors and the curvatures. Some algorithms also suffer from multiscale structure preservation, due to the assumption of global smoothness (Wang *et al.* 2018b).

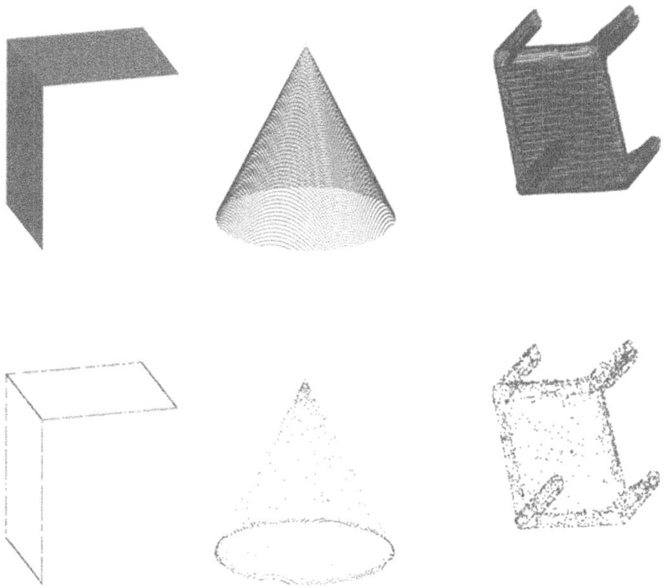

Figure 7.7. *Local variation based downsampling enhances the contour information (Chen* et al. *2018). For a color version of this figure, see www.iste.co.uk/cheung/graph.zip*

With the development of deep neural networks, most upsampling algorithms adopt the learning-based approach. PU-Net (Yu *et al.* 2018b) is the first end-to-end 3D point cloud upsampling network, which extracts multi-scale features based on PointNet++. The architecture is similar to latentGAN for 3D point cloud reconstruction, but reconstructs many more 3D points than the original 3D point cloud. The loss function includes a reconstruction loss and a repulsion loss, pushing a more uniform distribution for the generated points. Inspired by the recent success of neural-network-based image super-resolution, Wang *et al.* (2018b) proposed a patch-based progressive upsampling architecture for 3D point clouds. The multi-step upsampling strategy breaks an upsampling network into several sub-networks, where each sub-network focuses on a specific level of details. To emphasize edge preservation, EC-Net designs a novel edge-aware loss function (Yu *et al.* 2018a). During the reconstruction, EC-Net is able to attend to the sharp edges and provides more precise 3D reconstructions. Note that all those deep-neural-network-based methods are trained based on well-selected patches, which cover a rich variety of shapes.

7.4.3. *Datasets and evaluation metrics*

To evaluate the performance of the aforementioned tasks, here are several popular benchmarks, including:

– four MPEG models (*Longdress*, *Loot*, *Redandblack* and *Soldier*) from d'Eon *et al.* (2017);

– five Microsoft models (*Andrew9*, *David9*, *Phil9*, *Ricardo9* and *Sarah9*) from Loop *et al.* (2016);

– the Stanford 3D Scanning Repository (https://graphics.stanford.edu/data/3Dscanrep/) from Levoy *et al.* (1996);

– five models from the surface reconstruction *benchmark* (Berger *et al.* 2013);

– 100 models from the ShapeNetCore dataset (Chang *et al.* 2015);

– real-world noisy point cloud datasets, such as the *Face* model (Huang *et al.* 2009) and the *Iron Vise* model (Huang *et al.* 2013), which are raw scans from laser scanners.

Real holes exist in some of the datasets, such as the MPEG models and Microsoft models acquired from 3D sensors, which can be employed for inpainting. To evaluate point cloud restoration, such as denoising, with quantitative comparison, one may synthesize noisy point clouds by adding Gaussian noise to clean point clouds.

Popular evaluation metrics include the Earth mover's distance and the Chamfer distance. The Earth mover's distance is the objective function of a transportation problem, which moves one point set $\mathbf{S}$ to the other $\widehat{\mathbf{S}}$ with the lowest cost, i.e.

$$d_{\text{EMD}}(\mathbf{S}, \widehat{\mathbf{S}}) = \min_{\phi:\mathbf{S}\to\widehat{\mathbf{S}}} \sum_{\mathbf{x}\in\mathbf{S}} \|\mathbf{x} - \phi(\mathbf{x})\|_2, \quad [7.17]$$

where ϕ is a bijection. The Chamfer distance measures the total distance between each point in one set to its nearest neighbor in the other set, i.e.

$$d_{\mathrm{CH}}(\mathbf{S}, \widehat{\mathbf{S}}) = \frac{1}{N} \sum_{\mathbf{x} \in \mathbf{S}} \min_{\widehat{\mathbf{x}} \in \widehat{\mathbf{S}}} \|\mathbf{x} - \widehat{\mathbf{x}}\|_2 + \frac{1}{M} \sum_{\widehat{\mathbf{x}} \in \widehat{\mathbf{S}}} \min_{\mathbf{x} \in \mathbf{S}} \|\widehat{\mathbf{x}} - \mathbf{x}\|_2 . \qquad [7.18]$$

Both the Earth mover's distance and the Chamfer distance enforce the underlying manifold of the reconstruction to stay close to that of the original point cloud. Reconstruction with the Earth mover's distance usually outperforms that with the Chamfer distance; however, it is more efficient to compute the Chamfer distance.

Finally, a Chamfer distance is a measurement from a point to another point. When point clouds are considered as representations of a surface, a point to plane metric may be more appropriate. The point to point distance in equation [7.18] would be replaced by a projection to the associated normal vector. As proposed in Tian *et al.* (2017), the point to plane distance measures projected error vectors along normal directions rather than measuring the original error vectors directly. This imposes a larger penalty on errors that move more away from the local plane and promotes point cloud as a surface representation.

7.5. High-level point cloud understanding

In this section, we will present the benefits of GSP in high-level point cloud understanding from learning-based paradigms. We focus on representation learning methodologies of point clouds, especially in an unsupervised manner, aiming to capture the structure of the underlying manifold. The trained feature representations can adapt to downstream learning tasks on point clouds, such as classification, segmentation and generative learning. In particular, auto-encoders (AEs) are one of the most representative methods for unsupervised feature learning. Here, we introduce two types of AEs for point clouds. The first considers auto-encoding the signal (e.g. 3D coordinates) of a 3D point cloud (see Figure 7.8(a)). The second considers auto-encoding a transformation that could be applied to a 3D point cloud for transformation equivariant representation learning (see Figure 7.8(b)). For supervised point cloud learning (e.g. classification and segmentation) and generative learning on point clouds, please refer to Chapter 10. We also introduce popular benchmarks for high-level tasks of point clouds and commonly used evaluation metrics.

7.5.1. *Data auto-encoding for point clouds*

As a deep neural network framework, a deep AE is able to find compact representations of input data through self-representation. A common AE architecture consists of an encoder and a decoder. An encoder learns to compress input data from the input layer into a short code, and then a decoder decompresses that code into a reconstruction that closely matches the original data. This architecture has been generally applied in speech recognition (Lu *et al.* 2013) and image processing (Vincent *et al.* 2010). For 3D point clouds, an AE can learn from complicated geometric structures and find the corresponding compact representations without any supervision. Intuitively, an AE provides a bijective mapping between a 3D shape and a vector in the Euclidean space. In the following content, we will introduce the design of an encoder and a decoder for 3D point clouds, followed by the training of a deep AE.

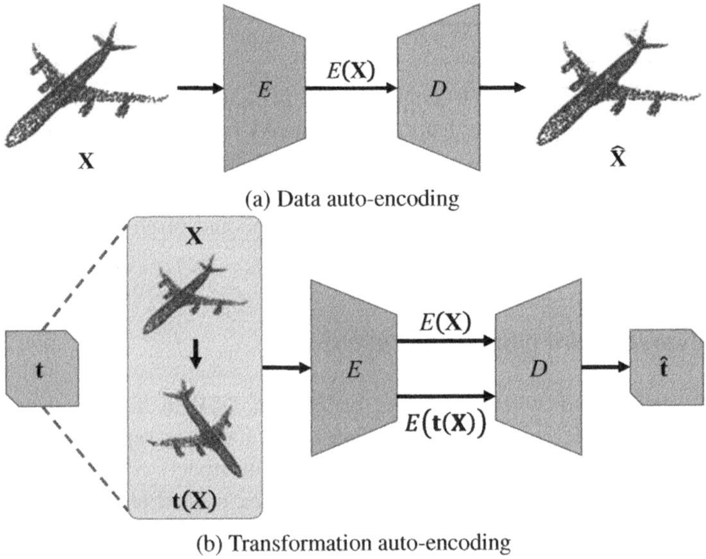

Figure 7.8. *Two auto-encoder frameworks for 3D point clouds. For a color version of this figure, see www.iste.co.uk/cheung/graph.zip*

7.5.1.1. *Encoder*

The functionality of an encoder $\Psi(\cdot)$ is to produce a low-dimensional code $\mathbf{c} \in \mathbb{R}^C$ to represent the original point cloud $\mathbf{X} \in \mathbb{R}^{N\times 3}$; that is, $\mathbf{c} = \Psi(\mathbf{X})$, where $C \ll 3N$, reflecting that $\mathbf{c}$ is a compact representation of the original point cloud.

There are many possibilities in designing an encoder. A straightforward architecture is to discretize a 3D point cloud into voxels and then use convolutional neural networks to design an encoder. The motivation of using convolutional neural networks is to leverage off-shelf deep learning tools. On the other hand, the voxel-based representation causes inevitable discretization error and is highly inefficient, as most voxels are often empty.

The pioneering work PointNet (Qi *et al.* 2017b) directly handles raw 3D points with deep neural networks without any discretization. Raw 3D point clouds are inherently unordered sets. When we permute the order of 3D points, we expect the low-dimensional code to remain the same; in other words, the compact representation of a 3D point cloud should be permutation-invariant. The key technical contribution of PointNet is to use a set of shared point-wise multi-layer perceptrons (MLPs) followed by global pooling to extract geometric features while ensuring the permutation-invariant property. Even though the architecture is simple, it has become a standard building block for numerous 3D point cloud learning algorithms.

Followed by the spirit of PointNet, we can design an encoder that includes a sequence of pointwise MLPs. For example, the first MLP maps a point from 3D space to a high-dimensional feature space. Since all 3D points share the same weights in the convolution, similar points will map to similar positions in the feature space. We next use the max-pooling to remove the point dimension, preserving global features. We finally use MLPs to map global features to codes. Mathematically, the encoder can be implemented as

$$\mathbf{h}_i = \text{MLP}^{(L_1)}\left(\mathbf{x}_i\right) \in \mathbb{R}^d, \quad \text{for } i = 1, \cdots, N, \tag{7.19a}$$

$$\mathbf{c}' = \text{maxpool}\left(\{\mathbf{h}_i\}_{i=1}^N\right) \in \mathbb{R}^d, \tag{7.19b}$$

$$\mathbf{c} = \text{MLP}^{(L_2)}\left(\mathbf{c}'\right) \in \mathbb{R}^C, \tag{7.19c}$$

where $\text{MLP}^{(\ell)}(\cdot)$ denotes ℓ layers of MLPs and $\mathbf{h}_i$ is the feature representation of the ith 3D point $\mathbf{x}_i$. Equation [7.19a] uses a cascade of MLPs to extract local features for each point; equation [7.19b] uses max-pooling to obtain global features; equation [7.19c] uses a cascade of MLPs to obtain the final code.

Intuitively, the first layer of MLPs proposes d representative geometric patterns and tests if those patterns appear around each 3D point. The max-pooling records the strongest response over all the 3D points for each pattern. Essentially, the global-feature vector $\mathbf{c}$ summarizes the activation level of d representative geometric patterns in a 3D point cloud, which can be used to recognize a 3D point cloud. Meanwhile, since each 3D point goes through the same MLPs separately and the max-pooling removes the point dimension, the entire computational block is permutation-invariant; that is, the ordering of 3D points does not influence the output of this block. To some extent, PointNet for 3D point cloud learning is similar to principal component analysis (PCA) for data analysis: it is simple, general and effective. Just like principal component analysis, PointNet extracts global features in a 3D point cloud.

PointNet treats each point individually and does not leverage the spatial relationships among 3D points, which is the characteristic of graphs. We thus can use graph-based operations to extract features from 3D point clouds. For example, we can use the edge convolution introduced in Chapter 4 to extract geometric features on a graph. The edge convolution exploits local neighborhood information and can be stacked to learn global geometric properties. A graph-based encoder can be implemented as

$$\begin{aligned} \mathbf{h}_i &= \square_{(i,j)\in\mathcal{E}}\, g(\mathbf{x}_i, \mathbf{x}_j) \in \mathbb{R}^d, \quad \text{for } i = 1, \cdots, N, \\ \mathbf{c} &= \text{maxpool}\left(\{\mathbf{h}_i\}_{i=1}^N\right) \in \mathbb{R}^d, \end{aligned}$$

where $\mathbf{h}_i \in \mathbb{R}^d$ is the feature for the ith point, $\mathcal{E}$ is the edge set, $g(\cdot,\cdot)$ is a generic mapping, implemented by MLPs and $\square$ is a generic aggregation function, which could be the summation or maximum operation. To some extent, the edge convolution extends PointNet by inputting a pair of neighboring points' features. The edge convolution is also similar to graph filtering: both aggregate neighboring information; however, the edge convolution specifically models each pairwise relationship with a nonparametric function.

7.5.1.2. *Decoder*

The functionality of a decoder $\Phi(\cdot)$ is to reconstruct a 3D point cloud that closely matches the original one; that is, the reconstructed 3D point cloud is

$$\widehat{\mathbf{X}} = \Phi(\mathbf{c}) \in \mathbb{R}^{M\times 3}, \tag{7.21}$$

which approximates the original 3D point cloud $\mathbf{X}$. Note that we do not restrict the number of points in the reconstructed 3D point cloud M to be the same as the number of points in the original 3D point cloud N.

A straightforward way to decode is to use fully connect layers that directly map a code to a 3D point cloud (Achlioptas *et al.* 2017); however, it does not explore any geometric property of 3D point clouds and requires a huge amount of training parameters. Since 3D point clouds are intrinsically sampled from 2D surfaces lying in the 3D space, it can mimic this process by considering the reconstruction as folding from a 2D surface to a 3D surface and the folding mechanism is determined by the code produced by the encoder. Since the decoding process involves folding a 2D lattice into a 3D point cloud, this decoder is called the FoldingNet (Yang *et al.* 2018).

Mathematically, let $\mathbf{Z} \in \mathbb{Z}^{M\times 2}$ be a matrix representation of nodes sampled uniformly from a fixed regular 2D lattice and the ith row vector $\mathbf{z}_i \in \mathbb{R}^2$ be the 2D coordinate of the ith node in the 2D lattice. Note that $\mathbf{Z}$ is fixed. It is used as a canonical base for the reconstruction and does not depend on the original point cloud.

The functionality of the folding module is to fold a 2D lattice into a surface in the 3D space. Since the code is trained in a data-driven manner, it preserves the details of the folding mechanism. We can thus concatenate the code with each 2D coordinate and then use MLPs to implement the folding process. Mathematically, the ith point after folding is

$$\widehat{\mathbf{x}}_i = f_{\mathbf{c}}(\mathbf{z}_i) = \text{MLP}\left(\left[\text{MLP}\left([\mathbf{z}_i, \mathbf{c}]\right), \mathbf{c}\right]\right) \in \mathbb{R}^3, \qquad [7.22]$$

where the code $\mathbf{c}$ is the output of the encoder and $[\cdot, \cdot]$ denotes the concatenation of two vectors. The folding function $f_{\mathbf{c}}(\cdot)$ consists of two-layer MLPs and the code is introduced in each layer to guide the folding process. We collect all the 3D points to form the reconstruction $\widehat{\mathbf{X}} \in \mathbb{R}^{M \times 3}$ with $\widehat{\mathbf{x}}_i$ the ith row.

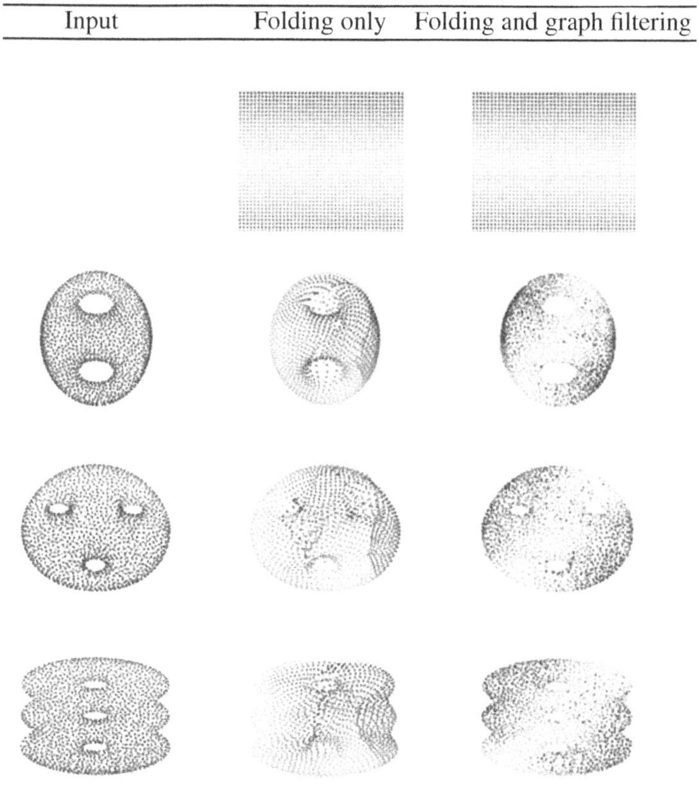

Figure 7.9. *Graph topology learning and filtering improves the reconstruction of a 3D point cloud (Chen* et al. *2019)*

The folding function $f_{\mathbf{c}}(\cdot)$ implicitly promotes smoothness; that is, when two nodes are close in the 2D lattice, their correspondence after folding is also close in the 3D space. The smoothness makes the networks easy to train; however, when the 3D surfaces have a lot of curvatures and complex shapes, the smoothness of the 2D to 3D mapping limits the representation power. Figure 7.9 shows that only training FoldingNet cannot reconstruct tori with a high-order genus. The color associated with each point indicates the correspondence between a node in the 2D lattice and a 3D point. The smoothness of the color transition reflects the difficulty of the folding process. We see that only training the FoldingNet cannot capture the holes in tori and the networks cannot find an appropriate way to fold. The reason behind this is that the folding process implies the spatial smoothness, which can hardly construct an arbitrarily complex shape.

To make the decoder more powerful, we can introduce a trainable graph structure to refine the reconstruction from the FoldingNet. This graph structure can flexibly regularize on the spatial relationships among 3D points and promotes complex 3D shapes. The learnted graph is initialized by the same 2D lattice used in the folding module. The nodes are fixed and the edges are updated during the training process. The initial graph adjacency matrix $\mathbf{W}^0 \in \mathbb{R}^{M \times M}$ is

$$\mathbf{W}^0_{ij} = \begin{cases} \frac{1}{Z_i} \exp(-\frac{\|\mathbf{z}_i - \mathbf{z}_j\|_2^2}{2\sigma^2}) & \text{if } \mathbf{z}_j \in \mathcal{N}_i, \\ 0 & \text{otherwise}, \end{cases} \qquad [7.23]$$

where $\mathbf{z}_i$ is the ith node in the lattice, the hyper-parameters σ reflect the decay rate, $\mathcal{N}_i$ represents the k-nearest neighboring nodes of node $\mathbf{z}_i$ and a normalization term $Z_i = \sum_j \exp(-\|\mathbf{z}_i - \mathbf{z}_j\|_2^2/(2\sigma^2))$ ensures that $\sum_j \mathbf{W}^0_{ij} = 1$.

Since the code produced by the encoder preserves information of the original point cloud, we concatenate the code with each row of $\mathbf{W}^0$, and then use MLPs to implement the graph topology inference. Mathematically, the ith row of the learnted graph adjacency matrix is obtained as

$$\mathbf{W}_i = g_{\mathbf{c}}(\mathbf{z}_i) = \text{softmax}\left(\text{MLP}\left(\left[\text{MLP}\left(\left[\mathbf{W}^0_i, \mathbf{c}\right]\right), \mathbf{c}\right]\right)\right) \in \mathbb{R}^M, \qquad [7.24]$$

where $\mathbf{W}_i \in \mathbb{R}^M$ is the ith row of $\mathbf{W}$ and $\mathbf{c}$ is the code produced by the encoder. The softmax operation promotes sparse connections, which can help reduce overfitting. Note that (i) the last layer of MLP uses a rectified linear unit (ReLU) as the nonlinear activation function, ensuring that all the edge weights are non-negative; (ii) the softmax ensures that the sum of each row in $\mathbf{W}$ is one, that is, $\mathbf{W}\mathbf{1} = \mathbf{1} \in \mathbb{R}^{\mathrm{M}}$. We also recognize this as a random-walk matrix (Newman 2010), whose element reflects the transition probability jumping from one node to another.

The intuition behind this is that the initial graph adjacent matrix provides initial pairwise relationships in the 2D lattice and the latent code provides the mechanism to adjust the edge weights in the graph adjacent matrix and to properly reflect the pairwise relationships. At the same time, during training, the graph topology inference module pushes the latent code to preserve the spatial relationships in the original 3D point cloud, guiding the evolvement of the graph adjacency matrix.

Based on the learnted graph adjacency matrix, we design a graph filter to filter the reconstruction from the FoldingNet and obtain the refined reconstruction. A graph filter allows each point to aggregate information from its neighbors to refine its 3D position. At the same time, the graph filtering pushes the networks to learn a graph topology that preserves smoothness on the graph for 3D points. Here, we consider an infinite impulse response graph filter based on the inversion of the graph Laplacian matrix. Let $\mathbf{L} = \widetilde{\mathbf{D}} - \widetilde{\mathbf{W}}$ be the graph Laplacian matrix, $\widetilde{\mathbf{W}} = \left(\mathbf{W} + \mathbf{W}^T\right)/2$ is the symmetric graph adjacency matrix and $\widetilde{\mathbf{D}} = \mathrm{diag}\left(\widetilde{\mathbf{W}}\right)$ is the degree matrix. We thus consider the graph filters as follows:

$$h(\mathbf{L}) = (\mu + \mathbf{L})^{-1} \in \mathbb{R}^{M \times M}, \qquad [7.25]$$

where $\mu > 0$ is a hyperparameter to avoid computational issues. The graph spectral representations of the graph filters are given as

$$h(\mathbf{L}) = \begin{bmatrix} \frac{1}{\mu} & 0 & \cdots & 0 \\ 0 & \frac{1}{\mu+\lambda_2} & \cdots & 0 \\ \vdots & \vdots & \ddots & \vdots \\ 0 & 0 & \cdots & \frac{1}{\mu+\lambda_M} \end{bmatrix}^T . \qquad [7.26]$$

We finally refine the reconstruction by applying a graph filter; that is,

$$\widehat{\mathbf{X}} \leftarrow h(\mathbf{L})\widehat{\mathbf{X}} \in \mathbb{R}^{M \times 3}. \qquad [7.27]$$

Through graph filtering, we conduct the refinement of reconstruction according to an irregular graph structure, enabling the reconstruction to capture complicated 3D shapes.

7.5.1.3. *Loss function*

To push the reconstructed 3D point cloud to match the original 3D point cloud, we aim to minimize their difference. Given a set of n 3D point clouds and a fixed code length C, the overall optimization problem is given as

$$\min_{\Psi(\cdot),\Phi(\cdot)} \sum_{i \in 1}^{n} d(\mathbf{X}^{(i)}, \widehat{\mathbf{X}}^{(i)}) \qquad [7.28]$$

$$\text{subject to } \mathbf{c}^{(i)} = \Psi\left(\mathbf{X}^{(i)}\right) \in \mathbb{R}^C, \quad \widehat{\mathbf{X}}^{(i)} = \Phi\left(\mathbf{c}^{(i)}\right),$$

where $\mathbf{X}^{(i)}$ is the ith 3D point cloud in the dataset and $d(\cdot,\cdot)$ is the distance metric that measures the difference between two point clouds. Here, we consider the augmented Chamfer distance,

$$d(\mathbf{X},\widehat{\mathbf{X}}) = \max\left\{\frac{1}{N}\sum_{i=1}^{N}\min_{j}\left\|\mathbf{x}_i - \widehat{\mathbf{x}}_j\right\|_2, \frac{1}{M}\sum_{i=1}^{M}\min_{j}\left\|\widehat{\mathbf{x}}_i - \mathbf{x}_j\right\|_2\right\}, \qquad [7.29]$$

where $\mathbf{x}_i \in \mathbb{R}^3$ is the ith row of $\mathbf{X}$. The first term $\min_{j=1,2,\cdots,M}\left\|\mathbf{x}_i - \widehat{\mathbf{x}}_j\right\|_2$ measures the ℓ_2 distance between each 3D point in the original point cloud and its correspondence in the reconstructed point cloud. The second term $\min_{j=1,2,\cdots,N}\left\|\widehat{\mathbf{x}}_i - \mathbf{x}_j\right\|_2$ measures the ℓ_2 distance between each 3D point in the reconstructed point cloud and its correspondence in the original point cloud. The maximum operation outside the bracket enforces the distance from the original point cloud to the reconstructed point cloud and the distance vice versa will be small simultaneously. This augmented Chamfer distance enforces the underlying manifold of the reconstruction to stay close to that of the original point cloud. Since we use the minimum and average operations to remove the influence from the number of points, the reconstructed 3D point cloud does not necessarily have the same number of points as in the original 3D point cloud.

We use stochastic gradient descent to solve equation [7.28]. Since we train the entire networks end-to-end, the performance of unsupervised learning depends on both the encoder and the decoder: the encoder extracts sufficient information such that the decoder is able to reconstruct; on the other hand, a decoder uses specific structures to push the encoder to extract specific information. We mainly consider the design of a decoder for now. Since we can reconstruct the original 3D point cloud, the code $\mathbf{c}$ preserves key features that describe 3D shapes of the original 3D point cloud. The code can thus be used in classification, matching and other related tasks.

7.5.2. *Transformation auto-encoding for point clouds*

Instead of reconstructing input data at the output end, which is referred to as auto-encoding data (AED) as discussed in section 7.5.1, we can learn unsupervised feature representations with auto-encoding transformations (AETs) rather than the data themselves. As demonstrated in Figure 7.8(b), by sampling some operators to transform point clouds (e.g. 3D rotation, translation and shearing), we seek to train AE that can directly reconstruct these operators from the learned feature representations between original and transformed point clouds. As long as the trained features are sufficiently informative, we can decode the transformations from the features that encode intrinsically morphable structures of point clouds well. Compared with the conventional paradigm of AED in Figure 7.8(a), AETs focus on exploring dynamics of feature representations under different transformations, thereby revealing not only static structures but also how they would change with the application of different transformations.

The paradigm of AET aims to learn transformation equivariant representations (TER), which assumes that representations equivarying to transformations are able to encode the intrinsic structures of data. Therefore, the transformations can be reconstructed from the representations before and after transformations (Qi 2019). Learning transformation equivariant representations has been advocated in Hinton's seminal work on learning transformation capsules (Hinton *et al.* 2011). Following this, a variety of approaches have been proposed to learn transformation equivariant representations (Kivinen and Williams 2011; Schmidt and Roth 2012; Sohn and Lee 2012; Skibbe 2013; Gens and Domingos 2014; Lenc and Vedaldi 2015; Dieleman *et al.* 2015, 2016; Qi *et al.* 2019; Zhang *et al.* 2019).

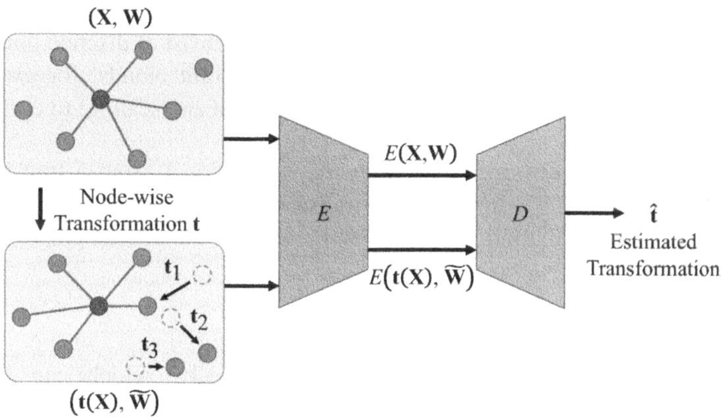

Figure 7.10. *An illustration of the GraphTER model for unsupervised feature learning (Gao* et al. *2020). For a color version of this figure, see www.iste.co.uk/cheung/graph.zip*

While most previous works focus on transformation equivariant representation learning of images that are Euclidean data, Gao *et al.* take a first step toward the extension of TER to graph data (e.g. point clouds) by formalizing Graph Transformation Equivariant Representation (GraphTER) learning via auto-encoding node-wise transformations in an unsupervised manner (Gao *et al.* 2020), as demonstrated in Figure 7.10. On the one hand, general *graph signal transformations* are defined as graph filtering on signals, where both the graph signal and graph structure could change. Then they present a graph-based AE architecture, which encodes the representations of the original and transformed graphs so that the graph signal transformations can be reconstructed from both representations. On the other hand, in contrast to the AET where global spatial transformations are applied to the *entire* input image, node-wise transformations are performed on each individual node. Representations of individual nodes are thus learned by decoding node-wise transformations to reveal the graph structures around it. This results in transformation equivariant representations to characterize the *intrinsically* morphable structures of graphs.

Next, we discuss graph signal transformation, the formulation and algorithm of GraphTER in detail.

7.5.2.1. *Graph signal transformation*

Unlike Euclidean data such as images, it is not straightforward to define transformations such as translation and rotation on graph signals. One intuitive way is to define a graph transformation on a graph signal $\mathbf{X}$ as *graph filtering* on $\mathbf{X}$. Graph filtering includes the low-pass graph filtering and high-pass filtering mentioned in sections 7.3.1 and 7.3.2. In particular, as point clouds consist of discrete points in the 3D space, it is easier to define transformations on point clouds, for example by applying affine transformations (e.g. translation, rotation and scaling) to each point.

Formally, suppose we sample a graph transformation $\mathbf{t}$ from a transformation distribution $\mathcal{T}_g$, i.e. $\mathbf{t} \sim \mathcal{T}_g$. Applying the transformation to graph signals $\mathbf{X}$ that are sampled from data distribution $\mathcal{X}_g$, i.e. $\mathbf{X} \sim \mathcal{X}_g$, leads to the filtered graph signal as

$$\tilde{\mathbf{X}} = \mathbf{t}(\mathbf{X}). \qquad [7.30]$$

The filter $\mathbf{t}$ is applied to each node individually, which can be either node-invariant or node-variant. In node-variant filtering, the transformation of each node signal associated with $\mathbf{t}$ can be different from each other. We call the graph transformation isotropic (anisotropic) if it is node-invariant (variant).

Suppose we construct the adjacency matrix as a linear or non-linear function $f(\cdot)$ applied to the graph signal $\mathbf{X}$, i.e. $\mathbf{W} = f(\mathbf{X})$. Consequently, the adjacency matrix of the transformed graph signal $\tilde{\mathbf{X}}$ equivaries:

$$\tilde{\mathbf{W}} = f(\tilde{\mathbf{X}}) = f(\mathbf{t}(\mathbf{X})), \qquad [7.31]$$

where the *graph structure* transforms implicitly, as edge weights are also filtered by $\mathbf{t}(\cdot)$.

Under this definition of graph filtering, there exists a wide spectrum of graph signal transformations. Examples include affine transformations (translation, rotation and shearing) on the location of nodes (e.g. 3D coordinates in point clouds), and graph filters such as low-pass filtering on graph signals by the adjacency matrix.

7.5.2.2. *The formulation of GraphTER*

Given a pair of graph signal and adjacency matrices $(\mathbf{X}, \mathbf{W})$, and a pair of *transformed* graph signal and adjacency matrices $(\tilde{\mathbf{X}}, \tilde{\mathbf{W}})$ by a node-wise graph transformation $\mathbf{t}$, a function $E(\cdot)$ is *transformation equivariant* if it satisfies

$$E(\tilde{\mathbf{X}}, \tilde{\mathbf{W}}) = E\left(\mathbf{t}(\mathbf{X}), f\left(\mathbf{t}(\mathbf{X})\right)\right) = \rho(\mathbf{t})\left[E(\mathbf{X}, \mathbf{W})\right], \qquad [7.32]$$

where $\rho(\mathbf{t})$ is a homomorphism of transformation $\mathbf{t}$ in the representation space.

The goal is to learn a function $E(\cdot)$, which extracts equivariant representations of graph signals $\mathbf{X}$. For this purpose, Gao *et al.* (2020) employed an encoder–decoder network: they learn a graph encoder $E : (\mathbf{X}, \mathbf{W}) \mapsto E(\mathbf{X}, \mathbf{W})$, which encodes the feature representations of individual nodes from the graph. To ensure the transformation equivariance of representations, they train a decoder $D : \left(E(\mathbf{X}, \mathbf{W}), E(\tilde{\mathbf{X}}, \tilde{\mathbf{W}})\right) \mapsto \hat{\mathbf{t}}$ to estimate the node-wise transformation $\hat{\mathbf{t}}$ from the representations of the original and transformed graph signals. Note that the node-wise transformation $\hat{\mathbf{t}}$ for point clouds could be a parametric matrix for affine transformations such as 3D translation and rotation. Hence, the learning problem of transformation equivariant representations is cast as the joint training of the representation encoder E and the transformation decoder D.

Further, instead of applying transformations to all the nodes, Gao *et al.* (2020) samples a subset of nodes $\mathbf{S}$ following a sampling distribution $\mathcal{S}_g$ from the original graph signal $\mathbf{X}$, locally or globally in order to reveal graph structures at various scales. Node-wise transformations are then performed on the subset $\mathbf{S}$ isotropically or anisotropically. In order to predict the node-wise transformation $\mathbf{t}$, a loss function $\ell_{\mathbf{S}}(\mathbf{t}, \hat{\mathbf{t}})$ can be chosen to quantify the distance between $\mathbf{t}$ and its estimate $\hat{\mathbf{t}}$ in terms of their parameters. Then the entire network is trained end-to-end by minimizing the loss

$$\min_{E,D} \underset{\mathbf{S}\sim\mathcal{S}_g}{\mathbb{E}} \underset{\substack{\mathbf{t}\sim\mathcal{T}_g \\ \mathbf{X}\sim\mathcal{X}_g}}{\mathbb{E}} \ell_{\mathbf{S}}(\mathbf{t}, \hat{\mathbf{t}}), \qquad [7.33]$$

where the expectation $\mathbb{E}$ is taken over the sampled graph signals and transformations, and the loss is taken over the (locally or globally) sampled subset $\mathbf{S}$ of nodes in each iteration of training.

In equation [7.33], the node-wise transformation $\hat{\mathbf{t}}$ is estimated from the decoder

$$\hat{\mathbf{t}} = D\left(E(\mathbf{X}, \mathbf{W}), E(\tilde{\mathbf{X}}, \tilde{\mathbf{W}})\right). \qquad [7.34]$$

Thus, the parameters in encoder E and decoder D are updated iteratively by backward propagation of the loss.

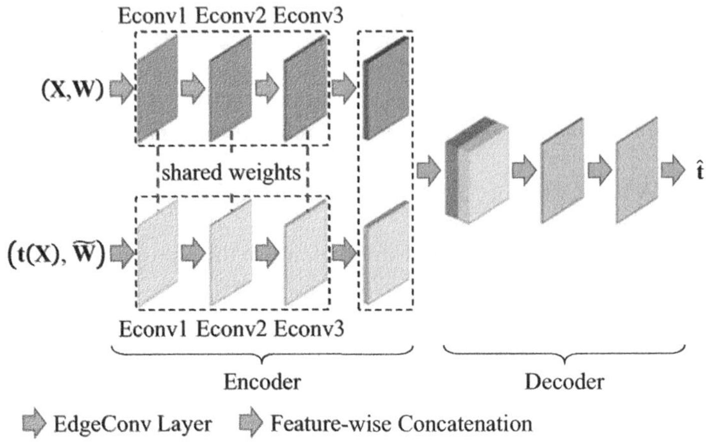

Figure 7.11. *The architecture of the unsupervised feature learning in GraphTER. The representation encoder and transformation decoder are jointly trained by minimizing equation [7.33]. For a color version of this figure, see www.iste.co.uk/cheung/graph.zip*

7.5.2.3. *The algorithm of GraphTER*

Given a point cloud $\mathbf{X}$ with 3D coordinates of each point as the feature, a subset of nodes $\mathbf{S}$ is *randomly sampled* either globally or locally in each iteration of training. Then a node-wise transformation $\mathbf{t}_i$ is drawn corresponding to each sample $\mathbf{x}_i$ of nodes in $\mathbf{S}$, either isotropically or anisotropically. Then, a kNN graph is constructed to make use of the connectivity between the nodes, whose matrix representation in $\mathbf{W}$ transforms equivariantly after applying the sampled node-wise transformations.

To learn the applied node-wise transformations, Gao *et al.* (2020) design a full graph-convolutional AE network as illustrated in Figure 7.11. Among various paradigms of GCNNs, they choose EdgeConv (Wang *et al.* 2019b) as a basic building block of the AE network, which efficiently learns node-wise representations by aggregating features of the neighborhood with the graph constructed dynamically at each layer, as discussed in section 7.3.3.

The representation encoder E takes the signals of an original point cloud $\mathbf{X}$ (i.e. 3D coordinates) and the transformed counterparts $\tilde{\mathbf{X}}$ as input, along with their corresponding graphs. E encodes node-wise features of $\mathbf{X}$ and $\tilde{\mathbf{X}}$ through a Siamese encoder network with shared weights, where multiple layers of regular edge convolutions (Wang *et al.* 2019b) are stacked to form the encoder for feature extraction.

Since the edge information of the underlying graph transforms with the transformations of individual nodes, edge convolution is able to extract higher level features from the original and transformed edge information. Also, as features of each node are learned via propagation from transformed and non-transformed nodes isotropically or anisotropically by both local or global sampling, the learned representation is able to capture intrinsic graph structures at multiple scales.

Node-wise features of the original and transformed graphs are then concatenated at each node, which are then fed into the transformation decoder. The decoder consists of several EdgeConv blocks to aggregate the representations of both the original and transformed graphs to predict the node-wise transformations $\mathbf{t}$. Based on the loss in equation [7.33], $\mathbf{t}$ is decoded by minimizing the mean squared error between the ground truth and estimated transformation parameters at each sampled node.

The general GraphTER model can be applied in classification, segmentation and other downstream tasks.

7.5.3. *Applications of GraphTER in point clouds*

The GraphTER model can be applied to graphs of point clouds on two representative tasks: point cloud classification and segmentation.

As presented in Gao *et al.* (2020), when evaluated on the ModelNet40 dataset (Wu *et al.* 2015b) for point cloud classification, the GraphTER model outperforms the state-of-the-art unsupervised methods. In particular, most of the compared unsupervised models map 3D point clouds to unsupervised representations by AED, such as MAP-VAE (Han *et al.* 2019) and L2G-AE (Liu *et al.* 2019a). Results in Table 7.1 show that the GraphTER model achieves significant improvement over these methods, showing the superiority of the auto-encoding transformation paradigm.

When applied to point cloud segmentation on the ShapeNet part dataset (Chang *et al.* 2015), the GraphTER model also significantly outperforms the state-of-the-art unsupervised method MAP-VAE (Han *et al.* 2019), as listed in Gao *et al.* (2020).

In Figure 7.12, we visualize the segmentation results of GraphTER and the state-of-the-art unsupervised method MAP-VAE (Han *et al.* 2019). The GraphTER model leads to more accurate segmentation results than MAP-VAE, especially in details such as the engines of planes and the legs of chairs.

7.5.4. *Datasets and evaluation metrics*

To evaluate the performance of high-level point cloud tasks, there are several popular benchmarks, including:

– ShapeNet part dataset from (Chang *et al.* 2015) for point cloud segmentation, which contains 16,881 shapes from 16 categories, annotated with 50 labels in total. Each 3D point cloud contains 2048 points, most of which are labeled with fewer than six parts;

– Stanford Large-Scale 3D Indoor Spaces Dataset (S3DIS) from Armeni *et al.* (2016) for point cloud segmentation, which includes 3D scan point clouds for six indoor areas including 272 rooms in total. Each point belongs to one of 13 semantic categories, e.g. board, bookcase, chair, ceiling and beam–plus clutter;

– ScanNet from Dai *et al.* (2017) for point cloud segmentation, which is an instance-level indoor RGB-D dataset that includes both 2D and 3D data;

– ModelNet40 from Wu *et al.* (2015b) for point cloud classification, which contains 12,311 meshed CAD models from 40 categories. Note that 9,843 models are used for training and 2,468 models are for testing. For each model, 1,024 points are sampled from the original mesh;

– KITTI Vision Benchmark Dataset since 2012 (Geiger *et al.* 2012), which is collected by an autonomous driving platform, Annieway. It is suitable for the following tasks: stereo, optical flow, visual odometry, 3D object detection and 3D tracking;

– WAYMO open dataset since 2019 (Sun *et al.* 2020), which is composed of high-resolution sensor data collected by Waymo self-driving cars in a wide variety of conditions.

Method	Year	Unsupervised	Accuracy
3D ShapeNets	2015	No	84.7
VoxNet	2015	No	85.9
PointNet	2017	No	89.2
PointNet++	2017	No	90.7
KD-Net	2017	No	90.6
PointCNN	2018	No	92.2
PCNN	2018	No	92.3
DGCNN	2019	No	92.9
RS-CNN	2019	No	93.6
T-L Network	2016	Yes	74.4
VConv-DAE	2016	Yes	75.5
3D-GAN	2016	Yes	83.3
LGAN	2018	Yes	85.7
MAP-VAE	2019	Yes	90.2
L2G-AE	2019	Yes	90.6
GraphTER	2020	Yes	**92.0**

Table 7.1. *Classification accuracy (%) on the ModelNet40 dataset*

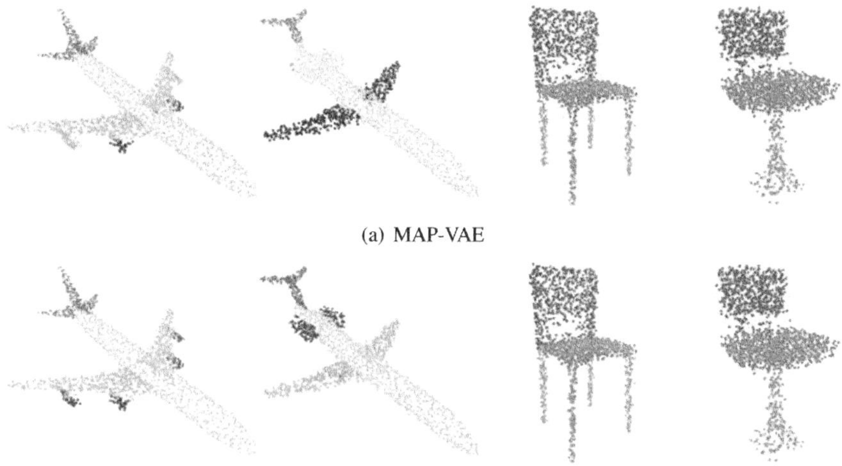

(a) MAP-VAE

(b) GraphTER

Figure 7.12. *Visual comparison (Gao* et al. *2020) of point cloud segmentation between GraphTER and MAP-VAE. For a color version of this figure, see www.iste.co.uk/cheung/graph.zip*

Different high-level tasks may employ various evaluation metrics. In unsupervised learning, popular metrics include the Earth mover's distance and the Chamfer distance, as discussed in section 7.4.3, as well as their variants. For example, an augmented Chamfer distance is considered in section 7.5.1 for data auto-encoding, while the prediction error of transformation parameters is employed as the loss in section 7.5.2 for transformation auto-encoding. In supervised learning such as point cloud classification and segmentation, the loss is often computed between ground truth labels and predicted ones. For example, point cloud segmentation is usually evaluated by the mean Intersection of Union (mIoU). IoU measures the ratio of the ground truth and prediction, while the mean IoU is the average of IoU for each label appearing in the model categories.

7.6. Summary and further reading

This chapter gives an overview of the recent progress in 3D point cloud processing via graph-based techniques. 3D points are irregularly scattered in the 3D space, which cannot be directly handled using traditional lattice-based methods. As a natural representation format, graphs bring spatial relationships for those points as well as characterize the manifold structure underlying the discrete point clouds, which is thus advantageous to point cloud processing.

We introduce two families of graph-based methodologies for point cloud processing, including GSP and geometric deep learning. Both families possess graph spectral interpretations to capture the unique characteristics of point clouds, such as piecewise smoothness of the manifold underlying a point cloud. Also, GSP enlightens the irregular kernel design and interpretation of geometric deep learning for point clouds. Finally, we present how these methods contribute in point cloud tasks, from low-level processing to high-level understanding.

Further, while most works focus on applications in static point clouds, some recent efforts exploit processing and understanding of dynamic 3D point clouds for time-varying scenarios such as autonomous vehicles, gaming and animation. Examples include dynamic point cloud denoising (Hu *et al.* 2020b) and inpainting (Fu *et al.* 2020), skeleton-based action recognition[3] (Gao *et al.* 2019; Li *et al.* 2019b, 2020) and so on. It still remains challenging to exploit the temporal correlation for processing and understanding of dynamic point clouds, because each frame is irregularly sampled with a possibly different number of points, and there is no explicit point-to-point temporal correspondence.

Open research issues in graph spectral point cloud processing are as follows:

- large-scale point cloud processing with efficient algorithms;
- processing and understanding of dynamic point clouds;
- applications in multi-modality scenarios.

7.7. References

Achlioptas, P., Diamanti, O., Mitliagkas, I., Guibas, L.J. (2017). Representation learning and adversarial generation of 3D point clouds. Paper, Stanford University, Department of Computer Science, Stanford, CA [Online]. Available at: https://www.arxiv-vanity.com/papers/1707.02392/.

Achlioptas, P., Diamanti, O., Mitliagkas, I., Guibas, L.J. (2018). Learning representations and generative models for 3D point clouds. *Proceedings of the 35th International Conference on Machine Learning, ICML 2018*, Stockholmsmässan, Stockholm, Sweden, July 10–15.

Alexa, M., Behr, J., Cohen-Or, D., Fleishman, S., Levin, D., Silva, C.T. (2003). Computing and rendering point set surfaces. *IEEE Transactions on Visualization and Computer Graphics*, 9(1), 3–15.

3 Skeleton data can be treated as sparse point clouds.

Armeni, I., Sener, O., Zamir, A.R., Jiang, H., Brilakis, I., Fischer, M., Savarese, S. (2016). 3D semantic parsing of large-scale indoor spaces. *Proceedings of the IEEE Conference on Computer Vision and Pattern Recognition*, 1534–1543.

Arthur, D. and Vassilvitskii, S. (2007). K-means++: The advantages of careful seeding. *SODA '07 Proceedings of the Eighteenth Annual ACM-SIAM Symposium on Discrete Algorithms*, New Orleans, Louisiana, 1027–1035.

Berger, M., Levine, J.A., Nonato, L.G., Taubin, G., Silva, C.T. (2013). A benchmark for surface reconstruction. *ACM Transactions on Graphics (TOG)*, 32(2), 20.

Bronstein, M.M., Bruna, J., LeCun, Y., Szlam, A., Vandergheynst, P. (2017). Geometric deep learning: Going beyond Euclidean data. *IEEE SPM*, 34(4), 18–42.

Chang, A.X., Funkhouser, T.A., Guibas, L.J., Hanrahan, P., Huang, Q., Li, Z., Savarese, S., Savva, M., Song, S., Su, H., Xiao, J., Yi, L., Yu, F. (2015). Shapenet: An information-rich 3D model repository. arXiv preprint: arXiv:1512.03012.

Chen, S., Tian, D., Feng, C., Vetro, A., Kovacevic, J. (2018). Fast resampling of three-dimensional point clouds via graphs. *IEEE Transactions on Signal Processing*, 66(3), 666–681.

Chen, S., Duan, C., Yang, Y., Li, D., Feng, C., Tian, D. (2019). Deep unsupervised learning of 3D point clouds via graph topology inference and filtering. *IEEE Transactions on Image Processing*, 29, 3183–3198.

Chen, S., Liu, B., Feng, C., Vallespi-Gonzalez, C., Wellington, C. (2021). 3D point cloud processing and learning for autonomous driving: Impacting map creation, localization, and perception. *IEEE Signal Processing Magazine*, 38(1), 68–86.

Dai, A., Chang, A.X., Savva, M., Halber, M., Funkhouser, T., Nießner, M. (2017). Scannet: Richly-annotated 3D reconstructions of indoor scenes. *Proceedings of the IEEE Conference on Computer Vision and Pattern Recognition*, 5828–5839.

Daniels, J., Ha, L.K., Ochotta, T., Silva, C.T. (2007). Robust smooth feature extraction from point clouds. *IEEE International Conference on Shape Modeling and Applications*, Lyon, France, 123–136.

Dieleman, S., Willett, K.W., Dambre, J. (2015). Rotation-invariant convolutional neural networks for galaxy morphology prediction. *Monthly Notices of the Royal Astronomical Society*, 450(2), 1441–1459.

Dieleman, S., De Fauw, J., Kavukcuoglu, K. (2016). Exploiting cyclic symmetry in convolutional neural networks. *Proceedings of the 33rd International Conference on Machine Learning (ICML)*, 1889–1898.

Dinesh, C., Cheung, G., Bajic, I.V. (2018). 3D point cloud denoising via bipartite graph approximation and reweighted graph Laplacian. arXiv preprint: arXiv:1812.07711.

Dovrat, O., Lang, I., Avidan, S. (2018). Learning to sample. arXiv preprint: arXiv:1812.01659.

Duan, C., Chen, S., Kovacevic, J. (2018). Weighted multi-projection: 3D point cloud denoising with estimated tangent planes. *2018 IEEE Global Conference on Signal and Information Processing, GlobalSIP 2018 – Proceedings*, Anaheim, CA, 725–729.

Duan, C., Chen, S., Kovacevic, J. (2019). 3D point cloud denoising via deep neural network based local surface estimation. *IEEE International Conference on Acoustics, Speech and Signal Processing, ICASSP 2019*, Brighton, United Kingdom, May 12–17, 8553–8557.

d'Eon, E., Bob Harrison, T.M., Chou, P.A. (2017). 8i voxelized full bodies, version 2 – A voxelized point cloud dataset. *ISO/IEC JTC1/SC29/WG11 m40059 ISO/IEC JTC1/SC29/WG1 M74006*, January.

Fu, Z., Hu, W., Guo, Z. (2020). 3D dynamic point cloud inpainting via temporal consistency on graphs. *IEEE International Conference on Multimedia and Expo*, London, 6–10 July.

Gao, X., Hu, W., Tang, J., Liu, J., Guo, Z. (2019). Optimized skeleton-based action recognition via sparsified graph regression. *Proceedings of the 27th ACM International Conference on Multimedia*, 601–610.

Gao, X., Hu, W., Qi, G.-J. (2020). Graphter: Unsupervised learning of graph transformation equivariant representations via auto-encoding node-wise transformations. *Proceedings of the IEEE/CVF Conferences on Computer Vision and Pattern Recognition (CVPR)*, Seattle, Washington, DC.

Gavili, A. and Zhang, X.-P. (2017). On the shift operator, graph frequency, and optimal filtering in graph signal processing. *IEEE Transactions on Signal Processing*, 65(23), 6303–6318.

Geiger, A., Lenz, P., Urtasun, R. (2012). Are we ready for autonomous driving? The KITTI vision benchmark suite. *IEEE CVPR*, 3354–3361.

Gens, R. and Domingos, P.M. (2014). Deep symmetry networks. *Advances in Neural Information Processing Systems (NIPS)*, 2537–2545.

Gomes, L., Bellon, O. R.P., Silva, L. (2014). 3D reconstruction methods for digital preservation of cultural heritage: A survey. *PRL*, 50, 3–14.

Han, Z., Wang, X., Liu, Y.-S., Zwicker, M. (2019). Multi-angle point cloud-VAE: Unsupervised feature learning for 3D point clouds from multiple angles by joint self-reconstruction and half-to-half prediction. *The IEEE International Conference on Computer Vision (ICCV)*. arXiv preprint: arXiv:1907.12704.

Hammond, D.K., Vandergheynst, P., Gribonval, R. (2011). Wavelets on graphs via spectral graph theory. *Applied and Computational Harmonic Analysis*, 30(2), 129–150.

Hein, M. (2006). Uniform convergence of adaptive graph-based regularization. *Proc. Int. Conf. Computa. Learn. Theory*, 50–64.

Hein, M., Audibert, J.-Y., Luxburg, U.V. (2007). Graph Laplacians and their convergence on random neighborhood graphs. *Journal of Machine Learning Research*, 8, 1325–1368.

Hinton, G.E., Krizhevsky, A., Wang, S.D. (2011). Transforming auto-encoders. *International Conference on Artificial Neural Networks (ICANN)*, Springer, 44–51.

Hornung, A., Wurm, K.M., Bennewitz, M., Stachniss, C., Burgard, W. (2013). Octomap: An efficient probabilistic 3D mapping framework based on octrees. *Autonomous Robots*, 34(3), 189–206.

Hu, W., Fu, Z., Guo, Z. (2019). Local frequency interpretation and non-local self-similarity on graph for point cloud inpainting. *IEEE Transactions on Image Processing*, 28(8), 4087–4100.

Hu, W., Gao, X., Cheung, G., Guo, Z. (2020a). Feature graph learning for 3D point cloud denoising. *IEEE Transactions on Signal Processing*, 68, 2841–2856.

Hu, W., Hu, Q., Zehua, W., Gao, X. (2020b). 3D dynamic point cloud denoising via spatial-temporal graph learning. arXiv preprint: arXiv:1904.12284.

Huang, H., Li, D., Zhang, H., Ascher, U., Cohen-Or, D. (2009). Consolidation of unorganized point clouds for surface reconstruction. *ACM Transactions on Graphics (TOG)*, 28, 176.

Huang, H., Wu, S., Gong, M., Cohen-Or, D., Ascher, U., Zhang, H. (2013). Edge-aware point set resampling. *ACM Transactions on Graphics (TOG)*, 32(1), 9.

Kivinen, J.J. and Williams, C.K. (2011). Transformation equivariant Boltzmann machines. *International Conference on Artificial Neural Networks (ICANN)*, Springer, 1–9.

Le Magoarou, L., Gribonval, R., Tremblay, N. (2017). Approximate fast graph Fourier transforms via multilayer sparse approximations. *IEEE Transactions on Signal and Information Processing over Networks*, 4(2), 407–420.

Lenc, K. and Vedaldi, A. (2015). Understanding image representations by measuring their equivariance and equivalence. *Proceedings of the IEEE Conference on Computer Vision and Pattern Recognition (CVPR)*, 991–999.

Levoy, H., Gerth, B.C., Pull, K. (1996). The stanford 3D scanning repository [Online]. https://graphics.stanford.edu/data/3Dscanrep/ [Accessed February 12th, 2021].

Li, G., Müller, M., Thabet, A.K., Ghanem, B. (2019a). DeepGCNs: Can GCNs go as deep as CNNs? *ICCV*, Seoul, South Korea.

Li, M., Chen, S., Chen, X., Zhang, Y., Wang, Y., Tian, Q. (2019b). Actional-structural graph convolutional networks for skeleton-based action recognition. *The IEEE Conference on Computer Vision and Pattern Recognition (CVPR)*, arXiv preprint: arXiv:1904.12659.

Li, M., Chen, S., Zhao, Y., Zhang, Y., Wang, Y., Tian, Q. (2020). Dynamic multiscale graph neural networks for 3D skeleton-based human motion prediction. *The IEEE Conference on Computer Vision and Pattern Recognition (CVPR)*, Seattle, WA, 13–19 June.

Liu, X., Han, Z., Wen, X., Liu, Y.-S., Zwicker, M. (2019a). L2g auto-encoder: Understanding point clouds by local-to-global reconstruction with hierarchical self-attention. *Proceedings of the 27th ACM International Conference on Multimedia (ACM MM)*, 989–997.

Liu, Y., Fan, B., Xiang, S., Pan, C. (2019b). Relation-shape convolutional neural network for point cloud analysis. arXiv preprint: arXiv:1904.07601.

Loop, C., Cai, Q., Escolano, S.O., Chou, P.A. (2016). Microsoft voxelized upper bodies – A voxelized point cloud dataset. Document, ISO/IEC JTC1/SC29 Joint WG11/WG1 (MPEG/JPEG) Input Document m38673/M72012, Microsoft.

Lu, X., Tsao, Y., Matsuda, S., Hori, C. (2013). Speech enhancement based on deep denoising autoencoder. *INTERSPEECH 2013, 14th Annual Conference of the International Speech Communication Association*, ISCA, Lyon, France, August 25–29, 436–440.

Mekuria, R., Blom, K., Cesar, P. (2016). Design, implementation, and evaluation of a point cloud codec for tele-immersive video. *IEEE T-CSVT*, 27(4), 828–842.

Newman, M. (2010). *Networks: An Introduction*. Oxford University Press, Oxford.

Nguyen, C.V., Izadi, S., Lovell, D. (2012). Modeling kinect sensor noise for improved 3D reconstruction and tracking. *International Conference on 3D Imaging, Modeling, Processing, Visualization and Transmission (3DIMPVT)*, 524–530.

Pan, L., Chew, C., Lee, G.H. (2019). PointAtrousGraph: Deep hierarchical encoder-decoder with atrous convolution for point clouds. arXiv preprint: arXiv:1907.09798.

Pang, J. and Cheung, G. (2017). Graph Laplacian regularization for image denoising: Analysis in the continuous domain. *IEEE Transactions on Image Processing*, 26(4), 1770–1785.

Peng, J. and Kuo, C.-C.J. (2005). Geometry-guided progressive lossless 3D mesh coding with octree (OT) decomposition. *ACM Transactions on Graphics Proceedings of ACM SIGGRAPH*, 24(3), 609–616.

Qi, G.-J. (2019). Learning generalized transformation equivariant representations via autoencoding transformations. arXiv preprint: arXiv:1906.08628.

Qi, C.R., Yi, L., Su, H., Guibas, L.J. (2017a). Pointnet++: Deep hierarchical feature learning on point sets in a metric space. *Advances in Neural Information Processing Systems 30: Annual Conference on Neural Information Processing Systems 2017*, 4–9 December, Long Beach, CA, USA. 5099–5108.

Qi, C., Su, H., Mo, K., Guibas, L.J. (2017b). Pointnet: Deep learning on point sets for 3D classification and segmentation. *Proceedings of IEEE Computer Vision and Pattern Recognition (CVPR)*, 1(2), 4.

Qi, G.-J., Zhang, L., Chen, C.W., Tian, Q. (2019). Avt: Unsupervised learning of transformation equivariant representations by autoencoding variational transformations. *Proceedings of the IEEE International Conference on Computer Vision (ICCV)*, Seoul, South Korea, 2 October–2 November.

Rosman, G., Dubrovina, A., Kimmel, R. (2013). Patch-collaborative spectral point-cloud denoising. *Computer Graphics Forum*, 32, 1–12.

Schmidt, U. and Roth, S. (2012). Learning rotation-aware features: From invariant priors to equivariant descriptors. *Proceedings of the IEEE Conference on Computer Vision and Pattern Recognition (CVPR)*, 2050–2057.

Schoenenberger, Y., Paratte, J., Vandergheynst, P. (2015). Graph-based denoising for time-varying point clouds. *3DTV-Conference: The True Vision–Capture, Transmission and Display of 3D Video (3DTV-CON)*, 1–4.

Skibbe, H. (2013). Spherical tensor algebra for biomedical image analysis, PhD thesis, Albert Ludwig University of Freiburg, Baden-Wurttemberg.

Sohn, K. and Lee, H. (2012). Learning invariant representations with local transformations. *Proceedings of the 29th International Conference on Machine Learning (ICML)*, 1339–1346.

Sun, X., Rosin, P.L., Martin, R.R., Langbein, F.C. (2008). Noise in 3D laser range scanner data. *IEEE International Conference on Shape Modeling and Applications*, 37–45.

Sun, P., Kretzschmar, H., Dotiwalla, X., Chouard, A., Patnaik, V., Tsui, P., Guo, J., Zhou, Y., Chai, Y., Caine, B., Vasudevan, V., Han, W., Ngiam, J., Zhao, H., Timofeev, A., Ettinger, S., Krivokon, M., Gao, A., Joshi, A., Zhang, Y., Shlens, J., Chen, Z., Anguelov, D. (2020). Scalability in perception for autonomous driving: Waymo open dataset. *Proceedings of the IEEE/CVF Conference on Computer Vision and Pattern Recognition*, 2446–2454.

Te, G., Hu, W., Zheng, A., Guo, Z. (2018). RGCNN: Regularized graph CNN for point cloud segmentation. *ACM MM*, 746–754.

Thomas, H., Qi, C.R., Deschaud, J., Marcotegui, B., Goulette, F., Guibas, L.J. (2019). Kpconv: Flexible and deformable convolution for point clouds. arXiv preprint: arXiv:1904.08889.

Tian, D., Ochimizu, H., Feng, C., Cohen, R., Vetro, A. (2017). Geometric distortion metrics for point cloud compression. *IEEE International Conference on Image Processing (ICIP)*, 3460–3464.

Ting, D., Huang, L., Jordan, M. (2010a). An analysis of the convergence of graph Laplacians. *Proceedings of the 27th International Conference on Machine Learning*, Haifa, Israel.

Ting, D., Huang, L., Jordan, M. (2010b). An analysis of the convergence of graph Laplacians. *International Conference on Machine Learning*, 1079–1086.

Vincent, P., Larochelle, H., Lajoie, I., Bengio, Y., Manzagol, P.-A. (2010). Stacked denoising autoencoders: Learning useful representations in a deep network with a local denoising criterion. *Journal of Machine Learning Research*. 11, 3371–3408.

Wang, C., Samari, B., Siddiqi, K. (2018a). Local spectral graph convolution for point set feature learning. *Computer Vision – ECCV 2018 – 15th European Conference: Proceedings Part IV*, Munich, Germany, September 8–14, 56–71.

Wang, Y., Wu, S., Huang, H., Cohen-Or, D., Sorkine-Hornung, O. (2018b). Patch-based progressive 3D point set upsampling. arXiv preprint: arXiv:1811.11286.

Wang, L., Huang, Y., Hou, Y., Zhang, S., Shan, J. (2019a). Graph attention convolution for point cloud segmentation. *Proceedings of the IEEE International Conference on Computer Vision Pattern Recognition*, 10296–10305.

Wang, Y., Sun, Y., Liu, Z., Sarma, S.E., Bronstein, M.M., Solomon, J.M. (2019b). Dynamic graph CNN for learning on point clouds. *ACM Transactions on Graphics (TOG)*, 38(5), 1–12.

Weber, C., Hahmann, S., Hagen, H. (2010). Sharp feature detection in point clouds. *Shape Modeling International Conference*, Aix en Provence, France, 175–186.

Wu, S., Huang, H., Gong, M., Zwicker, M., Cohen-Or, D. (2015a). Deep points consolidation. *ACM Transactions on Graphics (TOG)*, 34(6), 176.

Wu, Z., Song, S., Khosla, A., Yu, F., Zhang, L., Tang, X., Xiao, J. (2015b). 3D shapenets: A deep representation for volumetric shapes. *IEEE Conference on Computer Vision and Pattern Recognition, CVPR 2015*, Boston, MA, USA, June 7–12, 1912–1920.

Yang, Y., Feng, C., Shen, Y., Tian, D. (2018). Foldingnet: Point cloud auto-encoder via deep grid deformation. *IEEE Conference on Computer Vision and Pattern Recognition, CVPR 2018*, Salt Lake City, UT, USA, June 18–22, 206–215.

Yu, L., Li, X., Fu, C., Cohen-Or, D., Heng, P. (2018a). Ec-net: An edge-aware point set consolidation network. *Computer Vision – ECCV 2018 – 15th European Conference: Proceedings Part VII*, Munich, Germany, September 8–14, 398–414.

Yu, L., Li, X., Fu, C., Cohen-Or, D., Heng, P. (2018b). Pu-net: Point cloud upsampling network. *IEEE Conference on Computer Vision and Pattern Recognition, CVPR 2018*, Salt Lake City, UT, USA, June 18–22, 2790–2799.

Zeng, J., Cheung, G., Ng, M., Pang, J., Yang, C. (2019). 3D point cloud denoising using graph Laplacian regularization of a low dimensional manifold model. *IEEE Transactions on Image Processing*, 29, 3474–3489.

Zhang, L., Qi, G.-J., Wang, L., Luo, J. (2019). Aet vs. aed: Unsupervised representation learning by auto-encoding transformations rather than data. *Proceedings of the IEEE Conference on Computer Vision and Pattern Recognition (CVPR)*, 2547–2555.

Zhao, H., Jiang, L., Fu, C.-W., Jia, J. (2019). Pointweb: Enhancing local neighborhood features for point cloud processing. *Proceedings of the IEEE International Conference on Computer Vision and Pattern Recognition*, 5565–5573.

8

Graph Spectral Image Segmentation

Michael NG
The University of Hong Kong, China

8.1. Introduction

Image segmentation is an important and fundamental step in computer vision, image analysis and recognition (Gonzales and Woods 2018). It refers to partitioning an image into different regions, where each region has its own meaning or characteristic in the image (e.g. the same color, intensity or texture). In the literature, there are a large number of image segmentation methods, including threshold-based, edge-based, region-based and energy-based approaches; see the references in Peng *et al.* (2013).

These approaches have been applied to many image processing applications successfully, for example, in medical imaging, tracking and recognition. For image segmentation methods, the energy-based approach is to develop and study an energy function, which gives an optimum when the image is segmented into several regions, according to the objective function criteria. This approach includes several techniques, such as active contour (for example, Kass *et al.* (1988)) and graph cut (for example, Shi and Malik (2000), Boykov *et al.* (2001)). The main advantage of using graph cut is that the associated energy function can be globally optimized, whereas this may not be guaranteed in the other segmentation methods. In the graph cut segmentation, the energy function is constructed based on graphs where image pixels are mapped to graph vertices, and it can be optimized via graph-based algorithms and spectral graph theory results. By using the representation of graphs, morphological processing techniques can be applied to obtain many interesting image segmentation results. In this chapter, we focus on the concept of graph image segmentation methods.

8.2. Pixel membership functions

In this section, we discuss the constrained optimization model arising from the graph image segmentation problem. For detailed information, we refer to Ng and Yip (2010) and Law *et al.* (2012).

8.2.1. *Two-class problems*

Foreground–background segmentation has wide applications in computer vision (for example, scene analysis), computer graphics (for example, image editing) and medical imaging (for example, organ segmentation). The observed n-by-n image $\mathbf{U} = [u_{i,j}]_{i,j=1}^{n}$ is modeled as a convex combination of a foreground $\mathbf{F} = [f_{i,j}]_{i,j=1}^{n}$ and a background $\mathbf{B} = [b_{i,j}]_{i,j=1}^{n}$:

$$u_{i,j} = \alpha_{i,j} f_{i,j} + (1 - \alpha_{i,j}) b_{i,j}.$$

Here, (i, j) refers to the (i, j)th pixel location of a image, and $\alpha_{i,j}$ with $0 \leq \alpha_{i,j} \leq 1$ for each (i, j) indicates the degree of membership of each pixel to the foreground. With loss of generality, we assume the size of $\mathbf{U}$ is n-by-n, but their sizes are not necessarily the same. For simplicity, we denote

$$\mathcal{A} = \{(i, j) \ : \ 1 \leq i, j \leq n\}.$$

A foreground and a background in an image are, correspondingly, two classes. When $\alpha_{i,j} = 1$, the (i, j)th pixel is a certain foreground pixel. When $\alpha_{i,j} = 0$, the (i, j)-th pixel is a certain background pixel. In image segmentation, the problem is to estimate the membership function $[\alpha_{i,j}]_{i,j=1}^{n}$ from the given image $[u_{i,j}]_{i,j=1}^{n}$. In some practical settings, this can be done with the use of some sample pixels in which $u_{i,j} = f_{i,j}$ or $u_{i,j} = b_{i,j}$ are known and given. The sets $\mathcal{F}$ and $\mathcal{B}$ are pixels that are given and certain: foreground and background, respectively, supplied by the user. Therefore, the set of unknown membership functions is

$$\mathcal{G} = \mathcal{A} \setminus (\mathcal{F} \cup \mathcal{B}).$$

In the segmentation model, membership functions $[\alpha_{i,j}]_{i,j=1}^{n}$ are based on the similarity of both geometric and photometric neighbors. The geometric neighborhood of the (i, j)th pixel is defined as

$$\mathcal{N}_{i,j}^{g} := \{(k, \ell) \ : \ 0 < \|(i, j) - (k, \ell)\| \leq r_g, (k, \ell) \in \mathcal{G}\}$$

where $||\cdot||$ is the distance norm between two pixel locations and r_g is a positive number controlling the size of the neighborhood. For example, we often use a 3×3 or 5×5 window around a pixel as its geometric neighborhood (excluding the (i,j)th pixel itself). Such corresponds to $r_g = 1$ or $r_g = 3$ and $\cdot||$ is the vector infinity norm. For the photometric neighbor, it is based on the photometric features (for example color and texture) of neighborhood pixels. Let $\mathbf{p}_{i,j}$ be a photometric feature vector computed around the (i,j)th pixel. A natural way to define photometric neighborhood $\mathcal{N}^p_{i,j}$ of the (i,j)th pixel is those pixels whose photometric features are close to $\mathbf{p}_{i,j}$. But this may lead to a large neighborhood. Moreover, these neighbors may be computational intensive to be obtained. Instead, we can define the photometric neighborhood $\mathcal{N}^p_{i,j}$ to be those pixels that are top k photometrically closest pixels to (i,j) within a window around (i,j) (excluding (i,j) itself). The neighborhood $\mathcal{N}_{i,j}$ is defined to be $\mathcal{N}^p_{i,j} \cup \mathcal{N}^g_{i,j}$, which will be used to define membership functions. The geometric neighbor is a symmetric, i.e. $(k,\ell) \in \mathcal{N}^g_{i,j}$ if and only if $(i,j) \in \mathcal{N}^g_{k,\ell}$. However, the relationship in the photometric neighborhood is non-symmetric since it is defined based on k nearest neighbors. Now both geometric and photometric graphs on image pixels can be constructed and their edges are defined based on $\mathcal{N}^g_{i,j}$ and $\mathcal{N}^p_{i,j}$, respectively. Moreover, the similarity scores (edge weights) can be calculated based on $w^g_{(i,j),(k,\ell)}$ and $w^g_{(i,j),(k,\ell)}$ defined by

$$w^g_{(i,j),(k,\ell)} := \begin{cases} c^g_{i,j} e^{-||(i,j)-(k,\ell)||_2^2/\sigma_g^2}, & (k,\ell) \in \mathcal{N}^g_{i,j}, \\ 0, & \text{otherwise}; \end{cases} \qquad [8.1]$$

and

$$w^p_{(i,j),(k,\ell)} := \begin{cases} c^p_{i,j} e^{-||\mathbf{p}_{i,j}-\mathbf{p}_{k,\ell}||_2^2/\sigma_p^2}, & (k,\ell) \in \mathcal{N}^p_{i,j}, \\ 0, & \text{otherwise}. \end{cases} \qquad [8.2]$$

Here, σ_g is the variance related to the geometrical information within $\mathcal{N}^g_{i,j}$, σ_p is the variance related to the photometrical information within $\mathcal{N}^p_{i,j}$, and $c^g_{i,j}$ and $c^p_{i,j}$ are normalization constants so that

$$\sum_{(k,\ell)} w^g_{(i,j),(k,\ell)} + \lambda \sum_{(k,\ell)} w^p_{(i,j),(k,\ell)} = 1.$$

It is obvious that $w^g_{(i,j),(k,\ell)}$ and $w^p_{(i,j),(k,\ell)}$ measures the similarity between the geometric location and photometric features of the (i,j)th and (k,ℓ) pixels, respectively.

The basic idea of the optimization model is that the membership (to the foreground) of neighboring pixels based on geometric and photometric information should be similar. More precisely, the optimization model for the foreground–background image segmentation problem is given by

$$\min_{\alpha_{i,j}} \sum_{(i,j)\in\mathcal{G}} \left[\alpha_{i,j} - \sum_{(k,\ell)\in\mathcal{G}} (w^g_{(i,j),(k,\ell)} + \lambda w^p_{(i,j),(k,\ell)})\alpha_{k,\ell}\right]^2 \qquad [8.3]$$

subject to

$$0 \le \alpha_{i,j} \le 1, \quad (i,j) \in \mathcal{G} \qquad [8.4]$$

and the given boundary conditions:

$$\alpha_{i,j} = \begin{cases} 1, & u_{i,j} = f_{i,j} \in \mathcal{F}, \\ 0, & u_{i,j} = b_{i,j} \in \mathcal{B}. \end{cases} \qquad [8.5]$$

Here, λ is a positive number controlling the relative importance of the geometric and photometric similarity scores. Under the lexicographical ordering of pixels, the optimization problem in equation [8.3] can be expressed in the following matrix-vector notations:

$$\min_{\mathbf{0}\le\boldsymbol{\alpha}\le\mathbf{1}} \|\mathbf{D}_{\mathcal{G}}(\mathbf{I} - \mathbf{W})\boldsymbol{\alpha}\|_2^2 \qquad [8.6]$$

subject to equation [8.4] and [8.5]. Here, **0** and **1** are vectors of all entries of 0 and 1, respectively, the inequality of $\boldsymbol{\alpha}$ refers to the entry-wise inequality, $\boldsymbol{\alpha}$ is the n^2-vector containing the pixel membership functions, $\mathbf{D}_{\mathcal{G}}$ is an $|\mathcal{G}| \times n^2$ downsampling matrix from the n-by-n image domain to $\mathcal{G}$, $\mathbf{I}$ is the n^2-by-n^2 identity matrix and $\mathbf{W}$ is the n^2-by-n^2 matrix where its (x, y)th entry is given by

$$[\mathbf{W}]_{x,y} = w^g_{(i,j),(k,\ell)} + \lambda w^p_{(i,j),(k,\ell)}$$

with $x = (i-1)n + j$ and $y = (k-1)n + \ell$. We remark that the matrix $\mathbf{W}$ is non-symmetric because the relation in the photometric neighborhood is not symmetric. An example of $\mathbf{W}$ is displayed in Figure 8.1. We can see from the figure that the matrix is very sparse.

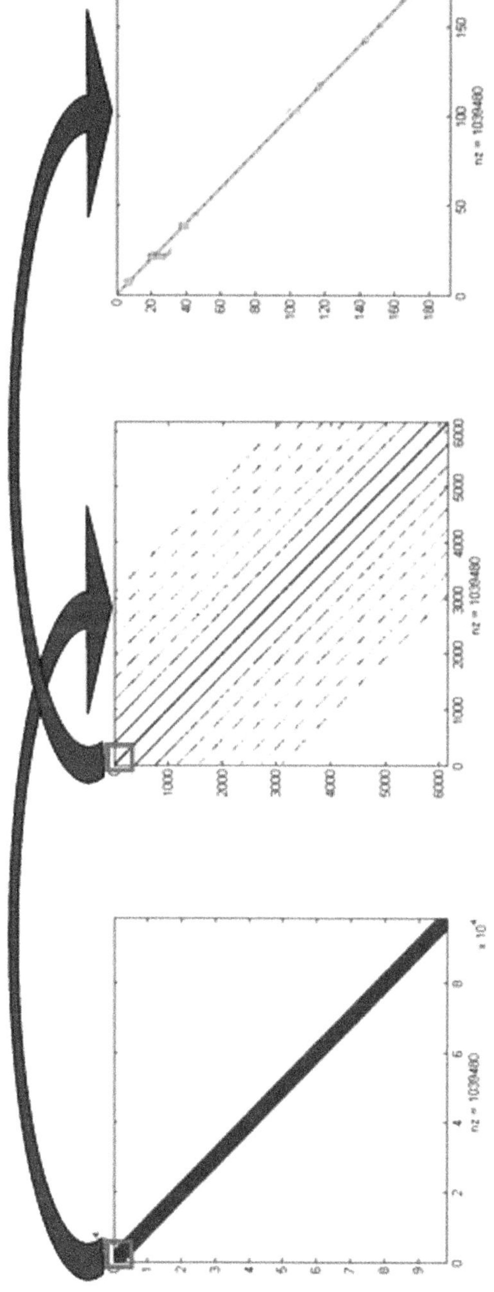

Figure 8.1. *An example of the matrix* **W** *. For a color version of this figure, see www.iste.co.uk/cheung/graph.zip*

8.2.2. *Multiple-class problems*

In this subsection, we generalize the two-class model to the case of multiple-class image segmentation. This multi-class model allows us to handle images with multiple segments. The image is now modeled as a convex combination of M images

$$U_{i,j} = \alpha_{i,j}^1 I_{i,j}^1 + \alpha_{i,j}^2 I_{i,j}^2 + \cdots + \alpha_{i,j}^M I_{i,j}^M,$$

where the M membership functions satisfy

$$0 \leq \alpha_{i,j}^m \leq 1,$$

for each m and each (i,j)th pixel, and

$$\sum_{m=1}^{M} \alpha_{i,j}^m, \quad (i,j) \in \mathcal{A},$$

for each (i,j)th pixel. In the multiple-class setting, M classes of known pixels are given, i.e.

$$\Gamma := \Gamma^1 \cup \Gamma^2 \cup \cdots \cup \Gamma^M,$$

where Γ^m is the set of pixels with certain membership to the mth class. Similar to the two-class model, $\mathcal{G}$ is the set of pixels at which the membership functions are unknown, i.e. $\mathcal{G} = \mathcal{A} \setminus \Gamma$.

The multiple-class optimization model is now given by

$$\min_{\mathbf{0} \leq \boldsymbol{\alpha}^1 \leq \mathbf{1}, \mathbf{0} \leq \boldsymbol{\alpha}^2 \leq \mathbf{1}, \cdots, \mathbf{0} \leq \boldsymbol{\alpha}^M \leq \mathbf{1}} \sum_{m=1}^{M} \|\mathbf{D}_{\mathcal{G}}(\mathbf{I} - \mathbf{W})\boldsymbol{\alpha}^m\|_2^2, \qquad [8.7]$$

subject to

$$\sum_{m=1}^{M} \boldsymbol{\alpha}^m = \mathbf{1}, \qquad [8.8]$$

and the given boundary conditions:

$$[\boldsymbol{\alpha}^m]_x = \begin{cases} 1, & (i,j) \in \Gamma^m, \\ 0, & (i,j) \in \Gamma \setminus \Gamma^m, \end{cases} \qquad [8.9]$$

with $x = (i-1)n + j$ for $m = 1, 2, \cdots, M$. Here, $\boldsymbol{\alpha}^m$ is the n^2-vector containing the pixel mth class membership functions. In Figure 8.2, we show an example of three-class image segmentation, given in Ng and Yip (2010).

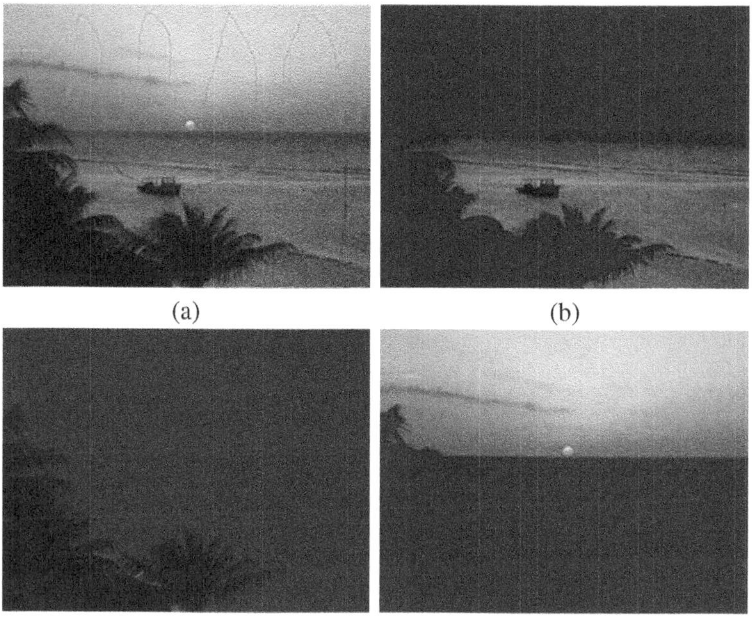

(a) (b) (c) (d)

Figure 8.2. *(a) The original 480 × 640 image with initial scribbles for three regions (blue, red and green). (b)–(d) The regions viewed against a uniform blue background, respectively. For a color version of this figure, see www.iste.co.uk/cheung/graph.zip*

8.2.3. *Multiple images*

In the last two subsections, the optimization models mentioned have been successful in segmenting single images (see Ng and Yip 2010; Law *et al.* 2012). In this subsection, we are interested in studying segmentation for multiple images. It is particularly useful to many applications where a collection of similar images is considered (see Figure 8.3). The basic idea is that some given pixels or objects in one or several images are specified once and then the segmentation models can be applied to the other images automatically. We expect that the computational time for segmentation may be reduced and human intervention can also be avoided.

For simplicity, we study two images in the following model. Let $\mathbf{U}^s = [u^s_{i,j}]^n_{i,j=1}$ be the sth given n-by-n images ($s = 1, 2$). Let $\mathcal{A}_s$ be the set of all pixels in the image $\mathbf{U}^s$. Let $\mathcal{G}_s$ be the set of all unlabeled pixels in the image $\mathbf{U}^s$. Let Γ_s be the set of pixels in the image $\mathbf{U}^s$ labeled to one of the M classes, i.e.

$$\Gamma_s = \Gamma^1_s \cup \Gamma^2_s \cup \cdots \cup \Gamma^M_s, \quad s = 1, 2.$$

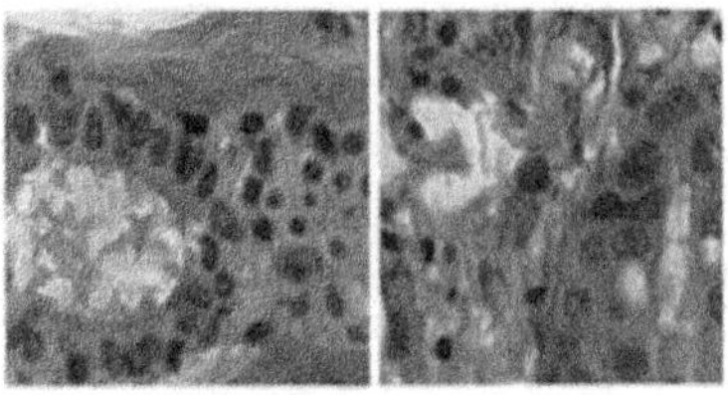

Figure 8.3. *An example of multiple images: two images of cells of benign and malignant types. For a color version of this figure, see www.iste.co.uk/cheung/graph.zip*

Note that Γ_s^m may be empty. Thus

$$\mathcal{A}_s = \mathcal{G}_s \cup \Gamma_s, \quad s = 1, 2.$$

As there are two images, we can compute similarity scores within an image and across two images. The similarity score based on geometric information is given as follows:

$$w^g_{s,(i,j),(k,\ell)} = \begin{cases} e^{-\|(i,j)-(k,\ell)\|_2^2/\sigma_g^2}, & (k,\ell) \in \mathcal{N}^g_{i,j}, \\ 0, & \text{otherwise}; \end{cases} \quad s = 1, 2.$$

Here, the geometric neighborhood $\mathcal{N}^g_{i,j}$ the (i,j)th pixel is the same as that used in section 8.2.1. Note that this similarity score is only dependent on geometric information. However, the similarity score based on photometric information is different, as we compare photometric features in the sth and the s'th images, i.e.

$$w^p_{(s,s'),(i,j),(k,\ell)} = \begin{cases} e^{-\|\mathbf{p}_{s,i,j}-\mathbf{p}_{s',k,\ell}\|_2^2/\sigma_p^2}, & (k,\ell) \in \mathcal{N}^p_{s,s',i,j}, \\ 0, & \text{otherwise}. \end{cases}$$

When $s = s'$ (the same image), the score is the same as in equation [8.2] and the photometric neighborhood of the (i,j)th pixel is the same (i.e. $\mathcal{N}^p_{s,s',i,j} = \mathcal{N}^p_{i,j}$). When $s \neq s'$ (different images), the score is calculated based on the comparison of photometric features at (i,j)th pixel of the sth image and at (k,ℓ)th pixel of the s'th image. The photometric neighborhood $\mathcal{N}^p_{s,s',i,j}$ refers to the set of pixels in other images. The combined within-image similarity score is given by

$$w^g_{s,(i,j),(k,\ell)} + \lambda_s w^p_{s,s,(i,j),(k,\ell)}, \quad s = 1, 2,$$

where λ_s is a positive number controlling the relative importance of the geometric and photometric similarity scores in the sth image. The across-image similarity score is given by

$$\sum_{s' \neq s} w^p_{s,s',(i,j),(k,\ell)}$$

and the resulting similar score is their weighted sum:

$$\sum_{(k,\ell)\in\mathcal{N}^g_{i,j}} w^g_{s,(i,j),(k,\ell)} + \lambda_s \sum_{(k,\ell)\in\mathcal{N}^p_{s,s,i,j}} w^p_{s,s,(i,j),(k,\ell)} +$$

$$\mu_{i,j} \sum_{s'\neq s} \sum_{(k,\ell)\in\mathcal{N}^p_{s,s',i,j}} w^p_{s,s',(i,j),(k,\ell)} = 1,$$

Here, $\mu_{i,j}$ is a normalization constant such that the above sum is equal to one.

The image segmentation model can be formulated by considering that the memberships of similar pixels should be similar. For each unlabeled pixel in $\mathcal{G}^s$, the membership function value to the mth class is inferred from its neighbors and is equal to the following weighted value:

$$\left[\sum_{(k,\ell)\in\mathcal{N}^g_{i,j}} w^g_{s,(i,j),(k,\ell)} + \lambda_s \sum_{(k,\ell)\in\mathcal{N}^p_{s,s,i,j}} w^p_{s,s,(i,j),(k,\ell)}\right] \alpha_{s,k,\ell} +$$

$$\left[\mu_{i,j} \sum_{s'\neq s} \sum_{(k,\ell)\in\mathcal{N}^p_{s,s',i,j}} w^p_{s,s',(i,j),(k,\ell)} \alpha_{s',k,\ell}\right].$$

The resulting optimization model is given by

$$\min_{\alpha^m_{s,i,j}} \sum_{m=1}^{M} \sum_{s=1}^{2} \sum_{(i,j)} \Bigg\| \alpha^m_{s,i,j}$$

$$- \left[\sum_{(k,\ell)\in\mathcal{N}^g_{i,j}} w^g_{s,(i,j),(k,\ell)} + \lambda_s \sum_{(k,\ell)\in\mathcal{N}^p_{s,s,i,j}} w^p_{s,s,(i,j),(k,\ell)}\right]$$

$$\alpha_{s,k,\ell} - \left[\mu_{i,j} \sum_{s'\neq s} \sum_{(k,\ell)\in\mathcal{N}^p_{s,s',i,j}} w^p_{s,s',(i,j),(k,\ell)} \alpha_{s',k,\ell}\right] \Bigg\|_2^2$$

subject to

$$0 \leq \alpha^m_{s,i,j} \leq 1, \quad \sum_{m=1}^{M} \alpha^m_{s,i,j} = 1$$

for each s and each (i,j)th pixel, and the given boundary conditions:

$$\alpha^m_{s,i,j} = \begin{cases} 1, & \Gamma^m_s, \\ 0, & \Gamma_s \setminus \Gamma^m_s, \end{cases}$$

for $m = 1, 2, \cdots, M$. In the vector notation, the kth entries of $\boldsymbol{\alpha}^m_s$ is $\alpha^m_{s,i,j}$ with $k = (i-1)n + j$. The optimization model can now be rewritten as follows:

$$\min_{0 \leq \boldsymbol{\alpha}^m_s \leq 1} \sum_{m=1}^{M} \left\| \begin{pmatrix} \mathbf{D}_{\mathcal{G}_1} & \mathbf{O} \\ \mathbf{O} & \mathbf{D}_{\mathcal{G}_2} \end{pmatrix} \begin{pmatrix} \mathbf{I} - \mathbf{W}_{1,1} & \mathbf{W}_{1,2} \\ \mathbf{W}_{2,1} & \mathbf{I} - \mathbf{W}_{2,2} \end{pmatrix} \begin{pmatrix} \boldsymbol{\alpha}^m_1 \\ \boldsymbol{\alpha}^m_2 \end{pmatrix} \right\|_2^2$$

subject to

$$\sum_{m=1}^{M} \boldsymbol{\alpha}^m_s = \mathbf{1}, \quad s = 1, 2, \qquad [8.10]$$

and the given boundary conditions:

$$[\boldsymbol{\alpha}^m_s]_x = \begin{cases} 1, & (i,j) \in \Gamma^m_s, \\ 0, & (i,j) \in \Gamma_s \setminus \Gamma^m_s, \end{cases} \qquad [8.11]$$

with $x = (i-1)n + j$ for $s = 1, 2$ and $m = 1, 2, \cdots, M$. Moreover, the (x, y)th entry of n^2-by-n^2 matrices $W_{s,s}$ and $W(s, s')$ $(s \neq s')$ are given by

$$[W_{s,s}]_{x,y} = w^g_{s,(i,j),(k,\ell)} + \lambda_s w^p_{s,s,(i,j),(k,\ell)},$$

and

$$[W_{s,s'}]_{x,y} = \mu_{i,j} w^p_{s,s',(i,j),(k,\ell)}$$

respectively, where $x = (i-1)n + j$ and $y = (k-1)n + \ell$ for $s = 1, 2$.

8.3. Matrix properties

In this section, we study the properties of $\mathbf{W}$ and the associated graph. For the sake of simplicity, we consider the optimality conditions for equation [8.6]. Let us define some notations. Let $\mathbf{A} = \mathbf{I} - \mathbf{W}$. Let $\mathbf{D}_{\mathcal{G}}$ and $\mathbf{D}_{\Gamma}$ be the downsampling matrices from the image domain to $\mathcal{G}$ and to Γ, respectively.

Let $\mathbf{A}_{\mathcal{G}\mathcal{G}} = \mathbf{D}_{\mathcal{G}} \mathbf{A} \mathbf{D}^T_{\mathcal{G}}$ and $\mathbf{A}_{\mathcal{G}\Gamma} = \mathbf{D}_{\mathcal{G}} \mathbf{A} \mathbf{D}^T_{\Gamma}$ be submatrices of $\mathbf{A}$ and let $\boldsymbol{\alpha}_{\mathbf{G}} = \mathbf{D}_{\mathcal{G}} \boldsymbol{\alpha}$ be a subvector of $\boldsymbol{\alpha}$. Here $\cdot^T$ denotes the transpose of the matrix. Other variables with subscripts $\mathcal{G}$ and/or Γ are defined similarly. Without a loss of generality, we assume that the pixels are ordered so that $\mathbf{A}$ and $\boldsymbol{\alpha}$ can be partitioned as follows:

$$\mathbf{A} = \begin{pmatrix} \mathbf{A}_{\mathcal{G}\mathcal{G}} & \mathbf{A}_{\mathcal{G}\Gamma} \\ \mathbf{A}_{\Gamma\mathcal{G}} & \mathbf{A}_{\Gamma\Gamma} \end{pmatrix} \quad \text{and} \quad \boldsymbol{\alpha} = \begin{pmatrix} \boldsymbol{\alpha}_{\mathcal{G}} \\ \boldsymbol{\alpha}_{\Gamma} \end{pmatrix}.$$

Here, we ignore the bilateral constraints $\mathbf{0} \le \boldsymbol{\alpha} \le \mathbf{1}$. We will see that an explicit imposition of such constraints is, indeed, unnecessary. The Lagrangian for equation [8.6] is now given by

$$L(\boldsymbol{\alpha}, \boldsymbol{\eta}) = \frac{1}{2}\|\mathbf{D}_{\mathcal{G}}\mathbf{A}\boldsymbol{\alpha}\|_2^2 - \langle \boldsymbol{\alpha}_{\Gamma} - \bar{\boldsymbol{\alpha}}_{\Gamma}, \boldsymbol{\eta}_{\Gamma} \rangle$$

where $\bar{\boldsymbol{\alpha}}_{\Gamma}$ are the given boundary data on Γ, and $\boldsymbol{\eta}_{\Gamma}$ are the Lagrange multipliers for the constraints $\boldsymbol{\alpha}_{\Gamma} = \bar{\boldsymbol{\alpha}}_{\Gamma}$. We have included the constant $1/2$ in the first term for convenience. The optimality condition is given by

$$\mathbf{A}^T\mathbf{D}_{\mathcal{G}}^T\mathbf{D}_{\mathcal{G}}\mathbf{A}\boldsymbol{\alpha} = \begin{pmatrix} \mathbf{0} \\ \bar{\boldsymbol{\alpha}}_{\Gamma} \end{pmatrix},$$

i.e.

$$\mathbf{A}_{\mathcal{G}\mathcal{G}}^T\boldsymbol{\alpha}_{\mathcal{G}} + \mathbf{A}_{\mathcal{G}\Gamma}\boldsymbol{\alpha}_{\Gamma} = \mathbf{0} \qquad [8.12]$$

$$\boldsymbol{\alpha}_{\Gamma} = \bar{\boldsymbol{\alpha}}_{\Gamma}. \qquad [8.13]$$

The coefficient matrix

$$\tilde{\mathbf{A}} = \begin{pmatrix} \mathbf{A}_{\mathcal{G}\mathcal{G}}^T & \mathbf{A}_{\mathcal{G}\Gamma} \\ \mathbf{O} & \mathbf{I} \end{pmatrix}$$

above is non-singular, and the linear system $\tilde{\mathbf{A}}\boldsymbol{\alpha} = \mathbf{b}$ above is solvable. It can be verified by the discrete maximum principle. We assume that the maximum is attained at an interior point $(i*, j*) \in \mathcal{G}$. Then $[\tilde{\mathbf{A}}\boldsymbol{\alpha}]_{x*} = 0$ with $x* = i*n + j*$ because $\mathbf{b}_{x*} = 0$. We find that the $x*$-th row $\tilde{\mathbf{A}}$ is the same as the $x*$-th row of $\mathbf{I} - \mathbf{W}$, i.e.

$$[\boldsymbol{\alpha}]_{x*} = \sum_{x=kn+\ell, (k,\ell)\in\mathcal{N}_{i*,j*}^g \cup \mathcal{N}_{i*,j*}^p} \left[w_{(i*,j*),(k,\ell)}^g + \lambda \sum_{(k,\ell)} w_{(i*,j*),(k,\ell)}^p \right] [\boldsymbol{\alpha}]_x.$$

It implies that the maximum value $[\boldsymbol{\alpha}]_{x*}$ is equal to the weighted average of $[\boldsymbol{\alpha}]_x$ over $\mathcal{N}_{i*,j*}^g \cup \mathcal{N}_{i*,j*}^p$. This is possible when $[\boldsymbol{\alpha}]_x = [\boldsymbol{\alpha}]_{x*}$. We can conclude that the values in $\boldsymbol{\alpha}_{\mathcal{G}}$ are the same, including the neighbor of $\mathcal{G}$ which covers Γ. Therefore,

$$0 = \min_{z=kn+\ell,(k,\ell)\in\Gamma} [\boldsymbol{\alpha}]_z \le [\boldsymbol{\alpha}]_x \le \max_{z=kn+\ell,(k,\ell)\in\Gamma} [\boldsymbol{\alpha}]_z = 1,$$

$$\forall x = in + j, (i,j) \in \mathcal{G}.$$

Since there are two classes in foreground–background image segmentation, $[\boldsymbol{\alpha}]_x$ is not constant in Γ, it follows that the bilateral constraints are strict in $\mathcal{G}$. Now it is obvious that $\tilde{\mathbf{A}}$ is non-singular.

Moreover, $\tilde{A}$ is an M-matrix. This property will be useful when we design some effective numerical methods for solving $\tilde{A}\boldsymbol{\alpha} = \mathbf{b}$ (see Ng and Yip 2010; Law *et al.* 2012). We recall that a matrix $\mathbf{M}$ is called an M-matrix if:

1) $[\mathbf{M}_{i,i} > 0$ for all i;

2) $[\mathbf{M}_{i,j} \leq 0$ for all $i \neq j$;

3) $\mathbf{M}$ is non-singular;

4) $\mathbf{M}^{-1} \geq 0$ (entrywise).

It can be shown that the third and the fourth conditions can be replaced by $\rho(\mathbf{I} - \mathbf{D}^{-1}\mathbf{M}) < 1$, where $\rho(\cdot)$ denotes the spectral radius (the largest eigenvalue in absolute value) and $\mathbf{D}$ is the diagonal part of $\mathbf{M}$ (see Saad (2003, p. 29)).

It is obvious that $\tilde{\mathbf{A}}$ satisfies the first three properties. We note that the diagonal part of $\tilde{\mathbf{A}}$ is the identity matrix. It is sufficient to consider $\mathbf{I} - \tilde{\mathbf{A}}$. Because the row sum of $\mathbf{W}$ is equal to 1, it is easy to show that $\|\mathbf{I} - \mathbf{A}\|_\infty = 1$, where $\|\cdot\|_\infty$ is the matrix infinity norm. Now we have $\rho(\mathbf{I} - \tilde{\mathbf{A}}) \leq 1$. Suppose $\mathbf{I} - \tilde{\mathbf{A}}$ has an eigenvalue λ with absolute value of 1, i.e.

$$(\mathbf{I} - \tilde{\mathbf{A}})\mathbf{z} = \lambda\mathbf{z}$$

with $\mathbf{z} \neq \mathbf{0}$, Because the entries of $(\mathbf{I} - \tilde{\mathbf{A}})$ are non-negative and the matrix row sum is equal to 1, we obtain

$$(\mathbf{I} - \tilde{\mathbf{A}})|\mathbf{z}| = |\mathbf{z}|$$

By using the discrete maximum principle, all the entries in $|\mathbf{z}|$ are the same. However, the entry of $(\mathbf{I} - \tilde{\mathbf{A}})\mathbf{z}$ at Γ is equal to zero, thus $\mathbf{z} = \mathbf{0}$. It implies that $\rho(\mathbf{I} - \tilde{\mathbf{A}}) < 1$.

We can note that the matrix properties above can be derived and shown for the cases of multiple classes and multiple images.

8.4. Graph cuts

We are given a graph $\mathcal{G} = (\mathcal{V}, \mathcal{E})$ composed of the vertex set $\mathcal{V}$ and the edge set $\mathcal{E} \subset \mathcal{V} \times \mathcal{V}$. The vertex set $\mathcal{V}$ contains the nodes of a two-dimensional or three-dimensional image pixels, together with two terminal vertices: the source vertex s and the sink vertex t. The edge set $\mathcal{E}$ contains two kinds of edges: (a) the edges $e = (i, j)$ where i and j are the image pixels, except the source and the sink vertices; and (b) the terminal edges $e_s = (s, i)$ and $e_t = (i, t)$ where i is the image pixel, except the source and the sink vertices. A cut on a graph is a partitioning of the vertices $\mathcal{V}$ into two disjoint and connected (through edges) sets $(\mathcal{V}_s, \mathcal{V}_t)$ such that $s \in \mathcal{V}_s$ and $t \in \mathcal{V}_t$. For each cut, the set of served edges $\mathcal{C}$ is defined as follows:

$$\mathcal{C}(\mathcal{V}_s, \mathcal{V}_t) = \{(i, j) \mid i \in \mathcal{V}_s, j \in \mathcal{V}_t \text{ and } (i, j) \in \mathcal{E}\}$$

We say that the graph cut uses the served edge (i, j) if (i, j) is contained in $\mathcal{C}$. Correspondingly, the cost of the cut is defined as follows:

$$\text{cost}(\mathcal{C}(\mathcal{V}_s, \mathcal{V}_t)) = \sum_{(i,j)\in\mathcal{C}(\mathcal{V}_s,\mathcal{V}_t)} w_{i,j}$$

In image segmentation, a cost function usually consists of the two terms: the region term and the boundary term (Boykov *et al.* 2001). The region term is used to give a cost function for a pixel assigned to a specific region. For example, the penalty can be referred to as the difference between the intensity value of a pixel and the intensity model of the region. This term is usually used for the cost of edges between the source/sink vertex and pixel vertices. The boundary term is used to give a cost function when two neighborhood pixels are assigned to two different regions (Shi and Malik 2000). This term is usually used for the cost of edges between neighborhood pixels. Basically, regional and edge information is used in graph cut. By incorporating shape information of the object into graph cut, image segmentation results can be improved. The main idea is to revise the region term and the boundary term in the cost function such that specific image segmentation results can be obtained. For instance, a distance function can be employed to represent some shapes for image segmentation (Kolmogorov and Zabih 2004) and surface segmentation (Boykov *et al.* 2006).

A minimum cut is the cut that has the minimum cost, also called min-cut (Ford and Fulkerson 1962). As an example, in foreground–background segmentation application, V_s contains vertices that corresponds to the foreground region in an image and V_t contains vertices that corresponds to the background region in an image. In the literature, we know that a minimum cut may favor giving regions with a small number of vertices (see, for instance, Wu and Leahy 1993; Shi and Malik 2000). To avoid such a situation for partitioning out small regions, the use of the normalized cut is proposed by Shi and Malik (2000). The cost of a cut is defined as a fraction of the total edge connections to all the vertices in the graph. In the literature, other cuts are proposed and studied for image segmentation, for instance, mean cut (Wang and Siskind 2001), ratio cut (Wang and Siskind 2003) and the ratio regions approach (Cox *et al.* 1986). In Peng *et al.* (2013), some comparisons are presented for different graph cut approaches. Recently, an exact l_1 relaxation of the Cheeger ratio cut problem for multi-class transductive learning is studied in Bresson *et al.* (2014). In general, the problem of finding a cut (min-cut, normalized cut, ratio cut, mean cut and ratio region) in an arbitrary graph is NP-hard. Definitely, efficient approximations to their solutions are required for image segmentation.

8.4.1. *The Mumford–Shah model*

Boykov and Funka-Lea (2006) showed an interesting connection between graph cuts and level sets (Sethian 1999), and discussed how combinatorial graph cut algorithms can be used for solving variational image segmentation problems, such as Mumford–Shah functionals (Mumford and Shah 1989). (Yuan *et al.* 2010) further investigated novel max-flow and min-cut models in the spatially continuous setting, and showed that the continuous max-flow models correspond to their respective continuous min-cut models as primal and dual problems.

The Mumford–Shah model is an image segmentation model with a wide range of applications in imaging sciences. Let f be the target image. We would like to seek a partition $\{\Omega_i\}_{i=1}^n$ of the image domain Ω, and an approximation image u which minimizes the functional

$$J(u, \{\Psi_i\}_{i=1}n) = \int_\Omega (u-f)^2 dx + \beta \int_{\Omega \setminus \cup_i \Psi_i} |\nabla u|^2 dx + \nu \sum_{i=1}^n \int_{\Psi_i} ds \qquad [8.14]$$

where $\{\Psi_i\}_{i=1}^n$ denotes the interphases between the regions $\{\Omega_i\}_{i=1}^n$. It is interesting to note that when u is assumed to be constant within each Ω_i, the second term in equation [8.14] disappears and the resulting functional is given as follows:

$$J(u, \{\Psi_i\}_{i=1}^n) = \int_\Omega (u-f)^2 dx + \nu \sum_{i=1}^n \int_{\Psi_i} ds, \qquad [8.15]$$

where

$$u = \sum_{i=1}^n c_i \xi_i \qquad [8.16]$$

and ξ_i is the characteristic function of Ω_i. Chan and Vese (2001) proposed using level set functions to represent the functional above and solve the resulting optimization problem via the gradient descent method. Piecewise constant level set functions are used in Lie *et al.* (2006):

$$\phi = i \quad \text{in } \Omega_i,\ 1 \leq i \leq n.$$

The relationship between the characteristic function and the level set function is given as follows:

$$\xi_i = \frac{1}{\alpha_i} \prod_{j=1, j\neq i} (\phi - j) \quad \text{with } \alpha_i = \prod_{k=1, k\neq i} (i - k).$$

The length term in equation [8.15] can be approximated by the total variation of the level set function itself. The resulting Mumford–Shah functional becomes

$$J(u, \phi) = \int_\Omega (u - f)^2 dx + \nu \int_\Omega |\nabla \phi| dx. \qquad [8.17]$$

8.4.2. *Graph cuts minimization*

Bae and Tai (2009) solved the minimization problem by graph cuts. They discretized the variational problem equation [8.17] on a grid, and the discrete energy function can be written as follows:

$$J_d(u, \phi) = \sum_i (u_i - f_i)^2 + \nu \sum_i \sum_{j\in\mathcal{N}(i)} d_{i,j} |\phi_i - \phi_j|, \qquad [8.18]$$

where i and j refer to the grid points, the weights $d_{i,j}$ are given by $d_{i,j} = \frac{1}{k\times distance(i,j)}$, $distance(i, j)$ is the distance between the two grid points i and j, and k refers to the neighborhood numbers in the discretization of different total variation forms. For fixed values of $\{c_i\}_{i=1}^n$, the minimizer of equation [8.18] can be solved by finding the minimum cut over a constructed graph. The work on graph cuts for the two regions with the Mumford–Shah model can be found in Darbon (2007) and Zehiry *et al.* (2007). For multiple regions, (Bae and Tai 2009) designed multiple layers to deal with multiple regions. We refer to Figure 8.4 for a one-dimensional example of five grid points and three regions segmentation for illustration. The graph consists of three layers referring to three regions segmentation ($n = 3$). Each layer contains five grid points as vertices (blue circles). The edges between grid points refer to their neighborhoods (blue arrowed lines). The cost of these edges is related to the total variation regularization term (or the boundary term, for the discontinuity of the two neighborhood grid points). The source vertex and the sink vertex are also constructed in the graph. The cost of the edges between the source vertex to the vertices in the top layer, and between the sink vertex to the vertices in the bottom layer, refer to the region penalty term. It was shown in Bae and Tai (2009) that, for any piecewise constant level set function ϕ taking values in $\{1, 2, \cdots, n\}$, there exists a unique admissible cut on the constructed graph. The level set function ϕ corresponds to a minimum cut in the constructed graph. After the level set function ϕ is determined, the values $\{c_i\}_{i=1}^n$ can be minimized by using the first term of equation [8.18], and they are given by

$$c_i = \frac{\sum_j f_j \xi_i(j)}{\sum_j \xi_i(j)}, \quad i = 1, 2, \cdots, n.$$

Numerical results in Darbon (2007), Zehiry *et al.* (2007) and Bae and Tai (2009) have shown that this graph cut approach for solving the Mumford–Shah segmentation model is superior in efficiency compared to the partial differential equation-based approach. Alternatively, convex approaches to segmentation with active contours have also been considered (see, for instance, Lézoray *et al.* 2012; Drakopoulos and Maragos 2012).

This graph cut approach can be used to address a class of multi-labeling problems over a spatially continuous image domain, where the data fitting term can be of any bounded function (see Ishikawa 2009; Yuan *et al.* 2010; Bae *et al.* 2014). It can also be extended to the convex relaxation of Potts' model (Potts 1952), describing a partition of the continuous domain into disjoint subdomains as the minimum of a weighted sum of data fitting and the length of the partition boundaries. Recent research development along this direction includes multi-class transductive learning based on L_1 relaxations of the Cheeger cut and the Mumford–Shah–Potts model (Bresson *et al.* 2014), and image segmentation by using the Ambrosio–Tortorelli functional and discrete calculus (Foare and Talbot 2016).

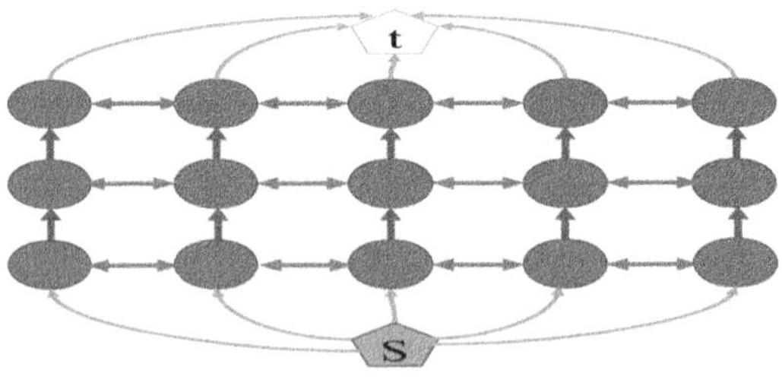

Figure 8.4. *A one-dimensional example of five grid points for three regions segmentation (blue circle: image pixel; blue arrow : an edge between two grid point vertices; brown arrow: an edge from the source vertex to a grid point vertex; green arrow: an edge from a grid point vertex to the sink vertex; red arrow: an edge from one region to another region. For a color version of this figure, see www.iste.co.uk/cheung/graph.zip*

8.5. Summary

For image segmentation methods, the energy-based approach is to develop and study an energy function that gives an optimum when the image is segmented into several regions according to the objective function criteria. The main advantage of using graph-based segmentation methods is that they can be performed and analyzed based on graph theory and associated matrix results. The membership functions model in equation [8.3] can be considered and applied to graph cuts minimization. The following objective function can be studied:

$$J_d(u, \phi) = \sum_i (u_i - f_i)^2 + \nu \sum_i \sum_{j \in \mathcal{N}(i)} w_{i,j} |\phi_i - \phi_j|, \tag{8.19}$$

where $w_{i,j}$ is the distance calculated based on geometric and photometric features in equations [8.1] and [8.2], respectively. As a future research work, it would be interesting to study such energy models for image processing applications.

8.6. References

Bae, E. and Tai, X.-C. (2009). Graph cut optimization for the piecewise constant level set method applied to multiphase image segmentation. In *Scale Space and Variational Methods in Computer Vision*, Tai, X.-C., Mørken, K., Lysaker, M., Lie, K.-A. (eds). Springer, Berlin/Heidelberg.

Bae, E., Yuan, J., Tai, X.-C., Boykov, Y. (2014). A fast continuous max-flow approach to non-convex multi-labeling problems. In *Efficient Algorithms for Global Optimization Methods in Computer Vision*, Bruhn, A., Pock, T., Tai X.-C. (eds). Springer, Berlin/Heidelberg.

Boykov, Y. and Funka-Lea, G. (2006). Graph cuts and efficient N-D image segmentation. *International Journal of Computer Vision*, 24, 109–131.

Boykov, Y., Veksler, O., Zabih, R. (2001). Fast approximate energy minimization via graph cuts. *IEEE Transactions on Pattern Analysis and Machine Intelligence*, 23, 1222–1239.

Boykov, Y., Kolmogorov, V., Cremers, D., Delong, A. (2006). An integral equation as surface evolution PDEs via geo-cuts. In *Computer Vision – ECCV 2006*, Leonardis, A., Bischof, H., Pinz, A. (eds). Springer, Berlin/Heidelberg.

Bresson, X., Tai, X.-C., Chan, T., Szlam, A. (2014). Multi-class transductive learning based on l1 relaxations of Cheeger cut and Mumford–Shah–Potts model. *Journal of Mathematical Imaging and Vision*, 49, 191–201.

Chan, T. and Vese, L. (2001). Active contours without edges. *IEEE Transactions on Image Processing*, 10, 266–277.

Cox, I., Rao, S., Zhong, Y. (1986). "Ratio regions": A technique for image segmentation. *IEEE Proceedings of the International Conference on Pattern Recognition*, 557–564.

Darbon, J. (2007). A note on the discrete binary Mumford–Shah model. In *Computer Vision/Computer Graphics Collaboration Techniques*, Gagalowicz, A., Philips, W. (eds). Springer, Berlin/Heidelberg.

Drakopoulos, K. and Maragos, P. (2012). Active contours on graphs: Multiscale morphology and graphcuts. *J. Sel. Topics Signal Processing*, 6, 780–794.

Ford, L. and Fulkerson, D. (1962). *Flows in Networks*. Princeton University Press, Princeton.

Gonzales, R. and Woods, R. (2018). *Digital Image Processing*. Pearson Prentice Hall, Upper Saddle River.

Ishikawa, H. (2009). Higher-order clique reduction in binary graph cut. *IEEE Conference on Computer Vision and Pattern Recognition*, 2993–3000.

Kass, M., Witkin, A., Terzopoulos, D. (1988). Snakes: Active contour models. *International Journal of Computer Vision*, 1(4), 321–331.

Kolmogorov, V. and Zabih, R. (2004). What energy functions can be minimized via graph cuts. *IEEE Transactions on Pattern Analysis and Machine Intelligence*, 26, 147–159.

Law, Y., Lee, H., Ng, M., Yip, A. (2012). A semi-supervised segmentation model for collections of images. *IEEE Transactions on Image Processing*, 21, 2955–2968.

Lézoray, O., Eimoataz, A., Ta, V. (2012). Nonlocal PdEs on graphs for active contours models with applications to image segmentation and data clustering. *IEEE International Conference on Acoustics, Speech, and Signal Processing*, 873–876.

Lie, J., Lysaker, M., Tai, X.-C. (2006). A variant of the level set method and applications to image segmentation. *Mathematics of Computation*, 75, 1155–1174.

Foare, J.L. and Talbot, H. (2016). Image restoration and segmentation using the Ambrosio-Tortorelli functional and discrete calculus. *IEEE Proceedings of the International Conference on Pattern Recognition*, 1418–1423.

Ng, G.Q. and Yip, A. (2010). Numerical methods for interactive multiple class image segmentation problems. *International Journal of Imaging Systems and Technology*, 20, 191–201.

Mumford, D. and Shah, J. (1989). Optimal approximations by piecewise smooth functions and associated variational problems. *Communications on Pure and Applied Mathematics*, 42, 577–685.

Peng, B., Zhang, L., Zhang, D. (2013). A survey of graph theoretical approaches to image segmentation. *Pattern Recognition*, 46, 1020–1038.

Potts, R. (1952). Some generalized order-disorder transformations. *Mathematical Proceedings of the Cambridge Philosophical Society*, 48(1), 106–109.

Saad, Y. (2003). *Iterative Methods for Sparse Linear Systems*, 2nd edition. Society for Industrial and Applied Mathematics, Philadelphia, 2003.

Sethian, J. (1999). *Level Set Methods and Fast Marching Methods*. Cambridge University Press, Cambridge.

Shi, J. and Malik, J. (2000). Normalized cuts and image segmentation. *IEEE Transactions on Pattern Analysis and Machine Intelligence*, 22, 888–905.

Wang, S. and Siskind, J. (2001). Image segementation with minimum mean cut. *IEEE International Conference on Computer Vision*, 517–524.

Wang, S. and Siskind, J. (2003). Image segmentation with ratio cut. *IEEE Transactions on Pattern Analysis and Machine Intelligence*, 25, 675–690.

Wu, Z. and Leahy, R. (1993). An optimal graph theoretic approach to data clustering. *IEEE Transactions on Pattern Analysis and Machine Intelligence*, 15, 1101–1113.

Yuan, J., Bae, E., Tai, X.-C., Boykov, Y. (2010). A continuous max-flow approach to Potts model. *Computer Vision – ECCV 2010*, Daniilidis, K., Maragos, P., Paragios, N. (eds). Springer, Berlin/Heidelberg.

Zehiry, N., Xu, S., Sahoo, P., Elmaghraby, A. (2007). Graph cut optimization for the Mumford–Shah model. In *VIIP 07: The Seventh IASTED International Conference on Visualization, Imaging and Image Processing*, Villanueva, J.J. (ed.). ACTA Press, Anaheim.

9

Graph Spectral Image Classification

Minxiang YE[1], Vladimir STANKOVIC[2], Lina STANKOVIC[2] and Gene CHEUNG[3]

[1] *Zhejiang Lab, Hangzhou, China*

[2] *University of Strathclyde, Glasgow, UK*

[3] *York University, Toronto, Canada*

The image classification problem is to categorize elements of an image dataset into two or more pre-defined classes based on inherent, detectable image features. Image classification falls within the wider data classification problem and is a fundamental image processing task, with applications including object recognition and tracking, image/video content analysis and remote sensing (Russakovsky *et al.* 2015; Jadhav *et al.* 2018; Fang *et al.* 2019).

Classification is typically posed in the context of *supervised* or *semi-supervised* learning (SSL). In supervised learning, a *training set* of labeled data (a set of data samples whose class labels are known) is inputted into an algorithm, to learn a mapping function from samples to their assigned labels. The function is subsequently used to classify new unlabeled data, called the *testing set*. On the other hand, SSL assumes that only a small sample subset, out of a large volume of data available for training, is labeled. The working assumption is that there are underlying relationships among data samples in the training set, so that information from the labeled data subset can be extrapolated to the unlabeled data for label assignment. This is in contrast to *unsupervised* learning, where only unlabeled data is provided to detect patterns or clusters in the data.

SSL is useful in scenarios where data with intrinsic inter-sample relationships is widely available, but labels are difficult or costly to obtain. Many imaging

Graph Spectral Image Processing,
coordinated by Gene CHEUNG and Enrico MAGLI.

applications fall into this category, including image retrieval (Song *et al.* 2016) and medical image classification (Liu *et al.* 2020). Moreover, a graph is a convenient abstraction to describe pairwise similarities among images (encoded as edge weights based on feature distances). Given a similarity graph, labels are computed for the unlabeled data samples via graph min-cuts (Blum and Chawla 2001), local and global consistency (Zhou *et al.* 2003) or harmonic energy minimization (Zhu *et al.* 2003). We focus on graph-based image classifiers in the SSL setting in this chapter.

We first formally define the problem of *binary classification*. Denote by $\mathcal{X}$ a dataset of N images, $\mathcal{X} = \{\mathbf{x}_1, \dots, \mathbf{x}_N\}$, where $\mathbf{x}_i \in \mathcal{R}^n$, $i = 1, \dots, N$, is the ith image containing n pixels. The task of binary classification is to learn a mapping function $f : \mathcal{R}^n \mapsto \mathcal{C}$ that maps each image $\mathbf{x}_i$ to a binary value $y_i \in \mathcal{C} = \{-1, 1\}$, called a *classification label*. Denote by $\mathbf{y} = [y_1, \dots, y_N]^\top \in \mathcal{C}^N$ a length-N vector containing labels of all N images in dataset $\mathcal{X}$. Using graph signal processing (GSP) terminologies, $\mathbf{y}$ is the target graph signal – signal on a similarity graph where the nodes are the N images – that we seek to reconstruct. The reason we focus on binary classification is as follows: reconstructing $\mathbf{y}$ as a graph signal for multi-class classification, with labels $\{1, \dots, L\}$, would imply a total order, where class l is closer to $l + 1$ in representation than class $l + 2$, which is not the typical case in multi-class classification problems. Using a binary classifier as a building block, a multi-class classifier can be built as a structured tree of binary classifiers (Sahu *et al.* 2015). However, this is outside the scope of this chapter and will not be discussed here.

Focusing on the SSL scenario, without loss of generality, denote by $\dot{\mathbf{y}} = [y_1, \dots, y_M]^\top$ a length-M vector, containing the observed labels of the first M images $\dot{\mathcal{X}} = \{\mathbf{x}_1, \dots, \mathbf{x}_M\}$ in $\mathcal{X}$. The goal of SSL image classification is thus to infer labels for the remaining $N - M$ images, given the first M labels in $\mathbf{y}$. Depending on the applications, the observed labels $\dot{\mathbf{y}}$ can be noise-corrupted (Tang *et al.* 2011; Frénay and Verleysen 2014; Wang *et al.* 2018) (e.g. when crowd-sourcing is employed for image labeling). In this case, the classification task is more complicated; instead of simply extrapolating label information from the first M images to the remaining $N - M$, the goal becomes a restoration task, where the first M labels require denoising, while the labels for the last $N - M$ images are inferred. We will discuss both the noiseless and noisy SSL image classification problems.

Often, the relevant image features – collected into a length-K *feature vector* $\mathbf{f}_i \in \mathbb{R}^K$ for each image i – can be computed *a priori* that characterize the image. Examples of image features include transform coefficients, color intensity, mean pixel values, image gradients, energy, etc. Thus, one can define a notion of *feature distance* $\delta_{i,j}$ between images i and j using the corresponding feature vectors $\mathbf{f}_i$ and $\mathbf{f}_j$. The weight $w_{i,j}$ of an edge in a graph that encodes the similarity between images i and j can thus be computed to be inversely proportional to $\delta_{i,j}$ (i.e. the larger the feature distance $\delta_{i,j}$, the smaller the edge weight $w_{i,j}$). Similarity graphs are therefore typically constructed based on feature distances. We will examine the importance of graph construction for SSL graph-based classification.

Instead of hand-crafted features, leveraging on the recent advances in deep learning (DL), one can alternatively learn the most salient features from images via state-of-the-art DL architectures, such as convolution neural networks (CNNs) (Hoffer and Ailon 2015). We will discuss DL-based image feature learning.

This chapter is organized as follows. We first discuss how SSL image classification can be mathematically formulated as optimization problems from a GSP perspective in section 9.1. In section 9.2, we discuss the critical issue of graph construction for SSL graph-based classifiers, assuming fixed and pre-defined features. In section 9.3, we discuss how DL techniques can be used to automatically learn relevant image features. Finally, we conclude in section 9.4.

9.1. Formulation of graph-based classification problems

In this section, we describe how SSL image classification can be formulated as graph signal restoration problems. We use the notations established in the Introduction.

Regularization on graphs has long been used to formulate well-posed SSL problems (Zhu *et al.* 2003; Belkin *et al.* 2004). The basic idea is to interpret the collection of classification labels $\mathbf{y}$ as a graph signal that is assumed to be "smooth" with respect to a constructed undirected similarity graph $\mathcal{G}(\mathcal{V}, \mathcal{E}, \mathbf{W})$, where each node $i \in \mathcal{V}$ represents an image $\mathbf{x}_i$. Obviously, many definitions of signal smoothness on graphs exist (Shuman *et al.* 2013; Sandryhaila and Moura 2013); we discuss them in detail in the SSL context here. The adjacency matrix $\mathbf{W}$ contains edge weights, where the typically non-negative weight $w_{i,j} \geq 0$ encodes the degree of similarity between images i and j. As discussed in Chapter 1, non-negative $\mathbf{W}$ implies that the graph Laplacian matrix $\mathbf{L}$ is a positive semi-definite (PSD), where eigenmodes of $\mathbf{L}$ can be interpreted as graph frequencies. We defer the important discussion on the proper selection of $\mathbf{W}$ to section 9.2; we assume here that graph $\mathcal{G}(\mathcal{V}, \mathcal{E}, \mathbf{W})$ is given, and we focus on the restoration of $\mathbf{y}$ using observed labels $\dot{\mathbf{y}}$.

9.1.1. *Graph spectral classifiers with noiseless labels*

We first assume an SSL scenario, where the observed labels are noiseless. We minimize the quadratic form of the graph smoothness, called the *graph Laplacian regularizer* (GLR) (see Introduction) as follows:

$$\min_{\mathbf{y}_2} \mathbf{y}^\top \mathbf{L} \mathbf{y} = \underbrace{\begin{bmatrix} \dot{\mathbf{y}}^\top & \mathbf{y}_2^\top \end{bmatrix}}_{\mathbf{y}^\top} \underbrace{\begin{bmatrix} \mathbf{L}_{1,1} & \mathbf{L}_{2,1}^\top \\ \mathbf{L}_{2,1} & \mathbf{L}_{2,2} \end{bmatrix}}_{\mathbf{L}} \underbrace{\begin{bmatrix} \dot{\mathbf{y}} \\ \mathbf{y}_2 \end{bmatrix}}_{\mathbf{y}} \qquad [9.1]$$

where the first M entries of $\mathbf{y}$, denoted by $\dot{\mathbf{y}}$, are known. The optimization problem in equation [9.1] is an unconstrained quadratic programming (QP) problem of variable $\mathbf{y}_2$ with a closed-form solution

$$\mathbf{y}_2^* = -\left(\mathbf{L}_{2,2}\right)^{-1}\mathbf{L}_{2,1}\dot{\mathbf{y}}, \tag{9.2}$$

where we assume that there exist edges from labeled nodes $1, \ldots M$ to unlabeled nodes $M+1, \ldots, N$, and $\mathbf{L}_{2,1} \neq \mathbf{0}$, $\mathbf{L}_{2,2}$ is a positive definite (PD) matrix, i.e. $\mathbf{L}_{2,2} \succ 0$, and is thus invertible.

One can prove $\mathbf{L}_{2,2}$ is PD as follows. Denote by $\mathbf{L}'$ the combinatorial graph Laplacian matrix corresponding to sub-graph $\mathcal{G}'$ of the last $N-M$ nodes. Suppose there exists just one edge from the first M nodes to the remaining $N-M$ nodes; specifically, an edge of weight $w > 0$ from nodes i to j, where $i \in \{1, \ldots, M\}$ and $j \in \{M+1, \ldots, N\}$. Then $\mathbf{L}_{2,2} = \mathbf{L}' + w\,\text{diag}(\mathbf{e}_{j-M})$, where $\mathbf{e}_k$ is a canonical vector of length $N-M$, with only the kth entry equal to 1, and $\text{diag}(\mathbf{v})$ is a diagonal matrix with vector $\mathbf{v}$ along its diagonal. We know that $\mathbf{L}'\mathbf{1} = \mathbf{0}$ and $\text{diag}(\mathbf{e}_{j-M})\mathbf{1} \neq \mathbf{0}$ for constant vector $\mathbf{1}$, which is the (unnormalized) first eigenvector of $\mathbf{L}'$. Thus, there is no vector $\mathbf{v}$ such that $\mathbf{L}'\mathbf{v} = \text{diag}(\mathbf{e}_{j-M})\mathbf{v} = \mathbf{0}$. Given that $\mathbf{L}'$ and $\text{diag}(\mathbf{e}_{j-M})$ are both PSD, we can conclude that $\mathbf{L}_{2,2} = \mathbf{L}' + w\,\text{diag}(\mathbf{e}_{j-M})$ is PD. Using Weyl's inequality, one can show that $\mathbf{L}_{2,2} \succ 0$ also holds true if there are more positive edges from the first M nodes to the remaining $N-M$ nodes.

Given $\mathbf{L}_{2,2}$ is real and symmetric, via the spectral theorem we can decompose it as $\mathbf{L}_{2,2} = \mathbf{U}\boldsymbol{\Lambda}\mathbf{U}^\top$, where $\mathbf{U}$ contains the $N-M$ eigenvectors of $\mathbf{L}_{2,2}$ as columns, and $\boldsymbol{\Lambda} = \text{diag}(\lambda_1, \ldots, \lambda_{N-M})$ is a diagonal matrix with eigenvalues λ_i of $\mathbf{L}_{2,2}$ along its diagonal. $\mathbf{L}_{2,2} \succ 0$ implies that $\lambda_i > 0, \forall i$. Solution $\mathbf{y}_2^*$ in equation [9.2] can now be expressed as:

$$\mathbf{y}_2^* = \underbrace{\left(\mathbf{U}\boldsymbol{\Lambda}^{-1}\mathbf{U}^\top\right)}_{f(\mathbf{L}_{2,2})}\underbrace{\left(-\mathbf{L}_{2,1}\dot{\mathbf{y}}\right)}_{\mathbf{z}_2} \tag{9.3}$$

We can interpret equation [9.3] as follows: after estimating a signal $\mathbf{z}_2$ from the observed labels $\dot{\mathbf{y}}$ using connectivity $-\mathbf{L}_{2,1} = \mathbf{W}_{2,1}$, we apply a *low-pass* (LP) filter $f(\mathbf{L}_{2,2})$, where $f(\mathbf{L}_{2,2})$ is a *spectral filter* of graph frequencies computed from graph operator $\mathbf{L}_{2,2}$. Specifically, we first transform input $\mathbf{z}_2$ to the frequency domain via the graph Fourier transform (GFT) $\mathbf{U}^\top$, scale each GFT coefficient i by $1/\lambda_i$, then transform the scaled GFT coefficients back to the graph signal domain via inverse GFT $\mathbf{U}$. The filter $f(\mathbf{L}_{2,2})$ is an LP filter because scalar $1/\lambda_i$ is smaller for the larger graph frequency λ_i.

Note that the obtained estimate $\mathbf{y}_2^*$ in equation [9.3] contains real numbers in general, and thus requires a conversion to binary $\mathcal{C} = \{-1, 1\}$, for example,

$\left[\mathbf{y}_2^b\right]_i = \text{sign}([\mathbf{y}_2^*]_i)$. An alternative approach is proposed by Zhu *et al.* (2003) to adjust class distribution based on a class prior. The approach assumes that it is possible to estimate the proportion of labels in each class, and then set the obtained soft values to match these proportions.

Using the same GLR, one can alternatively derive a *semi-definite programming* (SDP) formulation (Luo 2010) instead of equation [9.1]. We first formulate the following graph-based binary classification problem:

$$\min_{\mathbf{y}} \mathbf{y}^\top \mathbf{L} \mathbf{y}, \quad \text{s.t.} \quad \begin{cases} y_i^2 = 1, & \forall i \in \{1, \ldots, N\} \\ y_i = \dot{y}_i, & \forall i \in \{1, \ldots, M\} \end{cases} \qquad [9.4]$$

The first constraint is a binary constraint, requiring that $y_i \in \{-1, 1\}, \forall i$. The second constraint states that the first M entries of $\mathbf{y}$ should be $\dot{\mathbf{y}}$. The optimization problem in equation [9.4] is NP-hard due to the binary constraint.

We can relax the optimization problem to an SDP problem as follows. Suppose we define a matrix $\mathbf{Y} \in \mathbb{R}^{N \times N}$ as the outer product of $\mathbf{y}$, i.e. $\mathbf{Y} = \mathbf{y}\mathbf{y}^\top$. Define the next matrix $\mathbf{A} \in \mathbb{R}^{(N+1)\times(N+1)}$ as $\mathbf{A} = [\mathbf{Y}\ \mathbf{y}; \mathbf{y}^\top\ 1]$. Since 1 is PSD, and the Schur complement $\mathbf{A}/1 = \mathbf{Y} - \mathbf{y}\mathbf{y}^\top = 0$ is also PSD, the matrix $\mathbf{A}$ is also PSD. Hence $\mathbf{Y} = \mathbf{y}\mathbf{y}^\top$ implies $\mathbf{A} \succeq 0$; in other words, $\mathbf{A} \succeq 0$ is a necessary but not sufficient condition for $\mathbf{Y} = \mathbf{y}\mathbf{y}^\top$. We can thus formulate a relaxed version of equation [9.4]:

$$\min_{\mathbf{Y},\mathbf{y}} \text{Tr}(\mathbf{L}\mathbf{Y}) \quad \text{s.t.} \quad \begin{cases} Y_{ii} = 1, & \forall i \in \{1, \ldots, N\} \\ \begin{bmatrix} \mathbf{Y} & \mathbf{y} \\ \mathbf{y}^\top & 1 \end{bmatrix} \succeq 0 \\ y_i = \dot{y}_i, & \forall i \in \{1, \ldots, M\} \end{cases} \qquad [9.5]$$

Equation [9.5] has a linear objective and constraints *plus* a PSD cone constraint, and hence, it is an SDP problem. SDP problems can be solved in polynomial time using known methods, such as interior point methods (Boyd and Vandenberghe 2009), and with a tight performance bound (Wang *et al.* 2013). Compared to the earlier formulation in equation [9.1], however, the computation complexity of SDP is higher (Wang *et al.* 2013). The formulation in equation [9.5] was employed in Yang *et al.* (2018) as part of an alternating graph/classifier signal reconstruction strategy, similar, in spirit, to previous iterative image filtering (Milanfar 2013).

As discussed in the Introduction, instead of GLR in equation [9.1], one can interpret a "normalized" variant of adjacency matrix $\mathbf{W}^n = \mathbf{D}^{-1}\mathbf{W}$ – such that $\mathbf{W}^n$ is row-stochastic – as a graph shift operator and minimize graph shift variation (GSV) in equation [I.5], i.e.

$$\min_{\mathbf{y}_2} \|\mathbf{y} - \mathbf{W}^n\mathbf{y}\|_2^2 = \|(\mathbf{I} - \mathbf{W}^n)\mathbf{y}\|_2^2 = \left\| \begin{bmatrix} \mathbf{I} - \mathbf{W}_{1,1}^n & -\mathbf{W}_{1,2}^n \\ -\mathbf{W}_{2,1}^n & \mathbf{I} - \mathbf{W}_{2,2}^n \end{bmatrix} \begin{bmatrix} \dot{\mathbf{y}} \\ \mathbf{y}_2 \end{bmatrix} \right\|_2^2 \qquad [9.6]$$

where $\mathbf{I}$ is an appropriately sized identity matrix. Note that $\mathbf{W}^n$ is not symmetric, in general. Equation [9.6] also has a closed-form solution (Chen *et al.* 2015). Note that, unlike GLR, GSV can be used for directed graphs as well.

9.1.2. *Graph spectral classifiers with noisy labels*

Denote by $\mathbf{C} \in \{0,1\}^{M \times N}$ a 0/1 binary selection matrix that chooses the first M of N total entries in $\mathbf{y}$. Using a fidelity term $\|\dot{\mathbf{y}} - \mathbf{C}\mathbf{y}\|_2^2$ that evaluates the distance of the reconstructed signal $\mathbf{y}$ to the observed labels $\dot{\mathbf{y}}$, we can formulate an unconstrained QP problem together with GLR as follows:

$$\min_{\mathbf{y}} \|\dot{\mathbf{y}} - \mathbf{C}\mathbf{y}\|_2^2 + \gamma \mathbf{y}^\top \mathbf{L}\mathbf{y} \tag{9.7}$$

where $\gamma > 0$ is a parameter that trades off the fidelity term with GLR. Typically, γ is chosen to be a larger value if observed labels in $\dot{\mathbf{y}}$ are noisier.

Like equation [9.1], the optimization equation [9.7] also has a closed-form solution:

$$\mathbf{y}^* = \left(\mathbf{C}^\top \mathbf{C} + \gamma \mathbf{L}\right)^{-1} \mathbf{C}^\top \dot{\mathbf{y}} \tag{9.8}$$

where coefficient matrix $\mathcal{L} = \mathbf{C}^\top \mathbf{C} + \gamma \mathbf{L}$ is PD and is thus invertible. As done previously, we can decompose $\mathcal{L}$ as $\mathcal{L} = \mathbf{V}\mathbf{\Lambda}\mathbf{V}^\top$, where $\mathbf{\Lambda} = \text{diag}(\lambda_1, \ldots, \lambda_N)$, $\lambda_i > 0, \forall i$, and rewrite the solution in equation [9.8] as

$$\mathbf{y}^* = \underbrace{\left(\mathbf{V}\mathbf{\Lambda}^{-1}\mathbf{V}^\top\right)}_{f(\mathcal{L})} \underbrace{\left(\mathbf{C}^\top \dot{\mathbf{y}}\right)}_{\mathbf{z}} \tag{9.9}$$

In equation [9.9], M observed labels in $\dot{\mathbf{y}}$ are first mapped to corresponding entries in signal $\mathbf{z}$, then a graph spectral filter $f(\mathcal{L})$ is employed to low-pass filter $\mathbf{z}$. ($f(\mathcal{L})$ is again a low-pass filter, since a larger frequency component λ_i is scaled by a smaller scalar $1/\lambda_i$.) Compared to equation [9.3], where graph frequencies are computed using submatrix $\mathbf{L}_{2,2}$, frequencies, here they are computed using an operator $\mathcal{L}$ corresponding to the entire graph $\mathcal{G}$. This is reasonable, since the goal in equation [9.7] is to reconstruct labels for all N nodes, given the observed M labels are noise-corrupted.

Optimization in equation [9.7] can also be posed using GSV instead, resulting in

$$\min_{\mathbf{y}} \|\dot{\mathbf{y}} - \mathbf{C}\mathbf{y}\|_2^2 + \mu \|\mathbf{y} - \mathbf{W}^n \mathbf{y}\|_2^2, \tag{9.10}$$

where $\mu > 0$. Like equation [9.7], equation [9.10] also has a closed-form solution. For brevity, we omit the details here.

To achieve better predictive performance, one can iteratively perform graph-based regularization based on equation [9.7] or [9.10]. Particularly, having reconstructed graph signal $\mathbf{y}^*$ after the first iteration, we can refine the observed labels $\dot{\mathbf{y}}$ using a threshold τ as:

$$\dot{y}_i := \begin{cases} -1 & \text{if } y_i^* \leq -\tau \\ 1 & \text{if } y_i^* \geq \tau \\ y_i^* & \text{if } |y_i^*| < \tau \end{cases}, \qquad [9.11]$$

The refined soft labels $\dot{\mathbf{y}}$ would then be used for the next iteration until convergence (see section 4 in Zhu *et al.* (2003) and section VII.E in Cheung *et al.* (2018)).

9.2. Toward practical graph classifier implementation

In this section, we discuss how to design an appropriate similarity graph and how to define hyper-parameters. We examine the effect of the constructed graph and regularization weight parameter on classification performance, and experimentally demonstrate the potential of different graph classifiers in the presence of noisy labels.

9.2.1. *Graph construction*

As previously discussed, in graph-based SSL, a similarity graph is used to capture pairwise similarities between data samples. How to best construct the graph – how to define the notion of similarity and thus the edge weights for a particular SSL task – is not always obvious. We overview methods to construct a similarity graph connecting image samples, to maximize classification performance.

We focus again on binary semi-supervised image classification, where images need to be classified into two classes. The training dataset consists of a small number of labeled images, where class labels are known, and a large number of unlabeled images. Each image i is represented by a feature vector, $\mathbf{f}_i \in \mathbb{R}^K$, which is, in this section, assumed to be known and fixed. Feature vectors are used to compute edge weights in a similarity graph $\mathcal{G}(\mathcal{V}, \mathcal{E}, \mathbf{W})$, where node set $\mathcal{V}$ corresponds to images in the dataset, $\mathcal{E}$ is the set of edges connecting the nodes and matrix $\mathbf{W}$ defines the weights of the edges, i.e. $w_{i,j}$ is the weight of an edge connecting nodes i and j. To compute $w_{i,j}$, a *feature distance*, $\delta_{i,j}$, between the corresponding feature vectors $\mathbf{f}_i$ and $\mathbf{f}_j$, needs to be defined first.

The performance of an SSL graph classifier is strongly influenced by the constructed graph (de Sousa *et al.* 2013). It is common in many image processing tasks to define the feature distance $\delta_{i,j}$ between nodes i and j as the *Mahalanobis distance*, using the associated feature vectors $\mathbf{f}_i$ and $\mathbf{f}_j$, i.e.

$$\delta_{i,j} = (\mathbf{f}_i - \mathbf{f}_j)^\top \mathbf{M}(\mathbf{f}_i - \mathbf{f}_j) \qquad [9.12]$$

where $\mathbf{M} \in \mathbb{R}^{K \times K}$ is a PD metric matrix (Mahalanobis 1936). This implies $\delta_{i,j} > 0$ unless $\mathbf{f}_i = \mathbf{f}_j$, when $\delta_{i,j} = 0$. $\mathbf{M}$ enables the unequal weighting of features, achieving a form of feature selection. Subsequently, the edge weight $w_{i,j}$ is computed using a Gaussian kernel, i.e.

$$w_{i,j} = \exp\left\{-\delta_{i,j}\right\}. \qquad [9.13]$$

The popular *bilateral filter* (Tomasi and Manduchi 1998) in image filtering, discussed in equation [I.1] in section I.2, is one concrete example of equation [9.13], where the features are pixel locations and intensities, and metric $\mathbf{M}$ is a diagonal matrix $\text{diag}(1/\sigma_l^2, 1/\sigma_x^2)$.

In Yang *et al.* (2018), the off-diagonal elements of metric $\mathbf{M}$ were set to zero, assuming feature independence, and the diagonal elements $\mathbf{m} = [M_{11} \dots M_{KK}]^\top$ were optimized via alternating feature metric and classifier learning. Specifically, GLR was first used to perform semi-supervised binary classification (for fixed $\mathbf{M}$), as in equation [9.5], to obtain label signal $\tilde{\mathbf{y}}$. Then, the metric $\mathbf{M}$ is optimized using the same GLR-based smoothness objective, but with $\tilde{\mathbf{y}}$ fixed, and an additional constraint that the sum of non-negative $\mathbf{m}$ must sum to a total no larger than C, i.e.

$$\min_{\mathbf{m}} \tilde{\mathbf{y}}^\top \mathbf{L} \tilde{\mathbf{y}} = \min_{\mathbf{m}} \sum_{i,j} w_{ij} (\tilde{y}_i - \tilde{y}_j)^2, \quad \text{s.t..} \begin{cases} m_k \in [0, C], \ \forall k \\ \mathbf{1}^\top \mathbf{m} \leq C \end{cases} \qquad [9.14]$$

The convex optimization problem in equation [9.14] was solved in Yang *et al.* (2018) via an iterative proximal gradient descent algorithm. The two steps, classifier learning via equation [9.5] and metric learning, i.e. optimizing the diagonal entries in $\mathbf{M}$, via equation [9.14], are solved alternately until convergence.

Setting off-diagonal terms in $\mathbf{M}$ to zero has benefits in optimization complexity reduction, but can lead to performance degradation in some cases. For example, in Hu *et al.* (2020), the 3D Point Cloud denoising problem is considered, and both diagonal and off-diagonal terms in $\mathbf{M}$ were optimized alternately, in a similar proximal gradient procedure. In any case, armed with feature vectors $\mathbf{f}_i$ and metric $\mathbf{M}$, edge weights $w_{i,j}$ can be computed via equations [9.12] and [9.13]. Toward computation efficiency, a sparse graph is typically more desirable than a fully connected graph. For example, to solve equation [9.1], instead of computing the matrix inverse $\mathbf{L}_{2,2}^{-1}$ in equation [9.2], one can instead solve the system of linear equations for $\mathbf{y}_2^*$,

$$\mathbf{L}_{2,2} \, \mathbf{y}_2^* = -\mathbf{L}_{2,1} \dot{\mathbf{y}}, \qquad [9.15]$$

using fast numerical linear algebra methods such as *conjugate gradient* (CG) (Moller 1990), provided that the coefficient matrix $\mathbf{L}_{2,2}$ is symmetric, PD and sparse.

To construct a sparse graph, various graph *sparsification* methods are possible. A ϵ-neighborhood graph is obtained by setting all weights that are below a pre-determined threshold $\epsilon > 0$ to zero. This approach is very sensitive to the selection of ϵ, and can easily result in a disconnected graph (if ϵ is too large), or an almost fully connected graph (if ϵ is too small) (Wu *et al.* 2018). A more popular approach is a k-NN graph, where an edge $e_{i,j}$ is only kept if node j is one of the k closest neighbors (in terms of distance $\delta_{i,j}$) to node i. Thus, one can convert a fully connected graph into a k-NN graph, in a greedy way, by considering one vertex at the time, and by only keeping k highest weights for each vertex.

In general, the above approach will result in a graph whose vertices have degrees larger than or equal to k (instead of all degrees being equal to k). Thus, sparsification via b-matching was proposed (see Wang *et al.* (2013) and references therein), resulting in a regular k-NN graph, by replacing the greedy k-NN graph construction described above with an optimization procedure that adds as a constraint that the number of neighbors of each vertex must be exactly k. This optimization problem can be solved in polynomial time (Wang *et al.* 2013). A faster implementation is proposed by Huang and Jebara (2007) based on the Loopy Belief Propagation.

Note that the above graph construction methods fall in the general domain of *graph learning*, that is, inferring a graph topology that describes the underlying structure of the data. The main difference between the graph learning described in Chapter 2 and the methods commonly used to construct graphs for SSL image classification tasks is as follows. In the former case, multiple data observations generated from the same statistical model are available to compute pairwise correlations, which are used along with some prior data knowledge (e.g. sparsity, graph smoothness) to learn an appropriate graph. In the latter case, the multiple observations of the label signals on the same graph to compute pairwise correlations do not exist, and thus, one must rely on distance metrics to define edge weights between data samples.

9.2.2. *Experimental setup and analysis*

We next empirically analyze the effects of hyper-parameters on the performance of graph classifiers. First, we analyze how the size of the training set and graph sparsification parameter k affect performance in the noiseless label case. Then, we discuss the effect of regularization parameters γ and μ, introduced in section 9.1.2.

In all examples presented in this section, we use two classes from the ***MNIST*** dataset – handwritten digits "zero" and "one" and from the ***CIFAR10*** dataset – "airplane" and "ship". For the ***MNIST*** dataset, we construct a training set of 10,660 images, a validation set of 1186 images and a testing set of 2115 images. For the ***CIFAR10*** dataset, we construct a training set of 9,000 images, a validation set of 1000 images and a testing set of 2000 images. Note that we do stratified sampling for the training and validation sets, resulting in the same number of images for both classes. For each classifier, we randomly choose M labeled images from the training set and tested on the whole testing set in each experiment. To reduce the computational complexity, we split the testing set into batches, where each batch consists of $N - M$ images. For each batch, we construct a graph connecting M labeled images and $N - M$ testing images. Unless stated otherwise, we set M to be 7% of the total number of samples in the training set, and randomly choose 50 testing images per batch (i.e. $N = M + 50$).

The underlying N-node graph $\mathcal{G}(\mathcal{V}, \mathcal{E}, \mathbf{W})$ for the ***MNIST*** dataset is constructed, similarly to Sandryhaila and Moura (2014), by viewing each image as a point in a 28^2=784-dimensional feature space, with zero-mean normalization. For the ***CIFAR10*** dataset, following the study by Kornblith *et al.* (2019), we use a Residual Network with 152 layers (ResNet-152 pre-trained in ImageNet dataset) to extract 2,048 features.

We use a classification error rate, defined as the number of incorrectly classified samples divided by the total number of samples in the testing set, to evaluate the predictive performance of different classifiers in at least 50 experiments for each setting.

Classifiers based on the regularizations on graphs introduced in the previous section are used in all experiments, namely (1) classic k **nearest neighbor (k-NN)** classifier based on a directed k-NN graph, (2) **GLR** based on equation [9.3], (3) **GLR normlap** – the same as GLR, but using the normalized Laplacian matrix in equation [9.3] instead of a combinatorial Laplacian, (4) **GLR fidelity** based on equation [9.8], (5) **GLR normlap fidelity** – the same as GLR fidelity but using the normalized Laplacian matrix in equation [9.8], (6) **GSV** based on equation [9.6], (7) **GSV fidelity** based on equation [9.10], (8) **GLR fidelity iter** – the same GLR fidelity scheme but two iterations are performed based on equation [9.11], (9) **GLR normlap fidelity iter** – GLR normlap fidelity with two iterations, (10) **GSV fidelity iter** – GSV fidelity with two iterations.

All methods construct a non-symmetric k-NN graph using the greedy k-NN construction method explained in the previous subsection (see also Wu *et al.* 2018 and Wang *et al.* 2013). The weights are calculated based on the feature metric given in equation [9.12], with matrix $\mathbf{M}$ set to $\mathbf{I}$. As a result, a Boolean edge matrix $\mathbf{B}$ is constructed with element $b_{i,j}$ being 1 if nodes i and j are connected, and zero otherwise. The resulting graph is a directed graph, where each node i has exactly k outgoing edges corresponding to the positions of 1's in the i row of matrix $\mathbf{B}$. This directed graph is used by GSV-based classifiers. To construct an undirected graph (for the GLR and norm GLR classifiers), we symmetrize the Boolean edge matrix by performing an element-wise logic "or" operation between $\mathbf{B}$ and $\mathbf{B}^\top$.

The k-NN classifier, used for benchmarking, is based on the same directed graph used by the GSV classifier. The ith test sample is labeled by majority voting based on the class membership of all of its neighbours.

9.2.2.1. *Noiseless labels*

In this subsection, we analyze the case of clean labels. First, we look at designing the underlying sparse graph by setting the parameter k. Recall that, after calculating the edge weights $w_{i,j}$ using equation [9.13], only the k most significant edges are kept, leading to a potentially sparse k-NN graph, improving the smoothing performance.

Figure 9.1 shows the classification error rate as a function of the node degree k in the k-NN graph. We observe a significantly lower classification error rate obtained by constructing a sparse graph (small k), especially for the k-NN classifier. Moreover, GSV consistently outperforms GLR regardless of k, revealing a better underlying graph structure using a directed graph. However, by adopting the normalized graph Laplacian matrix in equation [9.3], the GLR normlap outperforms both, for all values of k. Further, the GLR normlap is much less sensitive to the choice of k than other classifiers.

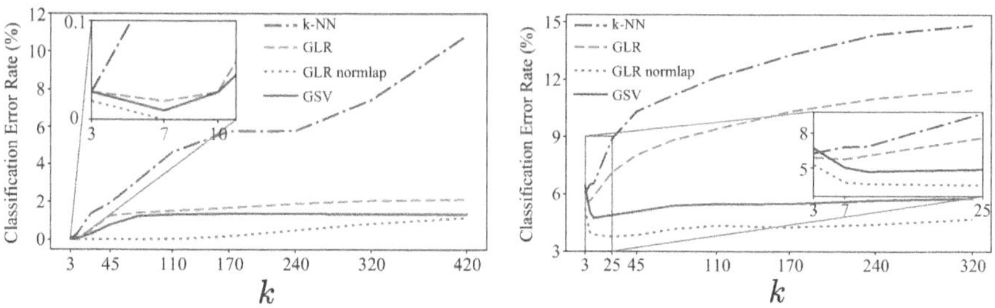

(a) Images of handwritten digits 0 and 1

(b) Images of airplane and ship

Figure 9.1. *Classification error rate (%) as a function of node degree k for the two datasets. For a color version of this figure, see www.iste.co.uk/cheung/graph.zip*

Recall that, graph regularization classifiers construct a similarity graph based purely on distance metrics, i.e. based on the similarity between individual image samples. Due to this, the graph regularization classifiers are expected to perform well, even when the training dataset is small. To illustrate this, in Figure 9.2, we show the classification error rate as a function of the labeling ratio, i.e. the ratio between M and the total size of the training set, to demonstrate the robustness of the graph-based classifiers to insufficient training samples.

We can observe that by increasing labeling ratio, the GSP-based classifiers achieve better predictive performance than the k-NN classifier. The GSV-based and

GSV-fidelity classifier leads to a relatively lower classification error rate than GLR and GLR fidelity, with better regularization performance on the directed graph. Using the normalized Laplacian matrix, the GLR normlap and GLR normlap fidelity significantly outperform others under a very low labeling ratio for both datasets. A detailed explanation for the superiority of the GLR normlap can be found in Johnson and Zhang (2007), and is beyond the scope of this book.

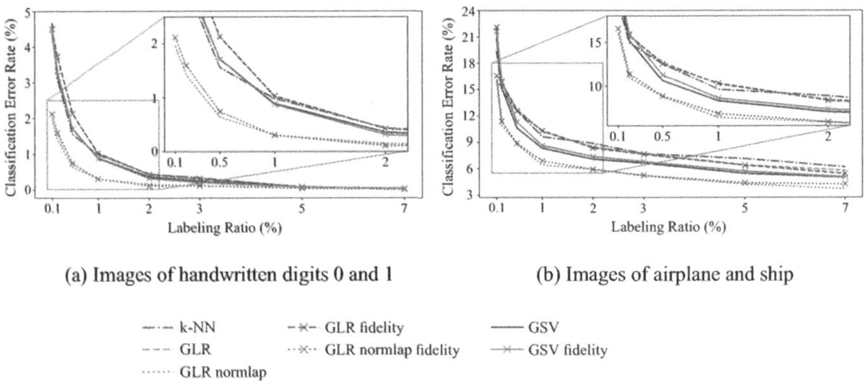

Figure 9.2. *Classification error rate (%) as a function of labeling ratio for the two datasets. For a color version of this figure, see www.iste.co.uk/cheung/graph.zip*

We can also observe that methods with the fidelity term constantly perform worse than the ones without the fidelity term. This is especially true when training samples become sufficient, as the predictive performance of the fidelity term-based classifiers drops quickly. Based on these findings, it is best to avoid using the fidelity term when training labels are clean, which resonates with the conclusions from the previous section.

9.2.2.2. *Noisy labels*

If noise is present in the labels, graph spectral classifiers with the fidelity term, introduced in section 9.1.2, may be more effective, in which case, based on the level of expected noise, one can trade-off the fidelity term and the regularization term differently, by putting more weight on the regularization term as the level of noise increases.

Figure 9.3 shows the classification error rate as a function of the parameter γ in equation [9.8], which trades off the fidelity and smoothness term. The results are given under uniform label noise levels of 0%, 10%, 20%, 30% and 40%.

One can see that there is an optimal value of γ for each method and each level of noise. As expected, this optimal γ to better regularize the prediction, increases with the level of label noise. We further observe that the GLR normlap achieves a low classification error rate for a small value of trade-off parameter γ, especially under a relatively higher noise level. Large γ leads to a performance drop for the GLR normlap due to over-smoothing the signal. However, one can also observe that $\gamma = 20$ achieves high performance across different noise levels.

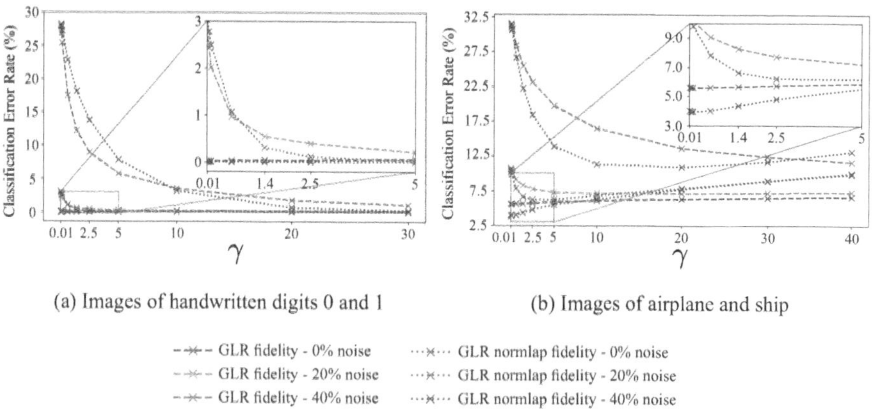

(a) Images of handwritten digits 0 and 1

(b) Images of airplane and ship

Figure 9.3. *Classification error rate (%) as a function of γ for the two datasets. For a color version of this figure, see www.iste.co.uk/cheung/graph.zip*

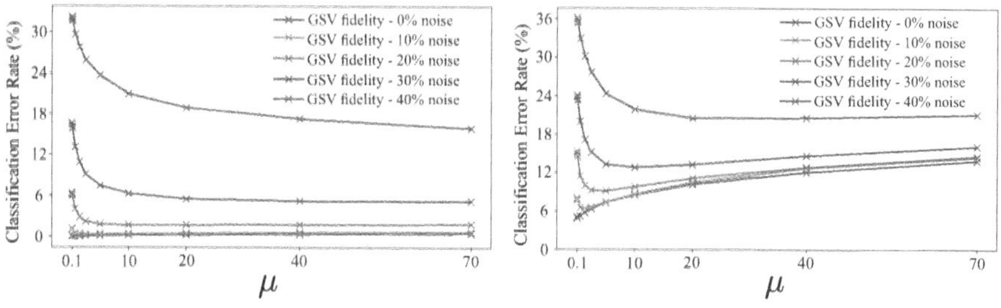

(a) Images of handwritten digits 0 and 1

(b) Images of airplane and ship

Figure 9.4. *Classification error rate (%) as a function of μ for the two datasets. For a color version of this figure, see www.iste.co.uk/cheung/graph.zip*

Figure 9.4 shows the classification error rate as a function of μ in equation [9.10] under different uniform label noise levels. Similar to the performance influence of γ in equation [9.8], a relatively large total variation smooth prior factor μ will potentially cause over-regularization effects. We can also observe that the optimal μ to achieve the lowest classification error rate is related to the level of label noise, which requires a clean validation set to tune.

The optimal smooth label signal, after the regularization, is a real-valued signal, which is then rounded to {-1, +1} values. After this step, it is possible to perform the regularization step again to potentially improve smoothing. This multi-step regularization could be beneficial in a noisy scenario, where it is reasonable to attempt to regularize the signal again after some noisy labels have been cleaned. Figure 9.5 shows the difference in classification error rate as a function of label noise and threshold τ in equation [9.11] when two iterations are performed, compared to the case without iterations.

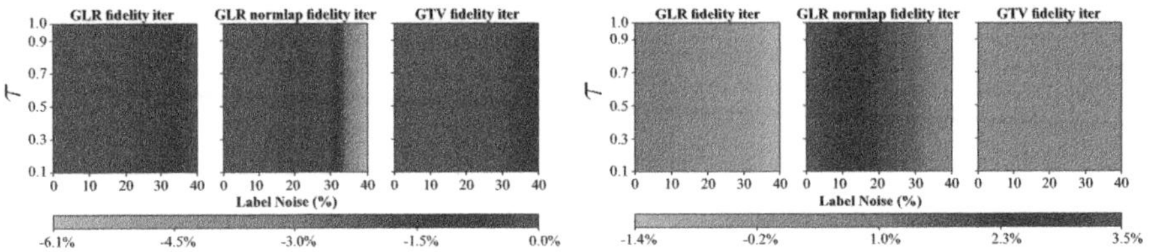

(a) Images of handwritten digits 0 and 1 (b) Images of airplane and ship

Figure 9.5. *Classification error rate (%) as a function of label noise and the threshold τ for (a)* ***MNIST*** *dataset and (b)* ***CIFAR10*** *dataset. We show classification error rate in color space as improvement when using two iterations, relative to the case with only one iteration; thus, a negative classification error rate means there is improvement due to the second iteration. For a color version of this figure, see www.iste.co.uk/cheung/graph.zip*

It can be seen from the figure that the threshold τ has very low impact on predictive performance. Furthermore, there is a significant improvement for the ***MNIST*** dataset by using two GLR normlap fidelity iterations under a high label noise level 40%, when the difference compared to no iterations exceeds 5%. For the ***CIFAR10*** dataset, there is negligible or no improvement. The results also show a predictive performance drop using the GLR normlap fidelity iter, causing over regularization under a relatively low label noise level.

Selecting appropriate values for γ and μ is important, as these parameters influence classification accuracy. If a reliable validation set is available and it is possible to estimate the level of noise, these should be used to tune the parameters. Otherwise, for a reasonably low level of noise, small γ and μ (below 1) is a safe choice; for a high level of noise, however, it is better to select a large γ and μ and rely more on the smoothness term. The graph regularization-based classifiers are much less sensitive to the node degree k than the k-NN classifier, with the GLR normlap scheme being the least sensitive. Using more than one regularization iteration, equation [9.11] provides negligible to no performance gain.

For all considered classifiers, Figure 9.6 shows the classification error rate as a function of the noise level in the training labels, evaluating the robustness to different levels of uniform label noise. For the fidelity term classifiers, the values of γ and μ are set based on the classification accuracy on the validation set using grid search.

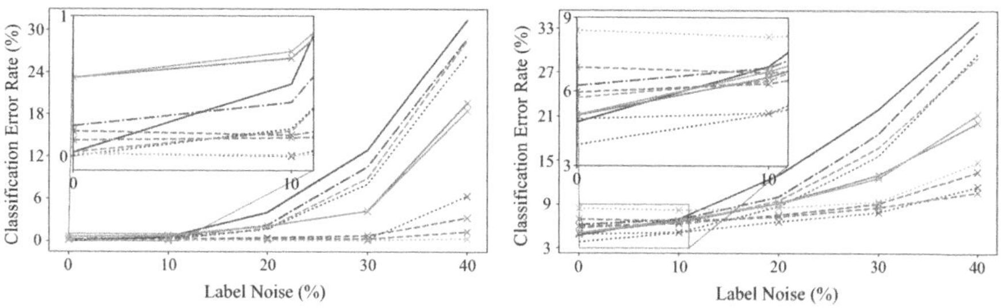

(a) Images of handwritten digits 0 and 1 (b) Images of airplane and ship

k-NN · GLR normlap fidelity · GLR fidelity iter · GLR · GSV · GLR normlap fidelity iter · GLR fidelity · GSV fidelity · GSV fidelity iter · GLR normlap

Figure 9.6. *Classification error rate (%) as a function of label noise for the two datasets. For a color version of this figure, see www.iste.co.uk/cheung/graph.zip*

From the figure, we can observe a significant predictive performance improvement made by using the fidelity term for better regularization effects. However, including the fidelity term tends to cause over-smoothing when the level of label noise is low. Moreover, when the normalized Laplacian matrix is used, the GLR normlap fidelity outperforms others across different label noise ratios. We can also find that the GLR fidelity iter, GLR normlap fidelity iter and GSV fidelity iter achieve better predictive performance under high label noise level for the ***MNIST*** dataset. However, for the ***CIFAR10*** dataset, we observe an over-regularization effect made by erforming the GLR normlap fidelity with two iterations, which leads to a predictive performance drop.

9.3. Feature learning via deep neural network

The previous section described several classifiers based on regularization on graphs, how to design the underlying similarity graph and how to define hyper-parameters. The potential of these classifiers is demonstrated when the training labels are corrupted or the training set is insufficient. In the previous section, however, it was assumed that features are pre-determined and fixed. In this section, the problem of learning discriminative features for the GLR-based classifiers is tackled. The aim is to better describe the correlation between data samples when constructing an underlying graph for SSL image classification.

As previously discussed, restoring corrupted training labels by representing them as smooth signals on graphs and applying a graph signal smoothness prior, in a similar manner to that used in image restoration tasks, is attractive. For example, to achieve robust binary classification, Cheung *et al.* (2018) introduce negative edge weights to separate nodes in two distinct clusters. Here, classifier learning is posed as *a Maximum a Posteriori* (MAP) problem, solved via the iterative re-weighted least squares strategy by Daubechies *et al.* (2010) and motivated by the signal restoration strategies by Zeng *et al.* (2019). The simulation results on various types of datasets demonstrate performance improvements compared to the positive graph edges of Belkin *et al.* (2004) and formulations of Chen *et al.* (2015) and Ekambaram *et al.* (2013), thus demonstrating the applicability of negative edge weights for robust graph-based classifier learning for small amounts of data, without learning feature representation.

Note that the methods proposed by Sandryhaila and Moura (2013); Chen *et al.* (2015); Cheung *et al.* (2018); Ekambaram *et al.* (2013), as well as the schemes described in the previous section, are pure model-based methods that construct similarity graphs to represent (i.e. to model) the relationship between data samples for fixed feature vectors, that is, these methods do not include feature learning. Constructing a similarity graph using *fixed, pre-determined, i.e. hand-crafted*, feature vectors can limit classification performance, either due to weak generalization ability of the hand-crafted features to describe the image samples or due to poor generalization of the feature distance $\delta_{i,j}$ (e.g. Euclidean distance, Mahalanobis distance, etc.) to quantify the node-to-node correlation. This is particularly pronounced in large-scale image classification tasks, when training labels are potentially noisy, or when the training dataset is small.

A promising alternative to using hand-crafted features is *feature learning*, which refers to methods that automatically discover, i.e. learn, feature vectors from the data. *Deep feature learning* offers improved feature representation using deep neural network (DNN) learning approaches. Deep feature learning methods can be categorized as:

1) unsupervised deep feature learning, where one learns an auto-encoder, without using training labels, to extract latent feature maps from the input image data. For example, in Liu *et al.* (2018), an unsupervised multiple layered sparse auto-encoder model is used to generate low-dimensional latent feature maps with the optimization goal of minimizing the reconstruction error. The learned latent feature maps are proven to be more effective than other classic dimensionality reduction methods for feature learning, in facilitating the further supervised classifier learning. A typical option to boost the feature representation capacity is to cooperate with other objectives, such as satisfying the similarity constraints on those images with similar input features sharing the same category label based on pre-clustering results (Chu and Cai 2017);

2) deep feature learning for classifier learning, where given input images (comprising the training set) and their corresponding class labels, one can train a DNN to learn the mapping function to assign labels to other unlabeled images. Typically, feature maps returned by one or more of the last DNN layers are good representatives and are used as features for consecutive classifier learning or clustering. This type of feature learning to classify unlabeled images is commonly referred to as transfer learning for classification – see Kornblith *et al.* (2019) for a review of transfer learning, where several supervised state-of-the-art large-scale CNNs are first trained on the ImageNet dataset (Russakovsky *et al.* 2015) for classification purposes. By dropping the last fully connected layer in those CNNs, the rest are used as fixed feature extractors and are proven to produce transferable "fixed" feature vectors, given a new set of images from other image datasets for further classifier learning;

3) deep feature learning for deep metric learning (DML): instead of learning the mapping function, i.e. classifier learning, as in (2), these methods optimize a feature metric-based distance function that discriminates feature vectors that correspond to opposite classes. Typically, N-tuples of labeled images are formed, by (randomly) sampling the training set, and then the results are averaged over all sampled N-tuples. For example, a 3-tuple (triplet) consists of images (i, j, v), where the ith image and jth image have the same label, while the vth image has the opposite label. To learn the features, a cost function is formed to promote small distance between the feature vectors of images i and j and large distance between the features of images i and v, and j and v.

Recent approaches to DML leverage on CNN to facilitate image classification or clustering tasks (Meyer *et al.* 2018; Song *et al.* 2017). Specifically, given a set of images without prior knowledge of the structure of their underlying similarity graph, DML methods learn a loss function that quantifies how well a pre-defined feature metric function measures the distances between subsets of images with and without the same labels.

Inspired by signature verification applications (Bromley *et al.* 1993), the main idea of DML is to learn a DNN-based metric function $d(a, b)$ (often called a similarity function), given two images a and b, based on a DNN mapping function. The structure of the DNNs and their similarity metric function d depends on the data scale and target applications. For example, in visual recognition applications (Yi *et al.* 2014; Song *et al.* 2016; Wang *et al.* 2017), flexible CNNs are developed to extract feature maps as the inputs to various distance functions, to assess the similarity. Typically used distance functions are Euclidean distance, cosine similarity, absolute difference and vector-concatenate as follows:

$$
\begin{aligned}
d(\mathcal{A}, \mathcal{B}) &= -\textstyle\sum_i (\mathcal{A}_i - \mathcal{B}_i), \\
d(\mathcal{A}, \mathcal{B}) &= \frac{\sum_i \mathcal{A}_i \mathcal{B}_i}{\sqrt{\sum_i \mathcal{A}_i \mathcal{A}_i \sum_i \mathcal{B}_i \mathcal{B}_i}},
\end{aligned}
$$

$$d(\mathcal{A},\mathcal{B}) = -\sum_i |\mathcal{A}_i - \mathcal{B}_i|,$$
$$d(\mathcal{A},\mathcal{B}) = \sum_i w_i[\mathcal{A};\mathcal{B}]_i, \qquad [9.16]$$

where the resulting DNN image feature maps for input images a and b are denoted as $\mathcal{A}$ and $\mathcal{B}$, respectively. Note that one can use pre-defined w_i to weight each element in the concatenated feature vector $[\mathcal{A};\mathcal{B}]_i$, in the vector concatenate-based distance function.

According to Yi *et al.* (2014), the cosine similarity distance measure, which is widely used in many pattern recognition problems, is invariant to the magnitude of the input feature maps. The resulting pairwise distance values are then used to compute the training cost via a loss function based on the training samples and their labels. Popular supervised loss functions for DML are as follows: triplet loss (Hoffer and Ailon 2015), N-pair loss (Sohn 2016), angular loss (Wang *et al.* 2017), clustering loss (Song *et al.* 2017), soft triple loss (Qian *et al.* 2019) and multi-similarity loss (Wang *et al.* 2019a).

To practically implement triplet loss-based DML for SSL tasks, a Boolean matrix that masks the distances of N^2 pairs over N samples is used. This operation is similar to defining edges during graph construction. To mitigate the effects of potentially overfitting the DNN for feature extraction, Ye *et al.* (2019) propose a regularized triplet loss correction function to mitigate the effects of insufficient training samples via a sparse k-NN graph construction and GLR.

Next, we describe how deep feature learning can be implemented, using CNNs and GLR, to restore noisy labels and perform classification. This is discussed in relation to other state-of-the-art semi-supervised classifiers. In the following subsections, the CNN-based DML problem is formulated. This is followed by iterative graph construction and an insight into how to implement such a connected multi-block system.

9.3.1. *Deep feature learning for graph construction*

Let $\dot{\mathbf{Y}}^0 = \{-1,1\}^M = \{y_1,\dots,y_M\} \subset \mathbf{Y}$, $0 < M < N$, be a set of known (possibly noisy) labels that correspond to the image instances $\dot{\mathbf{X}} = \{x_1,\dots,x_M\} \subset \mathbf{X}$ used for training. Let $\mathbf{Y}^0 = \{\dot{\mathbf{Y}}^0, \mathbf{0}^{N-M}\}$, where we set all $N-M$ unknown labels to zero (to be estimated during testing).

Given the set of images $\mathbf{X}$, the problem is to learn a robust mapping function to assign a classification label to each image observation $x \in \mathbf{X}$, when some classification labels $y \in \dot{\mathbf{Y}}^0$, used for training the model, are incorrect.

Let $\mathcal{G} = (\boldsymbol{\Psi}, \mathbf{E}, \mathbf{W})$ be an undirected graph, where $\boldsymbol{\Psi} = \{\psi_1, \ldots, \psi_N\}$ is a set of nodes, each corresponding to one instance in $\mathbf{X}$, $\mathbf{E} = \{e_{i,j}\}, i, j \in \{1, \ldots, N\}$, is a matrix representing the edge connectivity of $\mathcal{G}$; that is, $e_{i,j} = 1$ if there is an edge connecting the vertices i and j and $e_{i,j} = 0$ otherwise; and each entry $w_{i,j}$ in the weight matrix $\mathbf{W} = \{w_{i,j}\}, i, j \in \{1, \ldots, N\}$, where $w_{i,j}$ corresponds to the weight associated with the edge $e_{i,j}$. Then, $\mathbf{Y}^0$ is a graph signal where each sample of $\mathbf{Y}^0$ corresponds to a vertex in the graph $\mathcal{G}$. The combinatorial graph Laplacian matrix, as defined in section I.3, is given by $\mathbf{L} = \mathbf{D} - \mathbf{A}$, where $\mathbf{A}$ is a symmetric $N \times N$ adjacency matrix with each entry $a_{i,j} = \max(w_{i,j} \cdot e_{i,j}, w_{j,i} \cdot e_{j,i})$, and $\mathbf{D}$ is a degree matrix with entries $d_{i,i} = \sum_{j=1}^{N} a_{i,j}$, and $d_{i,j} = 0$ for $i \neq j$.

As in Hoffer and Ailon (2015), *triplets* are observations (x_a, x_p, x_n), $x_a, x_p, x_n \in \mathbf{X}$ corresponding to vertices $\psi_a, \psi_p, \psi_n \in \boldsymbol{\Psi}$, respectively, such that $y_a = y_p \neq y_n$, and $y_a, y_p, y_n \in \dot{\mathbf{Y}}$. Let $\mathbf{P}$ be a set of all edges $e_{a,p}$, such that $y_a = y_p$, and $\mathbf{Q}$ a set of all edges $e_{a,n}$, for which $y_a \neq y_n$, that is, $\mathbf{P}$ and $\mathbf{Q}$ are sets of all edges that connect nodes with the same and opposite labels, respectively.

Motivated by CNN's ability to extract discriminative features and GLR's ability to "clean" unreliable labels, graph-based classifier learning is formulated as a two-stage learning process: (1) *feature learning* – extracting feature maps, i.e. learning a deep neural network that returns the most discriminative feature maps, and then generating an initial graph by learning the underlying $\mathbf{E}$ to maximize/minimize the similarity between any two nodes in $\mathcal{G}$ that are indexed by the same/opposite labels; (2) *classifier learning* – iteratively refining the graph and effectively performing GLR to restore the corrupted classifier signal. In the following, these two stages are described.

Given the image set $\mathbf{X}$ and the corresponding, potentially noisy labels, $\dot{\mathbf{Y}}^0$, the first task is to learn a discriminative feature map $\mathcal{V}^0(\cdot)$ and generate an initial underlying graph for the learnt feature map.

Let

$$d_{i,j}(\mathcal{V}^0) = \|\mathcal{V}^0(x_i) - \mathcal{V}^0(x_j)\|_2^2$$

be the feature metric expressed as Euclidean distance between the feature maps $\mathcal{V}^0(x_i)$ and $\mathcal{V}^0(x_j)$, based on the top equation in equation [9.16]. Note that $d_{i,j}$ can also be set to one of the other metric functions in equation [9.16]. For a node $\psi_i \in \boldsymbol{\Psi}$, let $\mathcal{E}_i$ be a set containing all vertices, except ψ_i, in ascending order with respect to the metric $d_{i,k}(\mathcal{V}^0)$, $k = 1, \ldots, N-1, k \neq i$. Let $\mathcal{S}_i$ be a subset of $\mathcal{E}_i$ containing the first γ_i elements of $\mathcal{E}_i$, that is, the set $\mathcal{S}_i$ contains γ_i most correlated vertices to vertex ψ_i, according to metric $d_{i,k}(\mathcal{V}^0)$.

As discussed previously, to effectively perform GLR, the underlying similarity graph should be a sparsely connected graph (see also Cheung *et al.* 2018; Zeng *et al.*

2019). To perform sparsification of the resulting graph while maintaining connectivity, an indicator operator is used to minimize the number of $\mathbf{Q}$ edges. A typical option is a k-NN indicator that only keeps a maximum of γ_i edges for each individual node i and sets others to zero. That is, each graph edge $e_{i,j}$ is set to:

$$e_{i,j} = \begin{cases} 1, & \text{if } \psi_i \in \mathcal{S}_j \text{ or } \psi_j \in \mathcal{S}_i \\ 0, & otherwise. \end{cases} \qquad [9.17]$$

Once an optimal edge matrix $\mathbf{E}^0 = \{e^0_{i,j}\}$ is computed through $\mathcal{V}^0(\cdot)$ and equation [9.17], an initial undirected and unweighted graph $\mathcal{G}^0 = (\mathbf{\Psi}, \mathbf{E}^0, \mathbf{W}^0 = \mathbf{1})$ is obtained. The block diagram is shown in Figure 9.7. Note that, in the implementation, the learning starts with $\gamma_1 = \cdots = \gamma_N = \gamma^0$, that is learnt as explained in section 9.3.2.

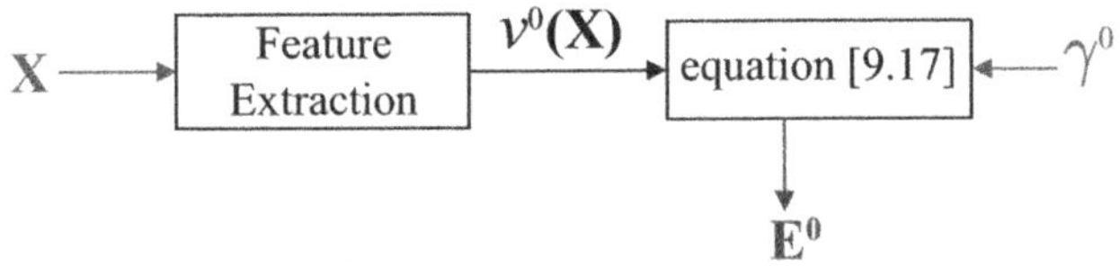

Figure 9.7. *The block diagram of the unweighted graph generation scheme. $\mathcal{V}^0(\cdot)$ is a feature map function that reflects the node-to-node correlation. For a color version of this figure, see www.iste.co.uk/cheung/graph.zip*

9.3.2. *Iterative graph construction*

If the noisy training labels are seen as a smooth graph signal, $\mathbf{Y}^0$, then one can iteratively perform GLR for denoising the labels and semi-supervised classification, while refining the set of deep feature maps and the underlying graph.

Let $r > 0$ be the iteration index, initialized to 1, and let $\mathcal{G}^r = (\mathbf{\Psi}, \mathbf{E}^{r-1}, \mathbf{W}^r)$ be the graph, with $\mathbf{W}^r$ to be learned, and $\mathbf{Y}^r$ the noisy labels in the rth iteration. Thus, $\mathcal{G}^1$ is an N-node graph with edges set by equation [9.17]. In the rth iteration, each vertex ψ_i is indexed by a label $y_i^{r-1} \in \mathbf{Y}^{r-1}$ (the graph signal), and is associated to a feature vector $\mathcal{V}^r(x_i)$.

As previously discussed, the edge weight is typically computed using a Gaussian kernel function, i.e. $\exp\left(-\frac{\|x_i - x_j\|_2^2}{2\sigma_x^2}\right)$, to quantify the node-to-node correlation. Instead of using a fixed σ_x like in Sandryhaila and Moura (2013); Cheung *et al.* (2018); Yang *et al.* (2018), motivated by Zelnik-Manor, Perona (2004), an auto-sigma Gaussian kernel function is used to assign edge weight $w^r_{i,j}$ in $\mathcal{G}^r$ by maximizing the margin between the edge weights assigned to **P**-edges and **Q**-edges, as:

$$\begin{aligned} \sigma^* &= \arg\max_{\sigma_x} \left[\exp\left(-\frac{\omega^2_{\{\psi_a,\psi_p\}}}{2\sigma_x^2} \right) - \exp\left(-\frac{\omega^2_{\{\psi_a,\psi_n\}}}{2\sigma_x^2} \right) \right] \\ w^r_{i,j} &= \exp\left(-\frac{\|\mathcal{V}^r(x_i) - \mathcal{V}^r(x_j)\|_2^2}{2\sigma^{*2}} \right), \end{aligned} \quad [9.18]$$

where $\omega_{\{\psi_a,\psi_p\}}$ and $\omega_{\{\psi_a,\psi_n\}}$ denote the mean Euclidean distance between the nodes connected by **P**-edges and **Q**-edges, respectively. By setting the first derivative to zero, the resulting optimal $\sigma^* = \sqrt{\frac{\omega^2_{\{\psi_a,\psi_n\}} - \omega^2_{\{\psi_a,\psi_p\}}}{2\log(\omega^2_{\{\psi_a,\psi_n\}}/\omega^2_{\{\psi_a,\psi_p\}})}}$, is obtained, which is used to assign the edge weights of the graph.

Then, the restored classifier signal is obtained by finding the smoothest graph signal $\mathbf{Y}^r$ as:

$$\mathbf{Y}^r = \arg\min_{\mathbf{B}} \left(\|\mathbf{Y}^{r-1} - \mathbf{B}\|_2^2 + \mu^r \mathbf{B}\mathbf{L}^r\mathbf{B}^T \right). \quad [9.19]$$

For more details, the reader can refer to (Cheung *et al.* 2018; Zeng *et al.* 2019). The minimization above finds a solution that is close to the observed set of labels in the previous iteration, $\mathbf{Y}^{r-1}$, while preserving smoothness. To guarantee that the solution $\mathbf{Y}^r$ to the QP problem (equation [9.19]) is numerically stable, theorem 1 from Zeng *et al.* (2019) can be adopted by setting an appropriate conditional number κ. The maximum value of the smoothness prior factor μ^r is then calculated as: $\mu^r_{\max} = (\kappa - 1)/(2d^r_{\max})$, where $d^r_{\max}$ is the maximum degree of the vertices in graph $\mathcal{G}^r$ (see Figure 9.8(a)).

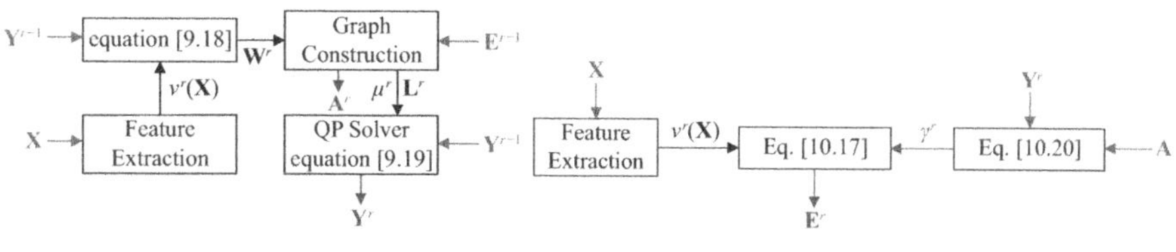

(a) The block diagram of the classifier scheme

(b) The block diagram of the graph update scheme

Figure 9.8. *The graph-based classifier and graph update scheme. The green and blue colors denote input and output, respectively. For a color version of this figure, see www.iste.co.uk/cheung/graph.zip*

COMMENT ON FIGURE 9.8.– (a) $\mathcal{V}^r(\cdot)$ *is a feature map function that reflects the node-to-node correlation. The edge matrix* $\mathbf{E}^{r-1}$ *is used as a mask when assigning edge weights to construct the adjacency matrix* $\mathbf{A}^r$. *We perform GLR to restore the corrupted classifier signal* $\mathbf{Y}^{r-1}$, *given the resulting sparse graph Laplacian* $\mathbf{L}^r$, *and apply the constrained smoothness prior factor* μ^r *to ensure the numerical stability of the QP solver. The implementation of* $\mathcal{V}^r(\cdot)$ *varies depending on the data scale and the dimension of the input observations. The output is the new set of "denoised" labels* $\mathbf{Y}^r$. (b) *Based on the adjacency matrix* $\mathbf{A}^r$ *and restored classifier signal* $\mathbf{Y}^r$, *we learn a feature map function to better refine the graph structure. The edge matrix* $\mathbf{E}^r$ *is updated via equation [9.20] and equation [9.17] based on both the previous restored classifier signal and the regularized feature map. The output of this block is the new edge matrix* $\mathbf{E}^r$ *that will be used in the next iteration.*

Between two iterations, a regularized feature map function is learnt to refine the underlying graph structure based on the denoised label signal, $\mathbf{Y}^{r-1}$, obtained in the previous iteration. See the illustration in Figure 9.8(b) for the graph update after the rth GLR iteration. The individual degree of vertex i is then updated as:

$$\begin{aligned} \mathring{e}_{i,j}^r &= \begin{cases} 1, & \text{if } e_{i,j}^r \in \mathbf{P}^r \,\&\, a_{i,j}^r > \beta \\ 0, & \text{if } e_{i,j}^r \in \mathbf{Q}^r \,\&\, a_{i,j}^r \leq \beta, \end{cases} \\ \gamma_i^r &= \sum_{j=1}^{N} \mathring{e}_{i,j}^r, \end{aligned} \qquad [9.20]$$

where $\mathbf{P}^r$ and $\mathbf{Q}^r$ sets are formed based on the denoised classifier signal $\mathbf{Y}^r$. The edge $e_{i,j}^r$ is removed if it connects vertices with opposite labels, or if the corresponding entry to adjacency matrix is less than β, which is heuristically set to 0.1.

9.3.3. *Toward practical implementation of deep feature learning*

Based on the key concepts described in the previous subsection, in this section, a typical algorithmic flow within a complex system of how deep feature learning can be implemented effectively with a deep neural network architecture is described. For illustration, we use the DynGLR-Net approach by Ye *et al.* (2020) as an example. Specifically, the overall implementation comprises multiple building blocks of CNNs, which interact with each other iteratively to learn the most discriminative features.

The block diagram of the DynGLR-Net approach is presented in Figure 9.9. This overall network consists of three sub-networks: (1) **G-Net** (graph generator network) used to learn a deep metric function to construct an undirected and unweighted k-NN graph $\mathcal{G}^0 = (\mathbf{\Psi}, \mathbf{E}^0, \mathbf{W}^0 = 1)$. (2) **W-Net** (graph weighting and classifier network) used to assign edge weights $\mathbf{W}^r$ for effectively performing GLR to restore the corrupted classifier signal $\mathbf{Y}^r$. (3) **U-Net** (graph update network) used to refine $\mathbf{E}^r$ to better reflect the node-to-node correlation based on the restored classifier signal in the previous iteration $\mathbf{Y}^{r-1}$. Each network is clarified in the following subsections.

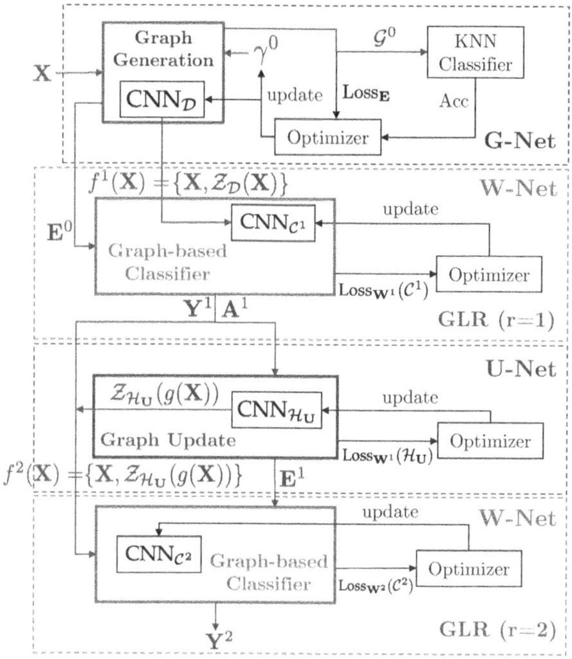

Figure 9.9. *The overall block diagram of the DynGLR-Net for* $r = 2$. *Given observations* $\mathbf{X}$, *G-Net first learns an initial undirected and unweighted* k*-NN-graph by minimizing* $\text{Loss}_{\mathbf{E}}$. *The resulting edge matrix* $\mathbf{E}^0$ *is used in the GLR iteration. The learnt shallow feature map* $f^1(X) = \{X, \mathcal{Z}_{\mathcal{D}}(X)\}$ *is then used as an input to learn a* $\text{CNN}_{\mathcal{C}^1}$ *network for assigning weights to the initial graph edges. Given a subset of potentially noisy labels,* $\dot{\mathbf{Y}}$, *GLR is performed on the constructed undirected and weighted graph to restore the labels. The resulting restored labels are used in the following GLR iterations. To assign the degree of freedom for refining graph connectivity, the graph edge sets are updated by minimizing* $\text{Loss}_{\mathbf{W}^1}(\mathcal{H}_{\mathrm{U}})$, *given neighbor information for each node, based on the resulting denoised classifier signal from the first GLR iteration. Then edge weights are reassigned to the updated graph edge sets to perform better node classification in the second GLR iteration. For a color version of this figure, see www.iste.co.uk/cheung/graph.zip*

In order to learn the optimal metric space, a CNN is used (as in Hoffer and Ailon (2015)), denoted by $\text{CNN}_{\mathcal{D}}$, to learn a mapping function $\mathcal{D}(\cdot)$ (see the architecture implementation of $\text{CNN}_{\mathcal{D}}$ in Ye *et al.* (2020)).

For a random observation image triplet (x_a, x_p, x_n), such that $e_{a,p} \in \mathbf{P}$ and $e_{a,n} \in \mathbf{Q}$, the following loss function is minimized to learn the feature map:

$$Loss_{\mathbf{E}} = \sum_{a,p,n} \left[\alpha_{\mathbf{E}} - \|\mathcal{D}(x_a) - \mathcal{D}(x_n)\|_2^2 + \|\mathcal{D}(x_a) - \mathcal{D}(x_p)\|_2^2\right]_+, \quad [9.21]$$

where $\mathcal{D}(\cdot)$ is a CNN-based feature map function, to be learnt, that returns a feature vector corresponding to the input observation, $\alpha_{\mathbf{E}}$ is the minimum margin, and operator $\left[\,\cdot\,\right]_+$ is a Rectified Linear Units (ReLU) activation function, which is equivalent to $\max(\cdot, 0)$. Let $\mathcal{Z}_{\mathcal{D}}(x)$ be the learnt feature map output at the second to the last layer of CNN$_{\mathcal{D}}$, obtained by minimizing the loss equation [9.21].

The loss function in equation [9.21] promotes a community structure graph that has relatively small Euclidean distance between the feature maps of vertices connected by the edges in $\mathbf{P}$, and a large distance between the vertices connected by the edges in $\mathbf{Q}$, while keeping a minimum margin $\alpha_{\mathbf{E}}$ between these two distances.

An alternative option to implement CNN$_{\mathcal{D}}$ is to use a pre-trained deep neural network on another dataset in the same domain, such as the pre-trained ResNet-152 (Kornblith *et al.* 2019), to extract the feature map.

For graph Laplacian estimation in graph learning, prior knowledge/assumptions about the graph structure is built into the choice of adjacency matrix. However, in practice, when the graph connectivity is unknown, then according to Egilmez *et al.* (2017), the adjacency matrix can first be set to represent a fully connected graph, and then be tuned until the desired level of sparsity is achieved. It is similar in tuning the parameter γ^0 for the graph generation scheme, where the level of sparsity in the edge matrix is tuned according to the KNN classifier, which uses max-vote for prediction. Since *a priori* knowledge of the connectivity of the nodes is not present in this example, the graph is initialized as a fully connected graph. A sparse $\mathbf{E}^0$ minimizes the number of $\mathbf{Q}$ edges by only keeping the connections with γ^0 neighbors per individual node. k-NN-graph construction is adopted based on equation [9.17], where optimal maximum number of neighbors γ^0 is obtained via a grid-search by evaluating the classification accuracy of the k-NN classifier (denoted by Acc in Figure 9.9), using the validation data with the same amount of noisy labels as the training dataset. Note that, due to a lack of any prior knowledge of the optimal maximum degree of each individual node, the initial setting is $\gamma^0 = \gamma_1 = \ldots = \gamma_N$. Once the optimal number of neighbors γ^0 is obtained, the resulting graph edges $\mathbf{E}^0$ are used in the following section for pruning the edge weights during edge weighting and are updated based on the regularized metric function and the difference between the classifier signal, before and after GLR.

To assign the edge weights $\mathbf{W}^r$ to the graph $\mathbf{G}^r$, a CNN is first employed, denoted by CNN$_{\mathcal{C}^r}$, to learn a deep metric function. To better learn the feature map $\mathcal{V}^r$, a robust graph-based triplet loss function is used:

$$Loss_{\mathbf{W}^r}(\mathcal{V}) = \sum_{\psi_a,\psi_p,\psi_n} \Big[\alpha_{\mathbf{W}} - \|\mathcal{V}^r(f^r(x_a)) - \mathcal{V}^r(f^r(x_n))\|_2^2 \cdot \pi_{(\psi_a,\psi_n|e_{a,n}\in\mathbf{Q})} + \|\mathcal{V}^r(f^r(x_a)) - \mathcal{V}^r(f^r(x_p))\|_2^2 \cdot \pi_{(\psi_a,\psi_p|e_{a,p}\in\mathbf{P})}\Big]_+ \qquad [9.22]$$

$$\Pi^r = \{\pi_{\psi_i,\psi_j}\} = \{\Theta(\dot{y}_i^r, \dot{y}_i^{r-1}, \dot{y}_j^r, \dot{y}_j^{r-1})\}.$$

Θ is an edge attention activation function (see equation [9.23] for the particular function used) that estimates how much attention should be given to each edge and $\mathbf{Y}^r = \{\dot{\mathbf{Y}}^r = [-1,1]^M, [-1,1]^{N-M}\}$ is the restored classifier signal obtained via equation [9.19], starting from the classifier signal in the previous iteration, $\mathbf{Y}^{r-1}$. π_{ψ_i,ψ_j} is the amount of attention, i.e. edge loss weights, assigned to the edge connecting vertices ψ_i and ψ_j. Note that $\mathcal{C}^r(\cdot)$ is the feature map learnt by minimizing equation [9.22]. The architectures for $r = 1$ and $r = 2$ are shown in Figure 9.10. Since, at the first iteration $r = 1$ many noisy labels are expected, the residual network architecture will be different to the $r > 1$ case, see He *et al.* (2016) for detailed arguments.

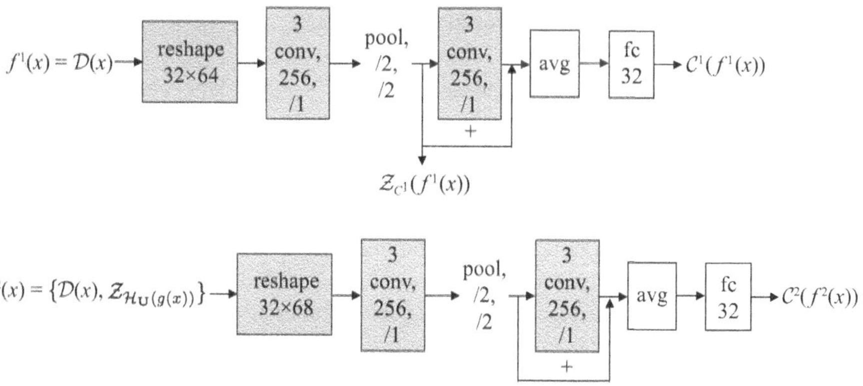

Figure 9.10. *$CNN_{\mathcal{C}^r}$ neural nets for **CIFAR10** dataset: "pool/q/w" refers to a max-pooling layer with q = pool size and w = stride size. "x conv y/z" refers to a 1D convolutional layer with y filters, each with kernel size x and stride size z. "fc x" refers to the fully connected layer with x = number of neurons. "reshape x" refers to a reshape layer to transform the size of the input to x. "avg" refers to the global average-pooling layer (Lin et al. 2013). For a color version of this figure, see www.iste.co.uk/cheung/graph.zip*

The architecture presented in Figure 9.10 (top) is used as the feature map $\mathcal{C}^1(\cdot)$, after G-Net, to construct the graph $\mathcal{G}^1 = (\mathbf{\Psi}, \mathbf{E}^0, \mathbf{W}^1)$ by minimizing $\mathrm{Loss}_{\mathbf{W}^1}(\mathcal{C}^1)$, taking the undirected graph $\mathcal{G}^0$ learned via G-Net as input. The input to $\text{CNN}_{\mathcal{C}^1}$ is the concatenated observations $\mathbf{X}$ and "shallow feature maps" learned via G-Net, i.e. the output of the second to last layer of $\text{CNN}_{\mathcal{D}}$, denoted by $\mathcal{Z}_{\mathcal{D}}(\mathbf{X})$.

The $r = 2$ architecture is shown in Figure 9.10 (bottom), with observations $\mathbf{X}$ and "shallow feature maps" learned via U-Net (described in the next subsection) to facilitate the regularization of $\text{CNN}_{\mathcal{C}^2}$ by minimizing $\text{Loss}_{\mathbf{W}^2}(\mathcal{C}^2)$, based on the denoised labels, convolution on both feature maps, denoised classifier signal and their differences across neighbors.

Note that unlike Hoffer and Ailon (2015), edge attention activation Θ is used in equation [9.22] to dropout some edges with relatively large changes between $\dot{\mathbf{Y}}^r$ and $\dot{\mathbf{Y}}^{r-1}$ via GLR. This helps to focus learning on edges with high confidence given noisy training labels. Therefore, the overall training performance is better than the standard dropout layer approach, which drops out random neuron units in the network. The edge attention activation Θ and Φ are implemented as:

$$\Phi(\dot{y}_i^{r-1}, \dot{y}_i^r) = \begin{cases} 1, & \text{if } |\dot{y}_i^{r-1} - \dot{y}_i^r| \leq \varepsilon^r \\ 0, & \text{if } |\dot{y}_i^{r-1} - \dot{y}_i^r| > \varepsilon^r \end{cases}$$
$$\Theta(\dot{y}_i^{r-1}, \dot{y}_i^r, \dot{y}_j^{r-1}, \dot{y}_j^r) = \min(\Phi(\dot{y}_i^{r-1}, \dot{y}_i^r), \Phi(\dot{y}_j^{r-1}, \dot{y}_j^r)), \qquad [9.23]$$

where the threshold ε^r is used to determine whether a node's label can be trusted and also helps to control the sparsity of edge attention matrix Π^r. That is, if the difference between the signal label in the previous and current iteration is large, this means that the label most likely changed sign (from -1 to +1 or vice versa) and is unreliable in this iteration. To reflect the fact that there might be many noisy (unreliable) labels at the start, heuristically ε^1 is initialized to 1 for the first GLR iteration. Since, after applying GLR, the classification signal is expected to be cleaner, heuristically threshold ε^2 is set to 0.01 for the second stacked W-Net during training, to ensure that CNNs are regularized to mitigate the over-fitting introduced by noisy labels.

Edge convolution has recently been proven to be a rich feature representation method (Wang *et al.* 2019b; Zhang *et al.* 2019). Thus, edge convolution is used in deeper feature map learning, i.e. after GLR $r = 1$. That is, given $\mathbf{A}^1$, from the first GLR iteration, each node's feature representation is enhanced by considering the observations of both $\mathbf{X}$ and the classifier signal $\mathbf{Y}^1$ from its six nearest neighbors (set heuristically), which are most likely to have the same label.

Incorporating an additional CNN, denoted by $\text{CNN}_{\mathcal{H}_\text{U}}$, shown in Figure 9.11, a richer feature representation $\mathcal{H}_\text{U}(g(x))$ is constructed to enhance the graph-based classifier learning with a single input to the network, $g(x)$, comprising x_i and $\{\mathbf{y}_i^1, \mathbf{y}_\mathcal{U}^1 - \mathbf{y}_i^1\}$, where $\mathbf{y}_i^1$ denotes a tuple $(y_i, 0)$, if $y_i > 0$ or $(0, y_i)$ otherwise. $\mathbf{y}_\mathcal{U}^1$ is a 6×2 matrix formed by concatenating $\mathbf{y}_i^1$ with the nearest six neighboring nodes. Finally, $\mathbf{y}_{\mathcal{U}}^1 - \mathbf{y}_i^1$ is obtained by subtracting each row of $\mathbf{y}_{\mathcal{U}}^1$ by $\mathbf{y}_i^1$.

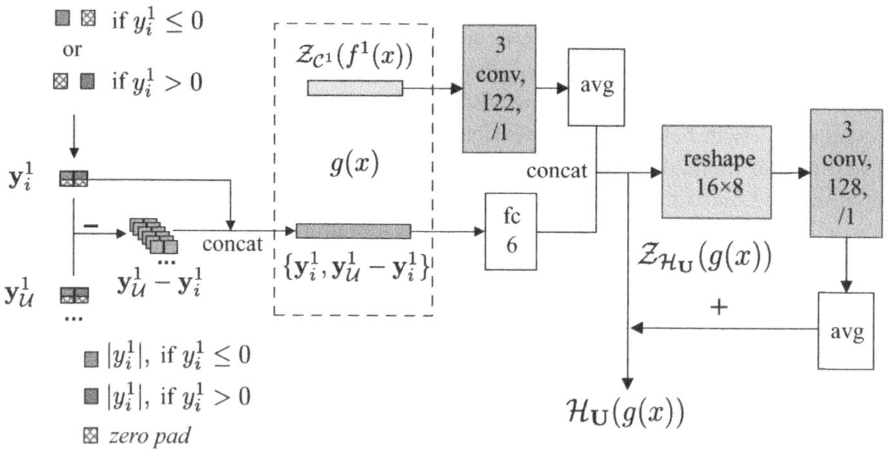

Figure 9.11. *CNN$_{\mathcal{H}_\mathbf{U}}$ neural nets for the **CIFAR10** dataset. For a color version of this figure, see www.iste.co.uk/cheung/graph.zip*

The graph edge $\mathbf{E}^{r-1}$ is updated by equation [9.20] based on the learnt regularized feature map $\mathcal{H}_\mathbf{U}(\cdot)$, in order to better reflect the node-to-node correlation. The new edge matrix $\mathbf{E}^1$ and the denoised classifier signal $\mathbf{Y}^1$ are then used in the second graph-based classifier iteration.

Though iterations between W-Net and U-Net can continue, only two iterations are performed in Ye *et al.* (2020) to reduce computational complexity, since it was observed that convergence was reached, i.e. there was no further improvement, after $r = 2$ iterations.

9.3.4. *Analysis on iterative graph construction for robust classification*

In SSL, given potentially noisy training labels, regularization on graphs helps to restore the corrupted training labels, as discussed in section 9.1.2. Based on these restored training labels, graph learning and GLR can be performed iteratively, as described in section 9.3.2. In this section, the benefits of this iterative graph construction process are analyzed based on the classification error rate and spectrum representation of the resulting underlying graphs per iteration.

To simplify the training process, the pre-trained ResNet-152 is used as CNN$_\mathcal{D}$ to extract the feature map for the ***CIFAR10*** dataset. The network described in the previous subsection is compared against the k-NN classifier using CNN-based DML (used CNN is the same as $\text{CNN}_{\mathcal{D}}$), as in Hoffer and Ailon (2015) (called DML-k-NN).

To evaluate the classification performance with insufficient training data, the labeling ratio of the training set is set to 0.5–10%, with balanced class distribution. For sufficient data with label noise, the experiments are designed by splitting each dataset into training, validation and testing sets with 40%, 20%, 40% of instances, respectively, with balanced class distribution. To evaluate the robustness of different classification methods against label noise (0–40%), the subsets of instances from both training and validation sets are randomly sampled and their labels are flipped. The classification error rates are measured by running at least 20 experiments per setting. The same random seed setting is used across all classification methods and all duplicated instances are removed to ensure a fair comparison. All CNNs are learnt by the ADAM optimizer. The hyper-parameters used for each experiment are obtained from the validation sets by using grid search.

To guarantee that the solution $\mathbf{Y}^r$ to equation [9.19] is numerically stable, the following parameters are initialised in all of the experiments as follows: $\kappa = 60$ and $\mu^r = 0.67\mu^r_{\text{max}}$. The distance margin $\alpha_{\mathbf{E}} = \alpha_{\mathbf{W}} = 10$ is used in both equations [9.21] and [9.22]. In each epoch, the batch size is 4, with each batch comprising 200 labeled instances from the training set and 50 unlabeled instances from the validation set; thus, $4 \cdot 250$ instances are randomly selected, resulting in four graphs to regularize training per epoch.

In order to provide insight into how each graph construction iteration improves graph representation and smoothing of the graph signals, the corresponding sub-architectures of the overall DynGLR-Net architecture are denoted by a DynGLR-G-number, where “G” refers to graph generation and the number = “1” refers to edge weighting, “2” to GLR and “3” to graph update. Particularly, the following variants of the scheme are compared: (1) DynGLR-G-2: the unweighted graph $\mathbf{G}^0$ generated by G-Net (see Figure 9.7) is imported into GLR for classification. (2) DynGLR-G-12: weights are assigned to the unweighted graph $\mathbf{G}^0$ via an adaptive Gaussian kernel function (see Equation [9.18]); the resulting undirected and weighted graph is then used to perform node classification via GLR. (3) DynGLR-G-1232: graph edge sets are updated by considering the neighbors of each node with denoised classifier signal and observed feature maps (see Figure 9.8); the resulting unweighted graph is then used for classification. (4) DynGLR-G-12312: weights are reassigned to the updated unweighted graph to effectively perform classification.

The effectiveness of the iterative graph update scheme can be shown, as in section 9.3.5, by visualizing the learnt underlying graph in the spectral domain. To analyze the impact of different components on the classification accuracy during the testing phase, an ablation study can be performed. This is discussed in sections 9.3.6 and 9.3.7 in context with other state-of-the-art schemes.

9.3.5. *Graph spectrum visualization*

GFT is another approach used to represent the smoothness and connectivity of an underlying graph. Following Shuman *et al.* (2013), the magnitude of the GFT coefficients can be visualized for iterations of the graph update, as shown in Figure 9.12.

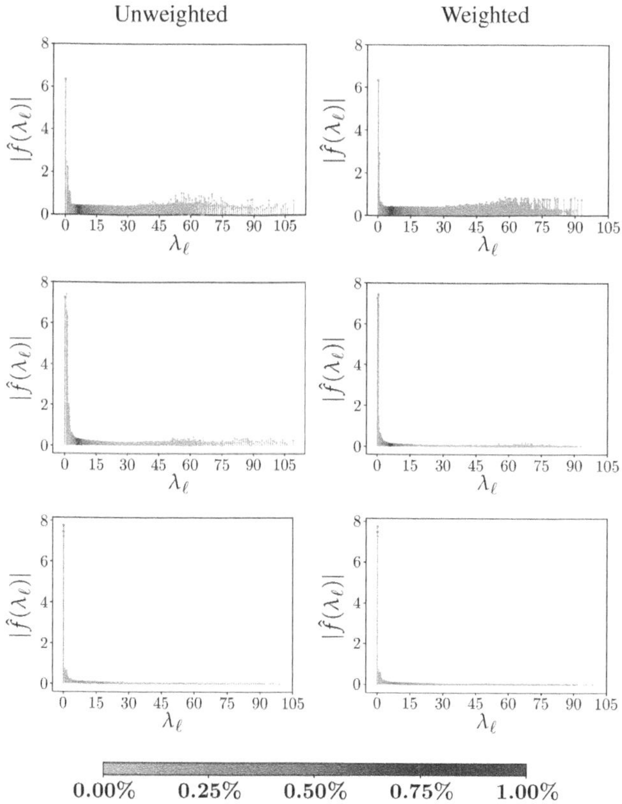

Figure 9.12. *The magnitude of the GFT coefficients for the* ***CIFAR10*** *dataset using sufficient data under 30% label noise level. The density of each eigenvalue λ_ℓ across all experiments on the testing sets is represented through color maps. The top row shows the result after initialization and before GLR (G-Net output) and the second and third row show the result after the first ($r = 1$) and the second iteration ($r = 2$), respectively. For a color version of this figure, see www.iste.co.uk/cheung/graph.zip*

In Figure 9.12 we can observe that the magnitude of the GFT coefficients decays more rapidly after a graph update, i.e. from iteration $r = 1$. Furthermore, comparing the left unweighted and right weighted columns for all iterations, we can observe that graph weighting via W-net smooths the graph data, with low frequency components becoming more prominent. Both observations are in accordance with (Shuman *et al.* 2013), where it is shown that the magnitude of the GFT coefficients decay rapidly for a smooth signal.

9.3.6. *Classification error rate comparison using insufficient training data*

Table 9.1 shows the comparison between the DynGLR networks and the benchmark, in terms of classification error rate using insufficient training data without label noise. Recall that the labeling ratio denotes the size of the training dataset relative to the total dataset used for training and testing.

Labeling ratio (%)	0.5	1.0	2.0	3.0	5.0	7.0	10.0
DML-k-NN	11.26	8.29	6.22	5.14	4.26	4.02	3.29
DynGLR-G-12	**10.88**	**7.32**	**5.92**	**4.77**	**4.20**	**3.92**	**3.03**

Table 9.1. *Classification error rate (%) for the* ***CIFAR10*** *dataset for different labeling ratios (%)*

From the table, it can be seen that DynGLR networks outperform the benchmark.

9.3.7. *Classification error rate comparison using sufficient training data with label noise*

Table 9.2 shows the comparison between the DynGLR networks and DML with K-NN instead of GLR, in terms of classification error rate using sufficient training data with label noise. From the table, it can be seen that the DynGLR networks with GLR always outperform classic DML with K-NN. The performance gap mainly increases with the level of label noise.

The DynGLR networks in Table 9.2 show the outcomes of the ablation study. Specifically:

– DynGLR-G-12 consistently outperforms DynGLR-G-2, showing the effect of edge weighting;

– the improvement due to graph update can be observed between DynGLR-G-12 and DynGLR-G-1232;

– by comparing DynGLR-G-1232 and DynGLR-G12312, small gains are observed due to iterative design, incorporating edge weighting;

– by comparing DynGLR-G-2 and DML-k-NN, performance improvements are obtained by replacing k-NN-based classification with GLR; larger gains can be observed as noise level increases.

% Label noise	0	10	20	35	40
DML-k-NN	4.02	7.91	15.52	22.70	30.50
DynGLR-G-2*	8.16	8.51	8.83	9.68	12.87
DynGLR-G-12*	3.92	4.49	5.26	6.22	7.41
DynGLR-G-1232*	3.43	3.82	4.79	5.15	7.32
DynGLR-G-12312*	**3.41**	**3.64**	**4.67**	**4.99**	**7.17**

Table 9.2. *Classification error rate (%) for the **CIFAR10** dataset using sufficient training data under different label noise levels*

Furthermore, the observations are that the importance of the following algorithmic steps, in order of most to least importance, on the performance can be summarized as: (I) iterative graph update (equation [9.20]) – disconnect $\mathbf{Q}$-edges and connect/reconnect $\mathbf{P}$-edges based on the restored labels after each GLR iteration to refine the graph structure, (II) edge convolution operation – performing feature and denoised label aggregation on neighboring nodes provides richer and smoother inputs and results in a spatially sparse graph, (III) edge attention function (equation [9.23]) – to better reflect node-to-node correlation, regularizing CNN training by weighting the edge loss based on classifier signal changes before and after GLR.

9.4. Conclusion

This chapter is focused on solving the problem of image classification when the training dataset is corrupted or insufficient. After providing an overview of popular model-driven and data-driven classifiers, we elaborate on how to build a robust graph for efficient classification, including designing the underlying similarity graph and defining hyper-parameters with practical implementation in mind. We compare several graph regularization-based classifiers and demonstrate their potential when the training labels are corrupted and the training set is insufficient, assuming fixed- and pre-defined features (i.e. without feature learning).

Then the problem of learning optimal features for the GSP-based classifiers is discussed to better describe the correlation between samples when constructing the underlying graph for SSL image classification. This is performed by deep feature learning, i.e. automatically learning features directly from the data, without the limitation of hand-crafted features and feature metrics. Having discussed the different approaches for deep feature learning, it is shown, by example, how this can be implemented by leveraging on CNN to learn a robust feature map to better reflect the underlying graph structure and find reliable training samples using a graph-based loss function.

9.5. References

Belkin, M., Matveeva, I., Niyogi, P. (2004). Regularization and semi-supervised learning on large graphs. *International Conference on Computational Learning Theory*, Springer, Heidelberg, Berlin, 624–638.

Blum, A. and Chawla, S. (2001). Learning from labeled and unlabeled data using graph mincuts. *Proceedings of the Eighteenth International Conference on Machine Learning*, 19–26.

Boyd, S. and Vandenberghe, L. (2009). *Convex Optimization*. Cambridge University Press, Cambridge.

Bromley, J., Guyon, I., LeCun, Y., Säckinger, E., Shah, R. (1993). Signature verification using a "siamese" time delay neural network. *NIPS', Morgan Kaufmann Publishers Inc.*, San Francisco, 737–744.

Chen, S., Sandryhaila, A., Moura, J., Kovacevic, J. (2015). Signal recovery on graphs: Variation minimization. *IEEE Transactions on Signal Processing*, 63(17), 4609–4624.

Cheung, G., Su, W.-T., Mao, Y., Lin, C.-W. (2018). Robust semisupervised graph classifier learning with negative edge weights. *IEEE Transactions on Signal and Information Processing over Networks*, 4(4), 712–726.

Chu, W. and Cai, D. (2017). Stacked similarity-aware autoencoders. *IJCAI*, 1561–1567.

Daubechies, I., DeVore, R., Fornasier, M., Güntürk, C.S. (2010). Iteratively reweighted least squares minimization for sparse recovery. *Communications on Pure and Applied Mathematics*, 63(1), 1–38.

Egilmez, H.E., Pavez, E., Ortega, A. (2017). Graph learning from data under Laplacian and structural constraints. *IEEE Journal of Selected Topics in Signal Processing*, 11(6), 825–841.

Ekambaram, V.N., Fanti, G., Ayazifar, B., Ramchandran, K. (2013). Wavelet-regularized graph semi-supervised learning. *IEEE Global Conference on Signal and Information Processing*, 423–426.

Fang, J., Yuan, Y., Lu, X., Feng, Y. (2019). Robust space-frequency joint representation for remote sensing image scene classification. *IEEE Transactions on Geoscience and Remote Sensing*, 57(10), 7492–7502.

Frénay, B. and Verleysen, M. (2014). Classification in the presence of label noise: A survey. *IEEE TNNLS*, 25(5), 845–869.

He, K., Zhang, X., Ren, S., Sun, J. (2016). Deep residual learning for image recognition. *IEEE CVPR*, 770–778.

Hoffer, E. and Ailon, N. (2015). Deep metric learning using triplet network. *International Workshop on Similarity-based Pattern Recognition*, Springer, Cham, 84–92.

Hu, W., Gao, X., Cheung, G., Guo, Z. (2020). Feature graph learning for 3D point cloud denoising. *IEEE Transactions on Signal Processing*, 68, 2841–2856.

Huang, B. and Jebara, T. (2007). Loopy belief propagation for bipartite maximum weight b-matching. *Artificial Intelligence and Statistics*, 195–202.

Jadhav, S.R., Das, R., Thepade, S.D., De, S. (2018). Applications of hybrid machine learning for improved content based image classification. *Fourth International Conference on Computing Communication Control and Automation (ICCUBEA)*, 1–6.

Johnson, R. and Zhang, T. (2007). On the effectiveness of Laplacian normalization for graph semi-supervised learning. *Journal of Machine Learning Research*, 8, 1489–1517.

Kornblith, S., Shlens, J., Le, Q.V. (2019). Do better imagenet models transfer better? *IEEE CVPR'*, 2661–2671.

Lin, M., Chen, Q., Yan, S. (2013). Network in network. arXiv:1312.4400.

Liu, J., Li, C., Yang, W. (2018). Supervised learning via unsupervised sparse autoencoder. *IEEE Access*, 6, 73802–73814.

Liu, Q., Yu, L., Luo, L., Dou, Q., Heng, P.A. (2020). Semi-supervised medical image classification with relation-driven self-ensembling model. *IEEE Transactions on Medical Imaging*, 39, 3429–3340.

Luo, Z.Q. (2010). Semidefinite relaxation of quadratic optimization problems. *IEEE Signal Processing Magazine*, 27(3)(April), 20–34.

Mahalanobis, P.C. (1936). On the generalized distance in statistics. *Proceedings of the National Institute of Science of India*, 2, 49–55.

Meyer, B.J., Harwood, B., Drummond, T. (2018). Deep metric learning and image classification with nearest neighbour Gaussian kernels. *IEEE ICIP*, 151–155.

Milanfar, P. (2013). A tour of modern image filtering. *IEEE Signal Processing Magazine*, 30(1), 106–128.

Moller, M.F. (1990). A scaled conjugate gradient algorithm for fast supervised learning. Paper, Computer Science Department, Aarhus University, Aarhus.

Qian, Q., Shang, L., Sun, B., Hu, J., Li, H., Jin, R. (2019). Softtriple loss: Deep metric learning without triplet sampling. *IEEE ICCV*. arXiv:1909.05235.

Russakovsky, O., Deng, J., Su, H., Krause, J., Satheesh, S., Ma, S., Huang, Z., Karpathy, A., Khosla, A., Bernstein, M., Berg, A.C., Fei-Fei, L. (2015). ImageNet Large Scale Visual Recognition Challenge. *International Journal of Computer Vision (IJCV)*, 115(3), 211–252.

Sahu, S.K., Pujari, A.K., Kagita, V.R., Kumar, V., Padmanabhan, V. (2015). Gp-svm: Tree structured multiclass svm with greedy partitioning. In *International Conference on Information Technology (ICIT)*, 142–147.

Sandryhaila, A. and Moura, J.M. (2013). Classification via regularization on graphs. In *2013 IEEE Global Conference on Signal and Information Processing*, 495–498.

Sandryhaila, A. and Moura, J.M. (2014). Discrete signal processing on graphs: Frequency analysis. *IEEE Transactions on Signal Processing*, 62(12), 3042–3054.

Shuman, D.I., Narang, S.K., Frossard, P., Ortega, A., Vandergheynst, P. (2013). The emerging field of signal processing on graphs: Extending high-dimensional data analysis to networks and other irregular domains. *IEEE signal processing magazine*, 30(3), 83–98.

Sohn, K. (2016). Improved deep metric learning with multi-class n-pair loss objective. In *Advances in Neural Information Processing Systems 29*, Lee, D.D., Sugiyama, M., Luxburg, U.V., Guyon, I., Garnett, R. (eds.), 1857–1865.

Song, H.O., Xiang, Y., Jegelka, S., Savarese, S. (2016). Deep metric learning via lifted structured feature embedding. *IEEE CVPR*.

Song, H.O., Jegelka, S., Rathod, V., Murphy, K. (2017). Deep metric learning via facility location. In *IEEE CVPR*, 2206–2214.

de Sousa, C.A.R., Rezende, S.O., Batista, G.E. (2013). Influence of graph construction on semi-supervised learning. *Joint European Conference on Machine Learning and Knowledge Discovery in Databases*, Springer, Berlin, Heidelberg, 160–175.

Tang, J., Hong, R., Yan, S., Chua, T.S., Qi, G.J., Jain, R. (2011). Image annotation by knn-sparse graph-based label propagation over noisily tagged web images. *ACM TIST*, 2(2), 14.

Tomasi, C. and Manduchi, R. (1998). Bilateral filtering for gray and color images. In *IEEE International Conference on Computer Vision*, 839–846.

Wang, J., Zhou, F., Wen, S., Liu, X., Lin, Y. (2017). Deep metric learning with angular loss. *IEEE ICCV*, 2612–2620.

Wang, P., Shen, C., Hengel, A.V.D. (2013). A fast semidefinite approach to solving binary quadratic problems. *Proceedings of the 2013 IEEE Conference on Computer Vision and Pattern Recognition*, 1312–1319.

Wang, Y., Liu, W., Ma, X., Bailey, J., Zha, H., Song, L., Xia, S.T. (2018). Iterative learning with open-set noisy labels. arXiv:1804.00092.

Wang, X., Han, X., Huang, W., Dong, D., Scott, M.R. (2019a). Multi-similarity loss with general pair weighting for deep metric learning. *IEEE CVPR*, 5022–5030.

Wang, Y., Sun, Y., Liu, Z., Sarma, S.E., Bronstein, M.M., Solomon, J.M. (2019b). Dynamic graph CNN for learning on point clouds. *ACM Transactions on Graphics (TOG)*, 38(5).

Wu, X., Zhao, L., Akoglu, L. (2018). A quest for structure: Jointly learning the graph structure and semi-supervised classification. *Proceedings of the 27th ACM International Conference on Information and Knowledge Management*, 87–96.

Yang, C., Cheung, G., Stankovic, V. (2018). Alternating binary classifier and graph learning from partial labels. *IEEE Asia-Pacific Signal and Information Processing Association Annual Summit and Conference (APSIPA ASC)*, 1137–1140.

Ye, M., Stankovic, V., Stankovic, L., Cheung, G. (2019). Deep graph regularized learning for binary classification. *IEEE International Conference on Acoustics, Speech and Signal Processing (ICASSP)*, 3537–3541.

Ye, M., Stankovic, V., Stankovic, L., Cheung, G. (2020). Robust deep graph based learning for binary classification. *IEEE Transactions on Signal and Information Processing over Networks*, November.

Yi, D., Lei, Z., Liao, S., Li, S.Z. (2014). Deep metric learning for person re-identification. *ICPR*, 34–39.

Zelnik-Manor, L. and Perona, P. (2004). Self-tuning spectral clustering. *Proceedings of the 17th International Conference on Neural Information Processing Systems*, NIPS, MIT Press, Cambridge, 1601–1608.

Zeng, J., Pang, J., Sun, W., Cheung, G. (2019). Deep graph Laplacian regularization for robust denoising of real images. *IEEE Conference on Computer Vision and Pattern Recognition Workshops*, June 16–17, Long Beach, CA.

Zhang, X., Xu, C., Tian, X., Tao, D. (2019). Graph edge convolutional neural networks for skeleton-based action recognition. *IEEE TNNLS*, 31, 1–14.

Zhou, D., Bousquet, O., Lal, T.N., Weston, J., Schölkopf, B. (2003). Learning with local and global consistency. *Proceedings of the 16th International Conference on Neural Information Processing Systems*, 321–328.

Zhu, X., Ghahramani, Z., Lafferty, J.D. (2003). Semi-supervised learning using Gaussian fields and harmonic functions. *Proceedings of the 20th International Conference on Machine Learning (ICML-03)*, 912–919.

10

Graph Neural Networks for Image Processing

Giulia FRACASTORO and Diego VALSESIA
Politecnico di Torino, Turin, Italy

10.1. Introduction

Processing visual data can be a daunting task due to the complexity and nuances of their representations. For this reason, decades of research have made great strides in defining more and more sophisticated models. Traditionally, such methods have been focused on images, which are still a major focus of the research in the field. However, new visual data types, such as 3D point clouds, are becoming increasingly relevant and extending image processing techniques to effectively process them provides a new set of challenges.

Recently, graph signal processing (GSP) has provided new powerful tools that are particularly suitable for visual data. On the one hand, traditional signal types, such as images, can benefit from richer representations induced by the graph structure (Liu *et al.* 2014; Kheradmand and Milanfar 2014; Bai *et al.* 2018; Pang and Cheung 2017). On the other hand, new data types with an irregular domain, such as point clouds, can now be effectively processed (Zeng *et al.* 2019; Chen *et al.* 2018; Thanou *et al.* 2016).

Concurrent to the emergence of GSP, data-driven solutions, based on neural networks have shown impressive performances in a variety of tasks, including low-level tasks, such as image restoration. The workhorse of data-driven methods is the convolutional neural network (CNN), which has been shown to capture highly complex features in images. However, two main shortcomings of CNNs have

Graph Spectral Image Processing,
coordinated by Gene CHEUNG and Enrico MAGLI. © ISTE Ltd 2021

recently emerged. First, CNNs are unable to process data that is defined on irregular domains. However, GSP can valuably contribute to fill this gap, which has led to the creation of graph convolutional neural networks (GCNNs), presented in Chapter 3. A second issue with CNNs concerns images: the local nature of convolutional layers cannot capture non-local self-similar patterns, which are well-known to be a powerful model of natural images (Dabov *et al.* 2007). In the last 2 years, the deep learning community has realized the importance of generalizing CNNs to include non-locality, albeit without any principled approach (Wang *et al.* 2018). Recent works have shown that GCNNs can elegantly and effectively exploit non-locality (Valsesia *et al.* 2019a; Valsesia *et al.* 2019b).

In this chapter, we present a few applications of GCNNs to visual data. We broadly subdivide them into two categories: problems solved by supervised learning approaches and problems solved with generative models, which are unsupervised. The former focuses on tasks where ground truth labels are available for the training procedure, in particular, we present point cloud classification and segmentation, as well as image denoising with additive white Gaussian noise. The latter category involves problems where the goal is to learn representations of the raw data points, in particular, we present generative models that can capture the structure of point clouds and effectively regularize inverse problems such as completing partial shapes. We note that, in this chapter, we focus only on methods based on GCNNs, and refer instead to Chapter 6 for a detailed description of image processing methods that incorporate graph spectral theory into standard CNNs.

10.2. Supervised learning problems

In this section, we provide an overview of the most relevant supervised methods based on graph neural networks. In the context of visual data processing, graph neural networks have been successfully used in point cloud processing, where they are considered the state-of-the art. In this section, we first focus on two supervised problems related to point cloud processing, namely, point cloud classification and point cloud segmentation. We then consider another supervised problem in this section, that of image denoising, where graph neural networks have been applied in order to exploit the non-local self-similarities of the image.

10.2.1. *Point cloud classification*

Point clouds are an important 3D data type that provide a geometrical representation of real-life objects and have widespread use in computer graphics and robotics. A point cloud is defined as an unordered set of points that are irregularly distributed in 3D space. Processing such a type of data poses many challenges, due to the irregular positioning of the points and the fact that any permutation of the points does not change the semantic meaning of the point cloud. For these reasons, it is not possible to use standard CNNs in point cloud processing. Some works have addressed this issue, either through voxelization (Maturana and Scherer 2015; Wu *et al.* 2015), where the irregular 3D structure of the point cloud is approximated with a 3D grid, or with architecture, such as PointNet (Qi *et al.* 2017a), where each point is processed independently and then a global symmetric aggregation function is applied. Recently, GCNNs have provided an elegant and effective way to overcome these limitations and have been successfully applied in point cloud processing.

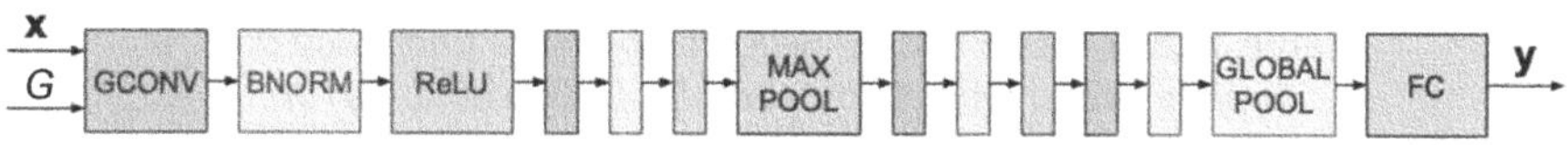

Figure 10.1. *Point cloud classification network (Simonovsky and Komodakis 2017). The network outputs the classification score vector* $\mathbf{y} \in \mathbb{R}^c$*, where* c *is the number of classes. GCONV: graph convolution operation; BNORM: batch normalization; FC: fully connected layer. For a color version of this figure, see www.iste.co.uk/cheung/graph.zip*

One of the most important applications of GCNNs is point cloud classification. Given a point cloud $\mathbf{x} \in \mathbb{R}^{N \times 3}$, we are interested in defining the object class of $\mathbf{x}$. We can define a graph $\mathcal{G}(\mathcal{V}, \mathcal{E})$, where the nodes of the graph are the 3D points and each node is connected to its neighboring nodes in order to capture the spatial arrangement of the points and their relative dependencies. The neighbors of a point can be selected as the k nearest nodes or the nodes whose distance from the point is lower than a fixed radius. The point cloud and its corresponding graph are then taken as input of a GCNN. Different architectures of GCNNs have been proposed for point cloud classification (see, for example, Simonovsky and Komodakis (2017), Wang *et al.* (2019)). In this section, we will focus on the network proposed in Simonovsky and Komodakis (2017), which is one of the most well-known architectures in the field. This network has the typical architecture of feed-forward networks, consisting of convolutional layers interlaced with pooling, followed by a global pooling and fully connected layers. An example of this architecture is shown in Figure 10.1. The graph is computed from the input point cloud, connecting each point to all the other points within a fixed radius. The graph convolutional layer uses the edge conditioned convolution (ECC) presented in Chapter 3. Each graph convolutional layer is followed by batch normalization and a Rectified Linear Unit (ReLU) function. The pooling layers aggregate the signals of neighboring nodes, obtaining, in this way, a new graph signal, defined on a coarsened graph, i.e. a graph with a lower number of nodes and edges. This means that a pyramid of progressively coarser graphs has to be constructed for each input graph. In Simonovsky and Komodakis (2017), the authors propose using the VoxelGrid algorithm (Rusu and Cousins 2011) to construct such a pyramid of graphs.

Model	Mean F1
Triangle+SVM (De Deuge *et al.* 2013)	67.1
GFH+SVM (Chen *et al.* 2014)	71.0
VoxNet (Maturana and Scherer 2015)	73.0
ORION (Sedaghat *et al.* 2016)	77.8
GCNN (Simonovsky and Komodakis 2017)	78.4

Table 10.1. *Mean F1 score weighted by class frequency on Sydney Urban Objects dataset (De Deuge* et al. *2013)*

Model	ModelNet10	ModelNet40
3DShapeNets (Wu *et al.* 2015)	83.5	77.3
MVCNN (Su *et al.* 2015)	-	90.1
VoxNet (Maturana and Scherer 2015)	92.0	83.0
ORION (Sedaghat *et al.* 2016)	93.8	-
SubvolumeSup (Qi *et al.* 2016)	-	86.0 (89.2)
GCNN (Simonovsky and Komodakis 2017)	90.0 (90.8)	83.2 (87.4)

Table 10.2. *Mean class accuracy (respectively, mean instance accuracy) on ModelNet datasets (Wu* et al. *2015)*

Table 10.1 and 10.2 show the results of the point cloud classification network described above, compared to methods based on volumetric CNNs. These results show that GCNNs outperform volumetric approaches on the Sydney Urban Objects dataset and reach a competitive level of performance on ModelNet datasets. The Sydney Urban Objects dataset consists of 588 objects, from 14 categories, manually extracted from 360° LiDAR scans. It represents the non-ideal sensing condition, with occlusions and a large variability in viewpoint. This makes object classification a very challenging task. On the other hand, the ModelNet10 (3991/908 train/test examples, in 10 categories) and ModelNet40 (9843/2468 train/test examples, in 40 categories) datasets are composed of synthetic point clouds that were created from meshes by uniformly sampling 1000 points from each mesh.

In the work presented in Simonovsky and Komodakis (2017), all the graph-convolutional layers use the same graph, which is computed at the input. Wang *et al.* (2019) instead proposed a different approach, where the graph is dynamically updated at each layer. In the following section, we will describe this method in detail.

10.2.2. *Point cloud segmentation*

Another important problem in point cloud processing is part segmentation. Different from point cloud classification, where we have to identify the semantic class of the entire point cloud, the task of part segmentation consists of associating a semantic label to each point of the point cloud. Since standard CNNs have the same shortcomings in this application as those encountered in point cloud classification, GCNNs can also provide an effective solution to overcome such limitations in this case.

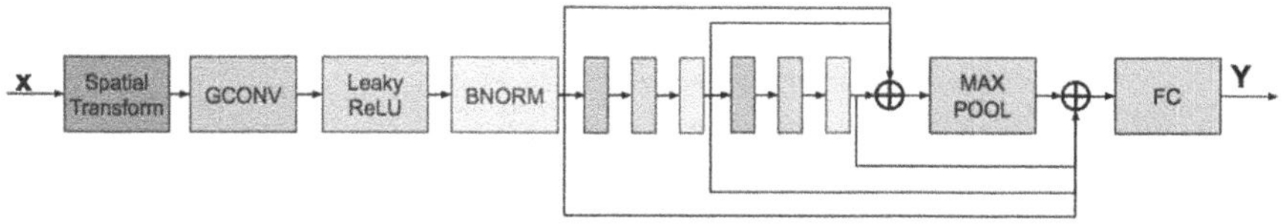

Figure 10.2. *Point cloud segmentation network (Wang* et al. *2019). The network outputs the per-point classification scores* $\mathbf{Y} \in \mathbb{R}^{N \times p}$ *for* p *semantic labels.* $\oplus$*: concatenation. For a color version of this figure, see www.iste.co.uk/cheung/graph.zip*

Model	mIoU (%)
PointNet (Qi *et al.* 2017a)	83.7
PointNet++ (Qi *et al.* 2017b)	85.1
Kd-Net (Klokov and Lempitsky 2017)	82.3
LocalFeatureNet (Shen *et al.* 2017)	84.3
PCNN (Atzmon *et al.* 2018)	85.1
PointCNN (Li *et al.* 2018)	86.1
GCNN (Wang *et al.* 2019)	85.2

Table 10.3. *Part segmentation results on ShapeNet part dataset (Yi* et al. *2016). The results show the mean computed for all classes; for more detailed results, we refer the reader to (Wang* et al. *2019)*

(Wang *et al.* 2019) propose a GCNN for part segmentation. An overview of the network architecture is shown in Figure 10.2. After a spatial transform, three graph convolutional layers are employed to extract geometrical features of the points. The output features of the convolutional layers are then aggregated globally to obtain a one-dimensional global descriptor. Finally, fully connected layers aggregate the global descriptor and the output features of the previous convolutional layers, in order to output the per-point classification scores. The graph convolutional layers use the DGCNN definition of graph convolution (see Chapter 3 for more details). It is important to note that the graph used in the convolutional operations is not fixed, but is updated at each layer. This is done by computing a k nearest neighbor graph between the feature vectors associated with each of the points, induced by the previous layer. Therefore, the network learns how to construct the graph from the feature space of each layer of the network, rather than considering the graph as an input of the network that is the same for all layers. Such a dynamic construction of the graph significantly improves the performance of the network. The reason behind such improvement can be found in the fact that the new graph captures dependencies among higher level semantic features.

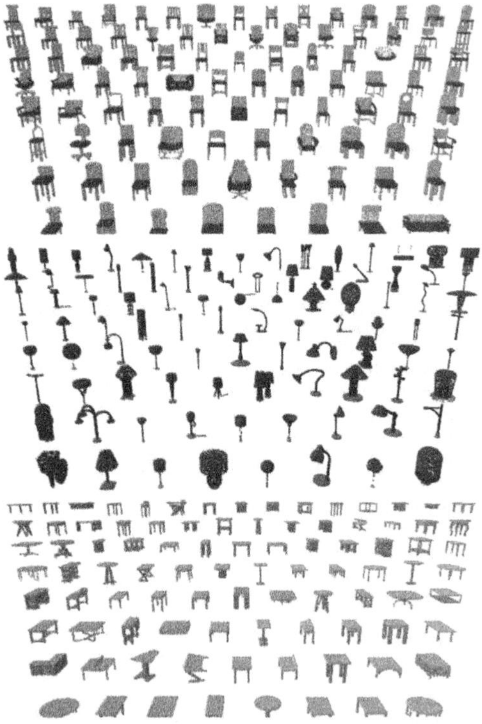

Figure 10.3. *Part segmentation results for chairs, lamps and tables. Figure from (Wang* et al. *2019). For a color version of this figure, see www.iste.co.uk/cheung/graph.zip*

Figure 10.3 shows some part segmentation results. Table 10.3 compares the results of the network described above with respect to alternative approaches. The results are expressed in terms of mean Intersection-over-Union (mIoU), where the IoU of a shape is computed by averaging the IoUs of all the different parts of that shape and the final mean IoU is obtained by averaging the IoUs of all the shapes.

10.2.3. *Image denoising*

While we have so far considered the new challenges and opportunities offered by the new data type of point clouds, this section deals with more traditional images. In particular, we discuss image denoising, which is a long-standing problem in image processing. The research on this topic has traditionally focused on developing increasingly sophisticated hand-crafted models of natural images. Among these techniques, methods based on non-local self-similarities, such as BM3D (Dabov *et al.* 2007) and WNNM (Gu *et al.* 2014), are those that achieve the best performance. Recently, approaches based on CNNs have shown to be able to capture highly complex image priors, outperforming model-based methods (Zhang *et al.* 2017). However, since these methods are based on convolutional operations, they can only capture local similarities, being unable to exploit non-local self-similarity patterns, which have proven to be very successful in model-based techniques. This represents a substantial shortcoming, which can limit the performance of CNN-based approaches. In order to overcome this issue, Valsesia *et al.* (2019a) proposes employing a GCNN, to exploit both local and non-local similarities. In fact, constructing a graph by looking at similarities between the feature vectors associated with pixels, it is possible to uncover non-local correlations.

An overview of the GCNN presented in Valsesia *et al.* (2019a) is shown in Figure 10.4. At a high level, we can observe that the network has a global input–output residual connection, learning to estimate the noise rather than the clean image. The network has, first, a preprocessing block, with standard 2D convolutions at different scales. This block creates a feature embedding with a receptive field larger than a single pixel to stabilize the graph construction operation, which otherwise would be affected by the input noise. After the preprocessing block, the remainder of the network is subdivided into an HPF block and multiple LPF blocks, which have been named after the analogy with high-pass and low-pass graph filters (for more details, we refer the reader to Valsesia *et al.* (2019a)). One of the main features of the network is the fact that the graph is dynamically computed from the features. Similarly as with a DGCNN, (Wang *et al.* 2019), the graph is updated by constructing a k-nearest neighbor graph from the feature vectors of every pixel produced by the hidden layers of the network. This results in a non-local receptive field, where pixels that are spatially distant, but have similar features, can be connected.

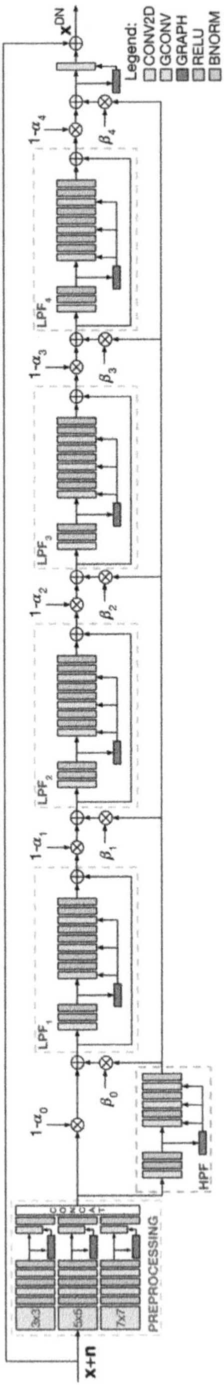

Figure 10.4. *Image denoising network. Figure from (Valsesia et al. 2019a). For a color version of this figure, see www.iste.co.uk/cheung/graph.zip*

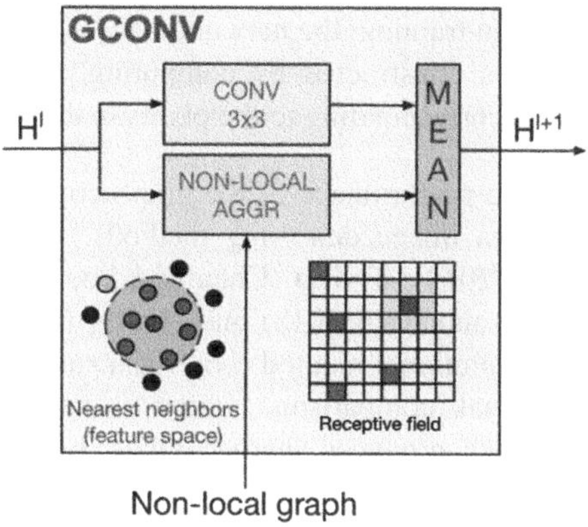

Figure 10.5. *Graph-convolutional layer. Figure from (Valsesia* et al. *2019a). For a color version of this figure, see www.iste.co.uk/cheung/graph.zip*

The graph-convolutional layer is the core of the network proposed by Valsesia *et al.* (2019a). Figure 10.5 shows a schematic representation of the operation performed by this layer. A standard 3×3 convolution processes the local neighbors. The non-local neighbors, instead which are selected as the k nearest neighbors in the feature space, are aggregated using the lightweight ECC described in Chapter 3. The features estimated by the standard 2D convolution are then combined with the features produced by the non-local aggregation to produce the output of the graph-convolutional layer.

Dataset	Noise σ	textbfBM3D	WNNM	TNRD	DnCNN	N^3Net	NLRN	GCNN
Set12	15	32.37	32.70	32.50	32.86	-	33.16	33.14
	25	29.97	30.28	30.06	30.44	30.55	30.80	30.78
	50	26.72	27.05	26.81	27.18	27.43	27.64	27.60
BSD68	15	31.07	31.37	31.42	31.73	-	31.88	31.83
	25	28.57	28.83	28.92	29.23	29.30	29.41	29.35
	50	25.62	25.87	25.97	26.23	26.39	26.47	26.38
Urban100	15	32.35	32.97	31.86	32.68	-	33.42	33.47
	25	29.70	30.39	29.25	29.97	30.19	30.88	30.95
	50	25.95	26.83	25.88	26.28	26.82	27.40	27.41

Table 10.4. *Natural image denoising results. The evaluation metric is PSNR (dB)*

It is worth mentioning that the proposed approach is fully convolutional in testing, as every pixel has a search window centered around it and its edges are defined by searching, in such a window, for the pixels with the most similar features. This is very important because it allows us to process images of any dimension, even very large images. Instead, in training, the network takes, as input, patches of a given dimension and the graph is constructed by computing all the pairwise distances among the features vectors corresponding to the pixels in the patch.

Table 10.4 compares the performance of the graph-convolutional approach with respect to other well-known image denoising methods, i.e. BM3D (Dabov *et al.* 2007), WNNM (Gu *et al.* 2014), TNRD (Chen and Pock 2016), DnCNN (Zhang *et al.* 2017), N^{3}Net (Plötz and Roth 2018) and NLRN (Liu *et al.* 2018). We can observe that the graph-convolutional method achieves a state-of-the-art performance. Figure 10.6 shows a visual comparison, where it can be observed that the graph-convolutional approach produces sharper edges. Finally, Figure 10.7 shows two examples of receptive field, where we can clearly see that the receptive field is adapted to the characteristics of the image.

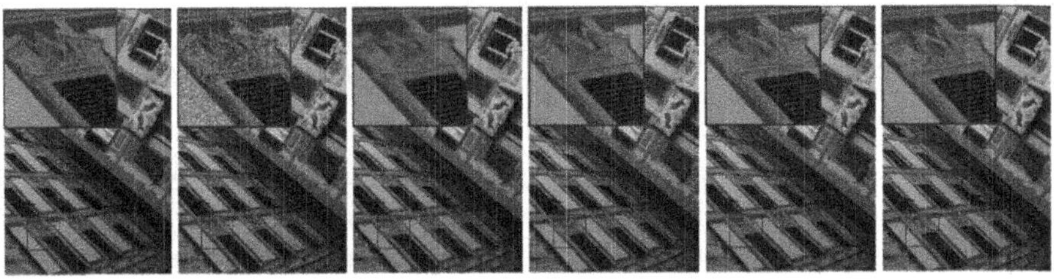

Figure 10.6. *Extract from Urban100 scene 13, $\sigma = 25$. Left to right: ground truth, noisy (20.16 dB), BM3D (30.40 dB), DnCNN (30.71 dB), NLRN (31.41 dB), GCNN (31.53 dB). Figure from (Valsesia* et al. *2019a)*

10.3. Generative models for point clouds

While supervised learning techniques, such as the ones presented in the previous section, can achieve impressive results on many tasks, they are fundamentally limited by the requirement of ground truth labels. Unsupervised learning techniques, such as generative models, are needed, in order to exploit the tremendous amount of unlabeled data. In this section, we present two graph-convolutional generative models that can capture complex representations of the data without requiring supervisory signals. Generative models aim at capturing the distribution of the data. Once this is learned, it can be used to sample new data or regularize inverse problems, such as shape completion.

10.3.1. *Point cloud generation*

Generative adversarial networks (GANs) (Goodfellow *et al.* 2014) are state-of-the-art generative models that can capture the data distribution with high fidelity. Figure 10.8 presents a high-level view of a GAN, which is composed of two neural networks: a generator, mapping a latent vector into a sample from the data distribution, and a discriminator (also known as a critic). The Wasserstein GAN (Arjovsky *et al.* 2017) addressed the stability problems of early GAN formulations by relying on a dual formulation of an optimal transport problem. In this method, the generator G and discriminator D are trained by solving the following optimization problem:

$$\min_G \max_{\|D\|_L \leq 1} E_{\mathbf{x} \sim p_{\text{data}}} [D(\mathbf{x})] - E_{\mathbf{z} \sim p_z} [D(G(\mathbf{z}))],$$

with p_{data} the distribution of training data and p_z a prior distribution on the latent vectors, typically a spherical Gaussian. The bound on the Lipschitz constant of the discriminator can be achieved by the gradient penalty method of (Gulrajani *et al.* 2017).

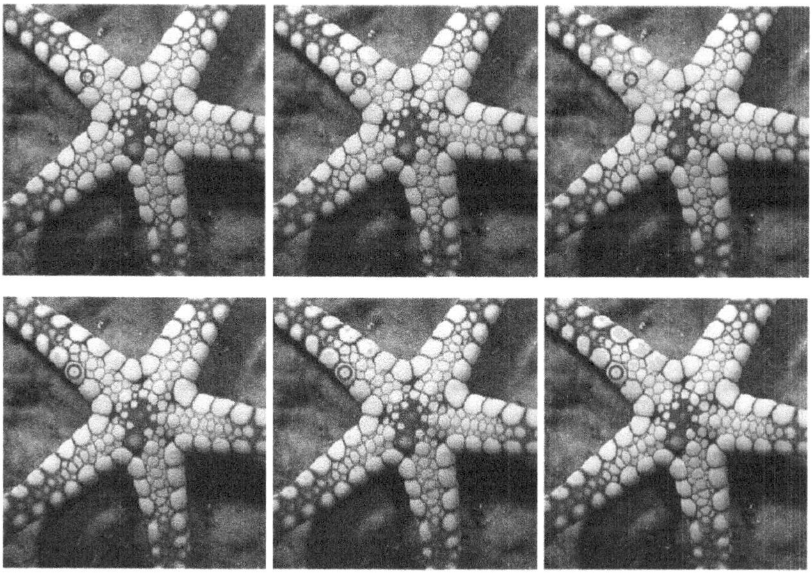

Figure 10.7. *Receptive field (green) of a single pixel (red, circled in purple) for the three graph-convolutional layers in the LPF_1 block with respect to the input of the first graph-convolutional layer in the block. Top row: gray pixel on an edge. Bottom row: white pixel in a uniform area. Figure from (Valsesia* et al. *2019a). For a color version of this figure, see www.iste.co.uk/cheung/graph.zip*

Extensive literature results show that GANs can generate very realistic-looking images, meaning that they can capture the complex data distribution in a non-trivial way. However, using GANs to learn the distribution of point cloud data can be challenging, due to their nature as unordered sets of points. Motivated by the results of graph-convolutional models in supervised tasks, one may seek to replicate their success in a GAN generator to exploit localized operators. However, a question immediately arises: how can one apply the graph convolution operation if the graph describing the connections between points is unknown in advance, as it is the output of the generator? Valsesia *et al.* (2019c) showed that a generator, built with graph-convolutional layers, where the graph is dynamically constructed at every layer from similarities between the feature vectors of the points, can successfully capture the distribution of point cloud data. Indeed, such a model learns to promote hidden features to be graph embeddings, i.e. high-dimensional representations of the local dependencies between a point and its neighbors. Furthermore, Valsesia *et al.* (2019c) addresses the problem of defining upsampling layers at the generator, so that hierarchical representations of the data can be exploited, where a point cloud with a high number of points can be created by leveraging representations with a lower number of points, e.g. by exploiting self-similarity.

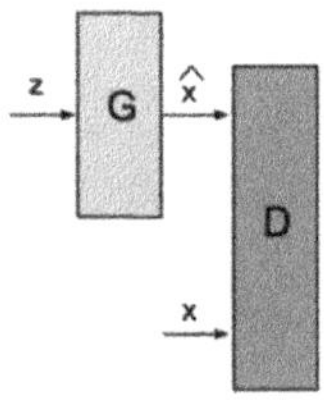

Figure 10.8. *Generative adversarial network. A generator G maps a latent vector $\mathbf{z}$ into a sample $\hat{\mathbf{x}}$ of the data distribution. The discriminator D can be interpreted as measuring the optimal transport cost between the true data and the generated data distributions. For a color version of this figure, see www.iste.co.uk/cheung/graph.zip*

Figure 10.9 shows the architecture of the GAN generator proposed by Valsesia *et al.* (2019c). The graph-convolutional layers use the edge-conditioned convolution (Simonovsky and Komodakis 2017), so the layer $l+1$ feature vector $\mathbf{H}_i^{l+1}$ is computed from the feature vectors in neighborhood $\mathcal{N}_i^l$ as:

$$\mathbf{H}_i^{l+1} = \sigma\left(\sum_{j\in\mathcal{N}_i^l} \frac{\mathcal{F}_{\mathbf{w}^l}^l\left(\mathbf{H}_j^l - \mathbf{H}_i^l\right)\mathbf{H}_j^l}{|\mathcal{N}_i^l|} + \mathbf{H}_i^l\mathbf{W}^l + \mathbf{b}^l\right)$$

$$= \sigma\left(\underbrace{\sum_{j\in\mathcal{N}_i^l} \frac{\mathbf{\Theta}^{l,ji}\mathbf{H}_j^l}{|\mathcal{N}_i^l|}}_{\text{neighborhood}} + \underbrace{\mathbf{H}_i^l\mathbf{W}^l}_{\text{node}} + \mathbf{b}^l\right).$$

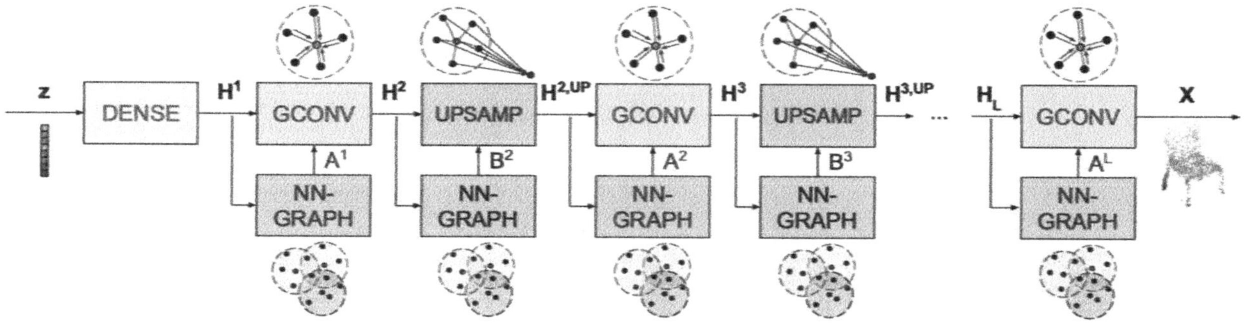

Figure 10.9. *Graph-convolutional GAN generator. The graph-convolutional and upsampling layers use a k nearest neighbor graph computed from the feature vectors at the input of the layer. For a color version of this figure, see www.iste.co.uk/cheung/graph.zip*

The upsampling layers also use a spatial aggregation that is similar to the ECC, but with a diagonal matrix in place of $\mathbf{\Theta}^{l,ji}$ and $\mathbf{W}^{l}$ to limit the number of trainable parameters. This can also be seen as treating each feature map independently, and is motivated by the desire to generate a new point in the same feature space, without mixing the features. The output set of points has the same cardinality as the input, and is concatenated to the input, thus doubling the total number of points.

Figure 10.10 shows some of the generated point clouds, and Table 10.5 reports some evaluation metrics to compare how well the graph-convolutional generator captures the data distribution with respect to alternative techniques. It can be noted that the lower values, in terms of Jensen Shannon divergence and minimum matching distance (MMD) (as defined by Achlioptas *et al.* (2017)), prove that the graph-convolutional generator provides a better approximation of the true data distribution. The baseline methods are represented by the r-GAN-dense (Achlioptas *et al.* 2017), which employs a fully connected generator and r-GAN-conv, which is a variant with single-point convolutions. Neither of these uses localized operations, limiting their representational capabilities.

Figure 10.10. *Generated point clouds. For a color version of this figure, see www.iste.co.uk/cheung/graph.zip*

Class	Model	JSD	MMD-CD	MMD-EMD
Chair	**r-GAN-dense**	0.238	**0.0029**	0.136
	r-GAN-conv	0.517	0.0030	0.223
	Graph GAN (no upsampling)	0.119	0.0033	0.104
	Graph GAN (upsampling)	**0.100**	**0.0029**	**0.097**
Airplane	**r-GAN-dense**	0.182	0.0009	0.094
	r-GAN-conv	0.350	**0.0008**	0.101
	Graph GAN (no upsampling)	0.164	0.0010	0.102
	Graph GAN (upsampling)	**0.083**	**0.0008**	**0.071**

Table 10.5. *Quantitative comparisons*

10.3.2. *Shape completion*

This section presents an approach to the shape completion problem, i.e. reconstruction of the missing parts of a 3D shape as a result of partial scans. This problem is tackled by means of variational autoencoders (VAEs) (Kingma and Welling 2013), which are generative models, often seen as alternatives to GANs. A VAE can be decomposed into an encoder function E, mapping the input data into a latent representation, and a decoder D, that reconstructs a data point from the latent variable. During training, the latent space is enforced to follow a distribution of choice, e.g. a Gaussian of a given mean and covariance.

Graph-convolutional VAEs are suitable for 3D models, extending convolution-like properties to meshes. (Litany *et al.* 2018) propose a graph-convolutional VAE for 3D shape completion based on graph-convolutional layers. Such a VAE is trained to create latent representations of the full 3D shapes. In the shape completion problem, a partial shape of the predefined topology is given and the task is to recreate the missing part. This inverse problem can be effectively regularized by means of the latent representations learned by the VAE.

Figure 10.11 shows a high level description of the architecture and its use in the shape completion task. The VAE is trained with a loss, which is the sum of two contributions:

$$L = \|D(E(\mathbf{X})) - \mathbf{X}\| + \lambda D_{\mathrm{KL}}(q(\mathbf{Z}|\mathbf{X})\|p(\mathbf{Z})),$$

where the first term measures the fidelity of the data reconstructed from the hidden representation, with $\mathbf{X} \in \mathbb{R}^{N\times 3}$ as the input shape, and the second term measures the Kullback–Leibler divergence (Kullback and Leibler 1951) between the distribution, estimated as the output of encoder ($q(\mathbf{Z}|\mathbf{X})$) and the prior distribution over the latent space ($p(\mathbf{Z}) = N(\mathbf{0}, \mathbf{I})$).

The graph-convolutional layers use the FeaStNet definition (Verma *et al.* 2018), thus computing the feature vector $\mathbf{H}_i^{l+1}$ of vertex i at layer $l+1$ as

$$\mathbf{H}_i^{l+1} = \mathbf{b} + \sum_{m=1}^{M} \exp\left(\boldsymbol{\omega}_{1,m}^T \mathbf{H}_i^l + \boldsymbol{\omega}_{2,m}^T \mathbf{H}_j^l + \omega_m\right) \boldsymbol{\Theta}^{(m)} \mathbf{H}_j^l,$$

for trainable weights $\boldsymbol{\omega}_{1,m}, \boldsymbol{\omega}_{2,m}, \omega_m, \boldsymbol{\Theta}^{(m)}, \mathbf{b}$. The graph is constructed as a vertex 2-ring.

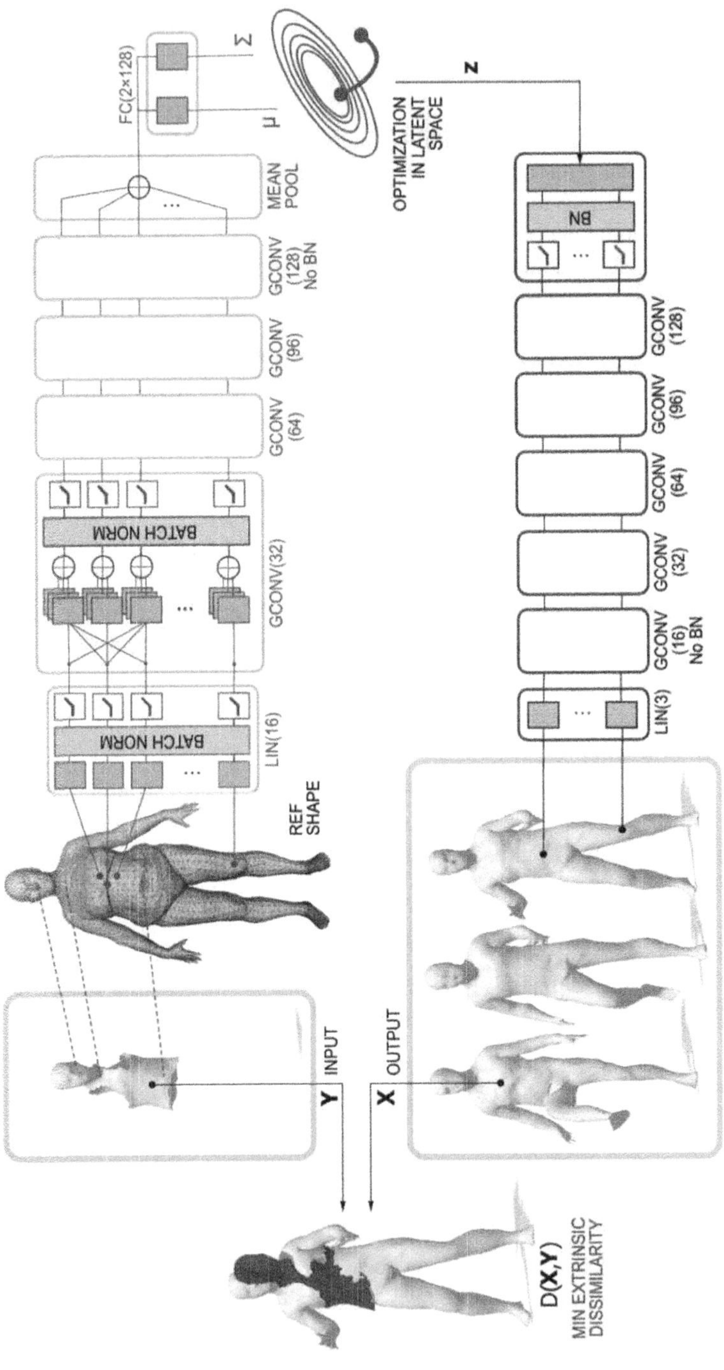

Figure 10.11. *Graph-convolutional variational autoencoder for shape completion. Figure from (Litany et al. 2018). For a color version of this figure, see www.iste.co.uk/cheung/graph.zip*

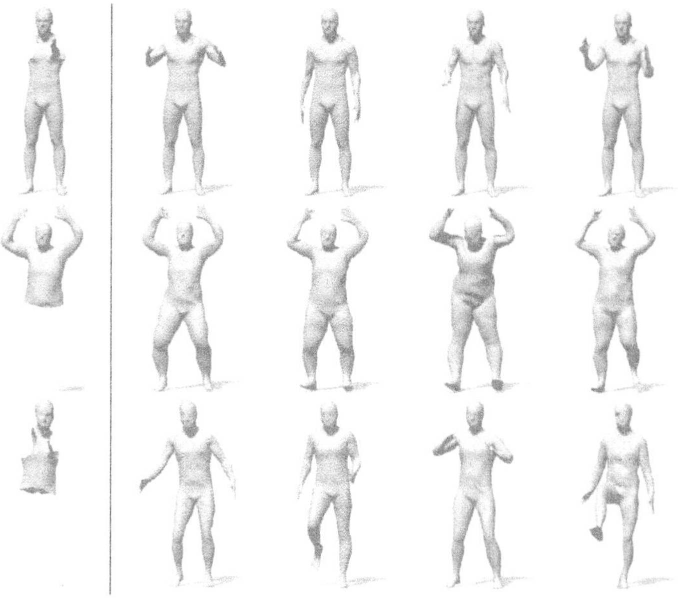

Figure 10.12. *Examples of completed shapes. Figure from (Litany* et al. *2018). For a color version of this figure, see www.iste.co.uk/cheung/graph.zip*

Once the VAE is trained on full shapes, partial shape completion only requires the use of the decoder network. In this problem, a partial shape $\mathbf{Y} \in \mathbb{R}^{N' \times 3}$ with a subset of vertices is given. Shape completion consists of finding the latent representation that, once decoded, yields the best match with the observed partial shape. Therefore, the full shape $\mathbf{Y}^*$ can be retrieved as

$$\mathbf{Y}^* = D(\mathbf{Z}^*)$$

$$\mathbf{Z}^*, \mathbf{T}^* = \arg \min_{\mathbf{Z}, \mathbf{T} \in \mathrm{SE}(3)} \|D(\mathbf{Z})\mathbf{\Pi} - \mathbf{T}\mathbf{Y}\|,$$

with $\mathbf{T}$ a rigid transformation and $\mathbf{\Pi}$ the partial permutation matrix describing the permutation and downsampling operations performed to get $\mathbf{Y}$ from a ground truth shape $\mathbf{X}$.

Figure 10.12 shows examples of completions from partial shapes. Note that, in general, the solution is not unique and multiple completions can fit the observed partial shape, especially when it is heavily subsampled. Finally, Table 10.6 shows the results achieved in terms of completion of synthetic range scans from the FAUST dataset (Bogo *et al.* 2014). The metrics include the mean Euclidean distance in centimeters and relative volumetric error for the missing region. It can be noted that the graph-convolutional VAE method can achieve lower errors.

	Euclidean distance	Volumetric error
(Kazhdan and Hoppe 2013)	7.3	24.8
(Dai *et al.* 2017)	4.43	89.7
(Litany *et al.* 2018)	3.40	12.51

Table 10.6. *Completion error for synthetic range scans.*

10.4. Concluding remarks

In this chapter, we present an overview of the most relevant methods for visual data processing based on GCNNs, considering both supervised and unsupervised approaches. The majority of the methods presented in this chapter focus on point cloud processing. Due to their irregular structure, point clouds cannot be processed with traditional CNNs and graph neural networks can represent a powerful solution to overcome such a limitation. Moreover, we also present an application of graph neural networks to image processing. In this case, a GCNN can help to use non-local self-similarities.

10.5. References

Achlioptas, P., Diamanti, O., Mitliagkas, I., Guibas, L.J. (2017). Representation learning and adversarial generation of 3D point clouds. arXiv preprint, arXiv:1707.02392.

Arjovsky, M., Chintala, S., Bottou, L. (2017). Wasserstein GAN. arXiv preprint, arXiv:1701.07875.

Atzmon, M., Maron, H., Lipman, Y. (2018). Point convolutional neural networks by extension operators. *ACM Transactions on Graphics (ToG)*, 37(4), 1–12.

Bai, Y., Cheung, G., Liu, X., Gao, W. (2018). Graph-based blind image deblurring from a single photograph. *IEEE Transactions on Image Processing*, 28(3), 1404–1418.

Bogo, F., Romero, J., Loper, M., Black, M.J. (2014). FAUST: Dataset and evaluation for 3D mesh registration. *Proceedings of the IEEE Conference on Computer Vision and Pattern Recognition*, 3794–3801.

Chen, Y. and Pock, T. (2016). Trainable nonlinear reaction diffusion: A flexible framework for fast and effective image restoration. *IEEE Transactions on Pattern Analysis and Machine Intelligence*, 39(6), 1256–1272.

Chen, T., Dai, B., Liu, D., Song, J. (2014). Performance of global descriptors for velodyne-based urban object recognition. *IEEE Intelligent Vehicles Symposium Proceedings*, 667–673.

Chen, S., Tian, D., Feng, C., Vetro, A., Kovacevic, J. (2018). Fast resampling of three-dimensional point clouds via graphs. *IEEE Transactions on Signal Processing*, 66(3), 666–681.

Dabov, K., Foi, A., Katkovnik, V., Egiazarian, K. (2007). Image denoising by sparse 3D transform-domain collaborative filtering. *IEEE Transactions on Image Processing*, 16(8), 2080–2095.

Dai, A., Ruizhongtai Qi, C., Nießner, M. (2017). Shape completion using 3D-encoder-predictor CNNS and shape synthesis. *Proceedings of the IEEE Conference on Computer Vision and Pattern Recognition*, 5868–5877.

De Deuge, M., Quadros, A., Hung, C., Douillard, B. (2013). Unsupervised feature learning for classification of outdoor 3D scans. *Australasian Conference on Robitics and Automation*, 2, 1.

Goodfellow, I., Pouget-Abadie, J., Mirza, M., Xu, B., Warde-Farley, D., Ozair, S., Courville, A., Bengio, Y. (2014). Generative adversarial nets. In *NIPS'14: Proceedings of the 27th International Conference on Neural Information Processing Systems – Volume 2*, Ghahramani, Z., Welling, M., Cortes, C., Lawrence, N.D., Weinberger, K.Q. (eds). MIT Press, Cambridge.

Gu, S., Zhang, L., Zuo, W., Feng, X. (2014). Weighted nuclear norm minimization with application to image denoising. *Proceedings of the IEEE Conference on Computer Vision and Pattern Recognition*, 2862–2869.

Gulrajani, I., Ahmed, F., Arjovsky, M., Dumoulin, V., Courville, A.C. (2017). Improved training of Wasserstein GANs. In *NIPS'17: Proceedings of the 31st International Conference on Neural Information Processing Systems*, von Luxburg, U., Guyon, I.M., Bengio, S., Wallach, H.M., Fergus, R. (eds). Curran Associates Inc, Red Hook.

Kazhdan, M. and Hoppe, H. (2013). Screened Poisson surface reconstruction. *ACM Transactions on Graphics (ToG)*, 32(3), 29.

Kheradmand, A. and Milanfar, P. (2014). A general framework for regularized, similarity-based image restoration. *IEEE Transactions on Image Processing*, 23(12), 5136–5151.

Kingma, D.P. and Welling, M. (2013). Auto-encoding variational Bayes. arXiv preprint, arXiv:1312.6114.

Klokov, R. and Lempitsky, V. (2017). Escape from cells: Deep Kd-networks for the recognition of 3D point cloud models. *Proceedings of the IEEE International Conference on Computer Vision*, 863–872.

Kullback, S. and Leibler, R.A. (1951). On information and sufficiency. *The Annals of Mathematical Statistics*, 22(1), 79–86 [Online]. Available at: https://doi.org/10.1214/aoms/1177729694.

Li, Y., Bu, R., Sun, M., Wu, W., Di, X., Chen, B. (2018). PointCNN: Convolution on X-transformed points. In *NIPS'18: Proceedings of the 32nd International Conference on Neural Information Processing Systems*, Bengio, S., Wallach, H.M., Larochelle, H., Grauman, K.L., Cesa-Bianchi, N. (eds). Curran Associates Inc, Red Hook.

Litany, O., Bronstein, A., Bronstein, M., Makadia, A. (2018). Deformable shape completion with graph convolutional autoencoders. *Proceedings of the IEEE Conference on Computer Vision and Pattern Recognition*, 1886–1895.

Liu, X., Zhai, D., Zhao, D., Zhai, G., Gao, W. (2014). Progressive image denoising through hybrid graph Laplacian regularization: A unified framework. *IEEE Transactions on Image Processing*, 23(4), 1491–1503.

Liu, D., Wen, B., Fan, Y., Loy, C.C., Huang, T.S. (2018). Non-local recurrent network for image restoration. In *NIPS'18: Proceedings of the 32nd International Conference on Neural Information Processing Systems*, Bengio, S., Wallach, H.M., Larochelle, H., Grauman, K.L., Cesa-Bianchi, N. (eds). Curran Associates Inc, Red Hook.

Maturana, D. and Scherer, S. (2015). Voxnet: A 3D convolutional neural network for realtime object recognition. *IEEE/RSJ International Conference on Intelligent Robots and Systems (IROS)*, 922–928.

Pang, J. and Cheung, G. (2017). Graph Laplacian regularization for image denoising: Analysis in the continuous domain. *IEEE Transactions on Image Processing*, 26(4), 1770–1785.

Plötz, T. and Roth, S. (2018). Neural nearest neighbors networks. In *NIPS'18: Proceedings of the 32nd International Conference on Neural Information Processing Systems*, Bengio, S., Wallach, H.M., Larochelle, H., Grauman, K.L., Cesa-Bianchi, N. (eds). Curran Associates Inc, Red Hook.

Qi, C.R., Su, H., Nießner, M., Dai, A., Yan, M., Guibas, L.J. (2016). Volumetric and multi-view CNNs for object classification on 3D data. *Proceedings of the IEEE Conference on Computer Vision and Pattern Recognition*, 5648–5656.

Qi, C., Su, H., Mo, K., Guibas, L.J. (2017a). PointNet: Deep learning on point sets for 3D classification and segmentation. *Proceedings of the IEEE Conference on Computer Vision and Pattern Recognition*, 1(2), 4.

Qi, C.R., Yi, L., Su, H., Guibas, L.J. (2017b). PointNet++: Deep hierarchical feature learning on point sets in a metric space. In *NIPS'17: Proceedings of the 31st International Conference on Neural Information Processing Systems*, von Luxburg, U., Guyon, I.M., Bengio, S., Wallach, H.M., Fergus, R. (eds). Curran Associates Inc, Red Hook.

Rusu, R.B. and Cousins, S. (2011). 3D is here: Point cloud library (PCL). *IEEE International Conference on Robotics and Automation*, 1–4.

Sedaghat, N., Zolfaghari, M., Amiri, E., Brox, T. (2016). Orientation-boosted voxel nets for 3D object recognition. arXiv preprint, arXiv:1604.03351.

Shen, Y., Feng, C., Yang, Y., Tian, D. (2017). Neighbors do help: Deeply exploiting local structures of point clouds. arXiv preprint, arXiv:1712.06760, 1(2).

Simonovsky, M. and Komodakis, N. (2017). Dynamic edge-conditioned filters in convolutional neural networks on graphs. *IEEE Conference on Computer Vision and Pattern Recognition*, 29–38.

Su, H., Maji, S., Kalogerakis, E., Learned-Miller, E. (2015), Multi-view convolutional neural networks for 3D shape recognition. *Proceedings of the IEEE International Conference on Computer Vision*, 945–953.

Thanou, D., Chou, P.A., Frossard, P. (2016). Graph-based compression of dynamic 3D point cloud sequences. *IEEE Transactions on Image Processing*, 25(4), 1765–1778.

Valsesia, D., Fracastoro, G., Magli, E. (2019a). Image denoising with graph-convolutional neural networks. *Proceedings of the IEEE International Conference on Image Processing*, 2399–2403.

Valsesia, D., Fracastoro, G., Magli, E. (2019b). Deep graph-convolutional image denoising, arXiv preprint, arXiv:1907.08448.

Valsesia, D., Fracastoro, G., Magli, E. (2019c). Learning localized generative models for 3D point clouds via graph convolution. *International Conference on Learning Representations*, New Orleans, USA, May 6–9.

Verma, N., Boyer, E., Verbeek, J. (2018). FeaStNet: Feature-steered graph convolutions for 3D shape analysis. *Proceedings of the IEEE Conference on Computer Vision and Pattern Recognition*, 2598–2606.

Wang, X., Girshick, R., Gupta, A., He, K. (2018). Non-local neural networks. *Proceedings of the IEEE Conference on Computer Vision and Pattern Recognition*, 7794–7803.

Wang, Y., Sun, Y., Liu, Z., Sarma, S.E., Bronstein, M.M., Solomon, J.M. (2019). Dynamic graph CNN for learning on point clouds. *ACM Transactions on Graphics (ToG)*, 38(5), 1–12.

Wu, Z., Song, S., Khosla, A., Yu, F., Zhang, L., Tang, X., Xiao, J. (2015). 3D ShapeNets: A deep representation for volumetric shapes. *Proceedings of the IEEE Conference on Computer Vision and Pattern Recognition*, 1912–1920.

Yi, L., Kim, V.G., Ceylan, D., Shen, I., Yan, M., Su, H., Lu, C., Huang, Q., Sheffer, A., Guibas, L.J. (2016). A scalable active framework for region annotation in 3D shape collections. *ACM Transactions on Graphics (ToG)*, 35(6), 1–12.

Zhang, K., Zuo, W., Chen, Y., Meng, D., Zhang, L. (2017). Beyond a Gaussian denoiser: Residual learning of deep CNN for image denoising. *IEEE Transactions on Image Processing*, 26(7), 3142–3155.

Zeng, J., Cheung, G., Ng, M., Pang, J., Yang, C. (2019). 3D point cloud denoising using graph Laplacian regularization of a low dimensional manifold model. *IEEE Transactions on Image Processing*, 29, 3474–3489.

List of Authors

Yung-Hsuan CHAO
Qualcomm Technologies Inc.
San Diego
USA

Siheng CHEN
Shanghai Jiao Tong University
China

Gene CHEUNG
York University
Toronto
Canada

Xiaowen DONG
University of Oxford
UK

Hilmi E. EGILMEZ
Qualcomm Technologies Inc.
San Diego
USA

Giulia FRACASTORO
Politecnico di Torino
Turin
Italy

Pascal FROSSARD
École polytechnique fédérale de Lausanne
Switzerland

Christine GUILLEMOT
Inria Rennes – Bretagne Atlantique
Rennes
France

Wei HU
Peking University
Beijing
China

Enrico MAGLI
Politecnico di Torino
Turin
Italy

Navid MAHMOUDIAN BIDGOLI
Inria Rennes – Bretagne Atlantique
Rennes
France

Thomas MAUGEY
Inria Rennes – Bretagne Atlantique
Rennes
France

Michael NG
The University of Hong Kong
China

Antonio ORTEGA
University of Southern California
Los Angeles
USA

Jiahao PANG
InterDigital Inc.
New York
USA

Michael RABBAT
Facebook
Montreal
Canada

Mira RIZKALLAH
École Centrale Nantes
France

Aline ROUMY
Inria Rennes – Bretagne Atlantique
Rennes
France

Lina STANKOVIC
University of Strathclyde
Glasgow
UK

Vladimir STANKOVIC
University of Strathclyde
Glasgow
UK

Yuichi TANAKA
Tokyo University of Agriculture and Technology
Japan

Dorina THANOU
École polytechnique fédérale de Lausanne
Switzerland

Dong TIAN
InterDigital Inc.
New York
USA

Diego VALSESIA
Politecnico di Torino
Turin
Italy

Minxiang YE
Zhejiang Lab
Hangzhou
China

Jin ZENG
SenseTime Research
Shenzhen
China

Index

Printed and bound by CPI Group (UK) Ltd, Croydon, CR0 4YY

13/09/2026

14961045-0003